Eugene Stefanovich
Elementary Particle Theory

De Gruyter Studies in Mathematical Physics

Volume 46

Eugene Stefanovich

Elementary Particle Theory

Volume 2: Quantum Electrodynamics

DE GRUYTER

Mathematics Subject Classification 2010
Primary: 81-02, 81V10, 81T15; Secondary: 47A40, 81T18

Author
Dr Eugene Stefanovich
San Jose, California
USA
eugene_stefanovich@usa.net

ISBN 978-3-11-049089-3
e-ISBN (PDF) 978-3-11-049320-7
e-ISBN (EPUB) 978-3-11-049143-2
ISSN 2194-3532

Library of Congress Control Number: 2018016481

Bibliographic information published by the Deutsche Nationalbibliothek
The Deutsche Nationalbibliothek lists this publication in the Deutsche Nationalbibliografie; detailed bibliographic data are available on the Internet at http://dnb.dnb.de.

Typesetting: VTeX UAB, Lithuania
Printing and binding: CPI books GmbH, Leck

www.degruyter.com

Contents

List of figures

https://doi.org/10.1515/9783110493207-201

List of tables

https://doi.org/10.1515/9783110493207-202

Postulates, statements, theorems

https://doi.org/10.1515/9783110493207-203

Conventional notation

See also Conventional notation of Volume 1.

Fock space

$\|\mathrm{vac}\rangle$	vacuum state vector (page 2)
$a^\dagger, a$	electron creation/annihilation operators (page 4)
$b^\dagger, b$	positron creation/annihilation operators (page 9)
$c^\dagger, c$	photon creation/annihilation operators (page 7)
$d^\dagger, d$	proton creation/annihilation operators (page 9)
$f^\dagger, f$	antiproton creation/annihilation operators (page 9)
$\alpha^\dagger, \alpha$	creation/annihilation operators of generic particles (page 10)
$V_{NM} \circ \zeta$	product of coefficient functions (page 18)
$\mathcal{E}_A$	energy function of operator A (page 19)
$\mathscr{H}$	Fock space (page 1)
$\mathscr{H}(i, j, k, l, m)$	sector in the Fock space (page 1)
$: abcd :$	normally ordered product (page 10)

Quantum fields

$\psi(\tilde{x})$	electron–positron quantum field (page 112)
$\overline{\psi}(\tilde{x})$	Dirac-conjugated field (page 113)
$\Psi(\tilde{x})$	proton–antiproton quantum field (page 113)
$\mathcal{A}_\mu(\tilde{x})$	photon quantum field (page 129)
$\mathcal{U}^\mu(\boldsymbol{p}'s'_z, \boldsymbol{p}s_z)$	$\equiv \overline{u}(\boldsymbol{p}', s'_z)\gamma^\mu u(\boldsymbol{p}, s_z)$ (page 118)
$\mathcal{W}^\mu(\boldsymbol{p}'s'_z, \boldsymbol{p}s_z)$	$\equiv \overline{w}(\boldsymbol{p}', s'_z)\gamma^\mu w(\boldsymbol{p}, s_z)$ (page 118)
H^n	naïve Hamiltonian of QED (page 61)
H^c	QED Hamiltonian with counterterms (page 42)
$\mathrm{j}^\mu_{\mathrm{ep}}(\tilde{x})$	electron–positron current density operator (page 139)
$\mathrm{j}^\mu_{\mathrm{pa}}(\tilde{x})$	proton–antiproton current density operator (page 139)
$\mathrm{j}^\mu(\tilde{x})$	total current density operator (page 139)
λ	infrared cutoff (page 80)
Λ	ultraviolet cutoff (page 80)

Miscellaneous

γ^μ	Dirac gamma matrices (page 108)
$a, b, c, \ldots$	Dirac indices (page 112)

https://doi.org/10.1515/9783110493207-204

$\mathcal{D}(\Lambda)$	Dirac representation of the Lorentz group (page B.27)
$\mathfrak{J}$, $\mathfrak{K}$	generators of the Dirac representation (page 109)
$\not{k}$	$\equiv k_\mu \gamma^\mu$ (page 109)
$\boldsymbol{\sigma}$, σ_0	Pauli matrices (page 107)
α	$= e^2/(4\pi\hbar c)$ fine structure constant (page 66)
∂^μ	4-gradient $\left(-\frac{1}{c}\frac{\partial}{\partial t}, \frac{\partial}{\partial x}, \frac{\partial}{\partial y}, \frac{\partial}{\partial z}\right)$ (page 83)
$\tilde{p}$	$\equiv (\omega_{\boldsymbol{p}}, cp_x, cp_y, cp_z)$ energy-momentum 4-vector (page 112)
$\tilde{x}$	$\equiv (t, x/c, y/c, z/c)$ time-position 4-vector (page 112)
$\tilde{p} \cdot \tilde{x}$	$\equiv \omega_{\boldsymbol{p}} t - \boldsymbol{p} \cdot \boldsymbol{x}$ (page 112)
$\boldsymbol{k}$	$\equiv \boldsymbol{q}' - \boldsymbol{q} = \boldsymbol{p} - \boldsymbol{p}'$ transferred 3-momentum (page 66)
$\tilde{k}$	$\equiv \tilde{q}' - \tilde{q} = \tilde{p} - \tilde{p}'$ transferred 4-momentum (page 66)

Preface

In a successful theory of elementary particles, at least three important conditions must be fulfilled:

(1) relativistic invariance in the instant form of dynamics;
(2) cluster separability of the interaction;
(3) description of processes involving creation and destruction of particles.

In the first volume of our book we discussed interacting quantum theories in Hilbert spaces with a fixed set of particles. We showed how it is possible to satisfy the first two requirements (relativistic invariance and cluster separability).[1] However, these theories were fundamentally incomplete, due to their inability to describe physical processes that change the types and/or number of particles in the system. Thus, condition 3 from our list was not fulfilled.

Familiar examples of the creation and annihilation processes are emission and absorption of light (photons), decays, neutrino oscillations, etc. Particles are produced especially intensively at high energies. This is due to the famous Einstein formula $E = mc^2$, which says, in particular, that if the system has sufficient energy E of relative motion, then this energy can be transformed into the mass m of newly created particles. Even in the simplest two-particle case, the energy of the relative motion of these reactants is unlimited. Therefore, there is no limit to the number of new particles that can be created in a collision.

To advance in the study of such processes, the first thing to do is to build a Hilbert space of states $\mathscr{H}$, which is capable of describing particle creation and annihilation. Such a space must include states with arbitrary numbers (from zero to infinity) of particles of all types. It is called the *Fock space*. This construction is rather simple. However, the next step – the definition of realistic interaction operators in the Fock space – is highly nontrivial. A big part of our third volume will be devoted to the solution of this problem. Here we will prepare ourselves to this task by starting with a more traditional approach, which is known as the renormalized relativistic *quantum field theory* (QFT). Our discussions in this book are limited to electromagnetic phenomena, so we will be interested in the simplest and most successful type of QFT – *quantum electrodynamics* (QED).

In Chapter 1, *Fock space*, we will describe the basic mathematical machinery of Fock spaces, including creation and annihilation operators, normal ordering and classification of interaction potentials.

A simple toy model with variable number of particles will be presented in Chapter 2, *Scattering in Fock space*. In this example, we will discuss such important ingredients of QFT as the S-matrix formalism, renormalization, diagram technique and

1 See, e. g., Subsection **1**-6.4.6.

https://doi.org/10.1515/9783110493207-205

cluster separability. Our first two chapters have a mostly technical character. They define our terminology and notation and prepare us for a more in-depth study of QED in the two following chapters.

In Chapter 3, *Quantum electrodynamics*, we introduce the important concept of the quantum field. This idea will be applied to systems of charged particles and photons in the formalism of QED. Here we will obtain an interacting theory, which satisfies the principles of relativistic invariance and cluster separability, where the number of particles is not fixed. However, the "naïve" version of QED presented here is unsatisfactory, since it cannot calculate scattering amplitudes beyond the lowest orders of perturbation theory.

Chapter 4, *Renormalization*, completes the second volume of the book. We will discuss the plague of ultraviolet divergences in the "naïve" QED and explain how they can be eliminated by adding counterterms to the Hamiltonian. As a result, we will get the traditional "renormalized" QED, which has proven itself in precision calculations of scattering cross sections and energy levels in systems of charged particles. However, this theory failed to provide a well-defined interacting Hamiltonian and the interacting time evolution (= dynamics). We will address these issues in the third volume of our book.

As in the first volume, here we refrain from criticism and unconventional interpretations, trying to keep in line with generally accepted approaches. The main purpose of this volume is to explain the basic concepts and terminology of QFT. For the most part, we will adhere to the logic of QFT formulated by Weinberg in the series of articles [19, 18, 20] and in the excellent textbook [21]. A critical discussion of the traditional approaches and a new look at the theory of relativity will be presented in Volume 3 [17].

References to Volume 1 [16] of this book will be prefixed with "**1**-". For example, (**1**-7.14) is formula (7.14) from Volume 1.

1 Fock space

There are more things in Heaven and on earth, dear Horacio, than are dreamed of in your philosophy.
Hamlet

In this chapter, we construct the Fock space $\mathscr{H}$ populated by particles of five types: electrons e^-, positrons e^+, protons p^+, antiprotons[1] p^- and photons γ. We will practice constructions of simple interaction operators and study their properties. In comparison with Volume 1, the main novelty is in working with operators that change the number of particles. This will prepare us for mastering a more realistic theory – *quantum electrodynamics* (QED) – in Chapters 3 and 4.

1.1 Creation and annihilation operators

Here we introduce the concepts of creation and annihilation operators. Though lacking autonomous physical meaning, these operators greatly simplify calculations in $\mathscr{H}$.

1.1.1 Sectors with fixed numbers of particles

The numbers of particles of each type are easily measurable in experiments, so we have the right to introduce in our theory five new observables, namely, the numbers of electrons (N_{el}), positrons (N_{po}), protons (N_{pr}), antiprotons (N_{an}) and photons (N_{ph}). Unlike in ordinary quantum mechanics from Volume 1, here we will not assume that the numbers of particles are fixed. We would like to treat these quantities on the same footing as other quantum observables. In particular, we will also take into account their quantum uncertainty. Then, in accordance with general quantum rules, these observables should be represented in the Hilbert space (= Fock space) $\mathscr{H}$ by five Hermitian operators. Obviously, their allowed values (spectra) are nonnegative integers $(0, 1, 2, \ldots)$. From part (II) of Postulate **1**-6.1, it follows that these observables are measurable simultaneously, so that the particle number operators commute with each other and have common eigensubspaces. Hence, the Fock space $\mathscr{H}$ splits into a direct sum of orthogonal subspaces, or *sectors*, $\mathscr{H}(i, j, k, l, m)$ containing i electrons,

1 In this book, protons and antiprotons are regarded as simple point charges. Their internal structure is ignored, as well as their participation in strong nuclear interactions.

https://doi.org/10.1515/9783110493207-001

j positrons, k protons, l antiprotons and m photons, so

$$\mathscr{H} = \bigoplus_{ijklm=0}^{\infty} \mathscr{H}(i,j,k,l,m), \tag{1.1}$$

where

$$\begin{aligned}
N_{\mathrm{el}}\mathscr{H}(i,j,k,l,m) &= i\mathscr{H}(i,j,k,l,m),\\
N_{\mathrm{po}}\mathscr{H}(i,j,k,l,m) &= j\mathscr{H}(i,j,k,l,m),\\
N_{\mathrm{pr}}\mathscr{H}(i,j,k,l,m) &= k\mathscr{H}(i,j,k,l,m),\\
N_{\mathrm{an}}\mathscr{H}(i,j,k,l,m) &= l\mathscr{H}(i,j,k,l,m),\\
N_{\mathrm{ph}}\mathscr{H}(i,j,k,l,m) &= m\mathscr{H}(i,j,k,l,m).
\end{aligned}$$

The one-dimensional subspace without particles $\mathscr{H}(0,0,0,0,0)$ is called the *vacuum subspace*. The *vacuum vector* $|\mathrm{vac}\rangle$ is defined in this subspace up to an unimportant phase factor.

Single-particle sectors are built according to the recipes from Chapter **1**-5. The subspaces $\mathscr{H}(1,0,0,0,0)$ and $\mathscr{H}(0,1,0,0,0)$ contain one electron and one positron, respectively. These subspaces carry unitary irreducible representations of the Poincaré group with mass $m_e = 0.511\,\mathrm{MeV}/c^2$ and spin 1/2 (see Table **1**-5.1). The subspaces $\mathscr{H}(0,0,1,0,0)$ and $\mathscr{H}(0,0,0,1,0)$ contain one proton and one antiproton, respectively. These particles have mass $m_p = 938.3\,\mathrm{MeV}/c^2$ and spin 1/2. The subspace $\mathscr{H}(0,0,0,0,1)$ contains one photon with zero mass. This subspace is the direct sum of two irreducible massless subspaces with helicities 1 and −1 (see Subsection **1**-5.4.4).

Sectors with two or more particles are constructed as (anti)symmetrized products of single-particle sectors.[2] For example, if $\mathscr{H}_{\mathrm{el}}$ is the one-electron sector and $\mathscr{H}_{\mathrm{ph}}$ is the one-photon sector, then sectors having only electrons and photons can be written as

$$\mathscr{H}(0,0,0,0,0) = |\mathrm{vac}\rangle, \tag{1.2}$$
$$\mathscr{H}(1,0,0,0,0) = \mathscr{H}_{\mathrm{el}}, \tag{1.3}$$
$$\mathscr{H}(0,0,0,0,1) = \mathscr{H}_{\mathrm{ph}}, \tag{1.4}$$
$$\mathscr{H}(1,0,0,0,1) = \mathscr{H}_{\mathrm{el}} \otimes \mathscr{H}_{\mathrm{ph}}, \tag{1.5}$$
$$\mathscr{H}(2,0,0,0,0) = \mathscr{H}_{\mathrm{el}} \otimes_{\mathrm{asym}} \mathscr{H}_{\mathrm{el}}, \tag{1.6}$$
$$\mathscr{H}(0,0,0,0,2) = \mathscr{H}_{\mathrm{ph}} \otimes_{\mathrm{sym}} \mathscr{H}_{\mathrm{ph}}, \tag{1.7}$$
$$\mathscr{H}(1,0,0,0,2) = \mathscr{H}_{\mathrm{el}} \otimes (\mathscr{H}_{\mathrm{ph}} \otimes_{\mathrm{sym}} \mathscr{H}_{\mathrm{ph}}), \tag{1.8}$$
$$\mathscr{H}(2,0,0,0,1) = (\mathscr{H}_{\mathrm{el}} \otimes_{\mathrm{asym}} \mathscr{H}_{\mathrm{el}}) \otimes \mathscr{H}_{\mathrm{ph}}, \tag{1.9}$$
$$\mathscr{H}(2,0,0,0,2) = (\mathscr{H}_{\mathrm{el}} \otimes_{\mathrm{asym}} \mathscr{H}_{\mathrm{el}}) \otimes (\mathscr{H}_{\mathrm{ph}} \otimes_{\mathrm{sym}} \mathscr{H}_{\mathrm{ph}}). \tag{1.10}$$

$$\ldots$$

2 See Subsection **1**-6.1.3. Note that electrons and protons are fermions, while photons are bosons.

1.1.2 Particle observables in Fock space

As explained in Subsection **1**-6.1.2, in each sector of the Fock space we can define observables of individual particles populating this sector, i. e., their positions, momenta, spins, etc. For example, in each (massive) one-particle subspace there is a Newton–Wigner operator describing measurements of the particle's position $\boldsymbol{r}$. In n-particle sectors, in addition to the center-of-energy position $\boldsymbol{R}$, positions $\boldsymbol{r}_i$ of individual particles are defined as well.

In each sector, we can choose a basis of common eigenvectors of a complete set of commuting one-particle observables. For further discussions it will be convenient to use the basis which diagonalizes momenta $\boldsymbol{p}$ and spin components s_z of massive particles or helicities τ of massless particles. For example, basis vectors in the two-electron sector $\mathscr{H}(2,0,0,0,0) = \mathscr{H}_{\mathrm{el}} \otimes_{\mathrm{asym}} \mathscr{H}_{\mathrm{el}}$ will be denoted as $|\boldsymbol{p}_1 s_{1z}, \boldsymbol{p}_2 s_{2z}\rangle$. Thus, in each sector one can define many-particle wave functions in the momentum–spin representation.

The arbitrary state $|\Psi\rangle$ in the Fock space can have components in many or all sectors.[3] So the number of particles in the state $|\Psi\rangle$ can be undefined, and a complete description of such a state requires the introduction of multi-sector state vectors, which can be expanded in the basis described above.

1.1.3 Noninteracting representation of Poincaré group

The construction given above gives us the Fock space $\mathscr{H}$, where many-particle states and observables of our theory live and where a convenient orthonormal basis is defined. To complete this formalism, we need to construct a realistic interacting representation U_g of the Poincaré group in $\mathscr{H}$. Let us first solve a simpler problem and define a noninteracting representation U_g^0 there.

From Subsection **1**-6.2.1 we already know how to build noninteracting representations of the Poincaré group in each separate sector of $\mathscr{H}$. This is done with the help of the tensor product[4] of one-particle irreducible representations corresponding to electrons U_g^{el}, photons U_g^{ph}, etc. Then, the *noninteracting* representation of the Poincaré group in the entire Fock space is formed as the direct sum of such sector representations. In accordance with the sector decomposition (1.2)–(1.10), we can write

$$U_g^0 = 1 \oplus U_g^{\mathrm{el}} \oplus U_g^{\mathrm{ph}} \oplus (U_g^{\mathrm{el}} \otimes U_g^{\mathrm{ph}}) \oplus (U_g^{\mathrm{el}} \otimes_{\mathrm{asym}} U_g^{\mathrm{el}}) \oplus \cdots . \tag{1.11}$$

3 *Superselection rules* forbid creating linear combinations of states with different charges. We will not discuss these rules here.

4 With the appropriate (anti)symmetrization.

The generators of this representation will be denoted by $\{H_0, \boldsymbol{P}_0, \boldsymbol{J}_0, \boldsymbol{K}_0\}$. In each sector, these generators are simply sums of single-particle generators.[5] As usual, we assume that the operators H_0, $\boldsymbol{P}_0$ and $\boldsymbol{J}_0$ represent the total energy, momentum and angular momentum, respectively.

Here we immediately notice a serious problem, which was not present in quantum mechanics with fixed particle content. For example, according to (1.11), a free Hamiltonian should be represented as a direct sum of sector components, i. e.,

$$H_0 = 0 \oplus H_0(1,0,0,0,0) \oplus H_0(0,0,0,0,1) \oplus H_0(1,0,0,0,1) \oplus \cdots.$$

It is tempting to use the notation from Section **1**-6.2 and express Hamiltonians in each sector through observables of individual particles there: $\boldsymbol{p}_1$, $\boldsymbol{p}_2$, etc. For example, in the one-electron sector $\mathscr{H}(1,0,0,0,0)$, the free Hamiltonian is equal to

$$H_0(1,0,0,0,0) = \sqrt{m_e^2c^4 + p^2c^2} \tag{1.12}$$

and the Hamiltonian in the sector $\mathscr{H}(2,0,0,0,2)$ is[6]

$$H_0(2,0,0,0,2) = p_1c + p_2c + \sqrt{m_e^2c^4 + p_3^2c^2} + \sqrt{m_e^2c^4 + p_4^2c^2}. \tag{1.13}$$

Obviously, such a notation is very cumbersome, because it does not give a single expression for the operator H_0 in the entire Fock space. Moreover, it is completely unclear how to use the single-particle observables for constructing operators of interactions that change the number of particles, i. e., moving state vectors across sector boundaries. We need to find a simple and universal method for writing operators in the Fock space. This problem is solved by introducing creation and annihilation operators.

1.1.4 Creation and annihilation operators for fermions

To begin with, it will be useful to consider the simpler case of a discrete spectrum of momentum. In theory, such a spectrum can be produced by standard methods of placing the system in an impenetrable box or using periodic boundary conditions. Then the eigenvalues of the momentum operator form a discrete three-dimensional lattice $\boldsymbol{p}_i$. In the limit of infinite box size, the usual continuous momentum spectrum is restored.

First turn to creation and annihilation operators for electrons. We define a (linear) *creation operator* $a^\dagger_{\boldsymbol{p}s_z}$ of an electron with momentum $\boldsymbol{p}$ and spin projection s_z by its action on basis vectors

$$|\boldsymbol{p}_1 s_{1z}, \boldsymbol{p}_2 s_{2z}, \ldots, \boldsymbol{p}_n s_{nz}\rangle. \tag{1.14}$$

5 For example, equations (**1**-6.10)–(**1**-6.13) are valid in each two-particle sector.

6 Two photons are labeled by indices 1 and 2, two electrons by indices 3 and 4.

in sectors with n electrons. We should distinguish two alternatives. In the first case, the created one-particle state $(\boldsymbol{p}s_z)$ is among the states present in (1.14), for example, $(\boldsymbol{p}s_z) = (\boldsymbol{p}_i s_{iz})$. Since electrons are fermions and two fermions cannot occupy the same state due to the Pauli principle, this action leads to the zero result, i. e.,

$$a^\dagger_{\boldsymbol{p}s_z} |\boldsymbol{p}_1 s_{1z}, \ldots, \boldsymbol{p}_{i-1} s_{(i-1)z}, \boldsymbol{p}_i s_{iz}, \boldsymbol{p}_{i+1} s_{(i+1)z}, \ldots, \boldsymbol{p}_n s_{nz}\rangle = 0. \tag{1.15}$$

In the second case, the created state $(\boldsymbol{p}s_z)$ is not among the single-particle states forming (1.14). Then, the creation operator $a^\dagger_{\boldsymbol{p}s_z}$ simply adds one electron to the beginning of the particle list, so

$$a^\dagger_{\boldsymbol{p}s_z} |\boldsymbol{p}_1 s_{1z}, \boldsymbol{p}_2 s_{2z}, \ldots, \boldsymbol{p}_n, s_{nz}\rangle \equiv |\boldsymbol{p}s_z, \boldsymbol{p}_1 s_{1z}, \boldsymbol{p}_2 s_{2z}, \ldots, \boldsymbol{p}_n s_{nz}\rangle. \tag{1.16}$$

In this case, the operator $a^\dagger_{\boldsymbol{p}s_z}$ converts a state with n electrons to a state with $n+1$ electrons. By repeatedly applying creation operators to the vacuum vector $|\text{vac}\rangle$, we can construct all basis vectors in the purely electronic part of the Fock space. For example,

$$\begin{aligned} a^\dagger_{\boldsymbol{p}s_z}|\text{vac}\rangle &= |\boldsymbol{p}s_z\rangle, \\ a^\dagger_{\boldsymbol{p}_1 s_{1z}} a^\dagger_{\boldsymbol{p}_2 s_{2z}}|\text{vac}\rangle &= |\boldsymbol{p}_1 s_{1z}, \boldsymbol{p}_2 s_{2z}\rangle \end{aligned} \tag{1.17}$$

are basis vectors in the one-electron and two-electron sectors.

We define the electron *annihilation operator* $a_{\boldsymbol{p}s_z}$ as a Hermitian conjugate to the creation operator $a^\dagger_{\boldsymbol{p}s_z}$. One can prove [21] that the action of $a_{\boldsymbol{p}s_z}$ on the n-electron state (1.14) is as follows. If the one-electron state with parameters $(\boldsymbol{p}s_z)$ is already occupied, for example, $(\boldsymbol{p}s_z) = (\boldsymbol{p}_i s_{iz})$, then this state is "annihilated" and the number of particles in the system decreases by one, i. e.,

$$\begin{aligned} & a_{\boldsymbol{p}s_z} |\boldsymbol{p}_1 s_{1z}, \ldots, \boldsymbol{p}_{i-1} s_{(i-1)z}, \boldsymbol{p}_i s_{iz}, \boldsymbol{p}_{i+1} s_{(i+1)z}, \ldots, \boldsymbol{p}_n s_{nz}\rangle \\ & \quad = (-1)^{\mathcal{P}} |\boldsymbol{p}_1 s_{1z}, \ldots, \boldsymbol{p}_{i-1} s_{(i-1)z}, \boldsymbol{p}_{i+1} s_{(i+1)z}, \ldots, \boldsymbol{p}_n s_{nz}\rangle. \end{aligned} \tag{1.18}$$

Here $\mathcal{P}$ is the number of permutations of neighboring particles, which is necessary to move the annihilated one-particle state i to the first place in the list. If the state $(\boldsymbol{p}s_z)$ is absent, i. e., $(\boldsymbol{p}s_z) \neq (\boldsymbol{p}_i s_{iz})$ for all i, then

$$a_{\boldsymbol{p}s_z} |\boldsymbol{p}_1 s_{1z}, \boldsymbol{p}_2 s_{2z}, \ldots, \boldsymbol{p}_n s_{nz}\rangle = 0. \tag{1.19}$$

Acting on the vacuum state, annihilation operators always yield zero, i. e.,

$$a_{\boldsymbol{p}s_z}|\text{vac}\rangle = 0. \tag{1.20}$$

The above formulas define the actions of creation and annihilation operators on the basis vectors in purely electronic sectors. These rules do not change in the presence of other particles, and they extend to any linear combinations of basis vectors by linearity. Creation and annihilation operators for other fermions – positrons, protons and antiprotons – are defined similarly.

For brevity, we will call the creation and annihilation operators jointly *particle operators*. In this way we will distinguish them from *particle observables*, such as momentum $\boldsymbol{p}_i$, position $\boldsymbol{r}_i$, energy h_i, etc. It should be emphasized that the (creation and annihilation) particle operators are not intended to directly describe any physical process or quantity. They are only formal mathematical objects intended to simplify the notation for working with other operators having direct physical meanings. Some examples will be provided in Subsection 1.1.10.

1.1.5 Anticommutators of particle operators

In practical calculations, we often encounter *anticommutators* of fermion particle operators. First we consider the case of annihilation/creation of unequal states of particles, such as $(\boldsymbol{p}s_z) \neq (\boldsymbol{p}'s'_z)$. In this case, the anticommutator is

$$\{a_{\boldsymbol{p}'s'_z}, a^\dagger_{\boldsymbol{p}s_z}\} \equiv a^\dagger_{\boldsymbol{p}s_z} a_{\boldsymbol{p}'s'_z} + a_{\boldsymbol{p}'s'_z} a^\dagger_{\boldsymbol{p}s_z}.$$

Acting by this operator on a one-particle state $|\boldsymbol{p}''s''_z\rangle$, which differs from both $|\boldsymbol{p}s_z\rangle$ and $|\boldsymbol{p}'s'_z\rangle$, we get

$$(a^\dagger_{\boldsymbol{p}s_z} a_{\boldsymbol{p}'s'_z} + a_{\boldsymbol{p}'s'_z} a^\dagger_{\boldsymbol{p}s_z})|\boldsymbol{p}''s''_z\rangle = a_{\boldsymbol{p}'s'_z}|\boldsymbol{p}s_z, \boldsymbol{p}''s''_z\rangle = 0.$$

Similarly we obtain

$$(a^\dagger_{\boldsymbol{p}s_z} a_{\boldsymbol{p}'s'_z} + a_{\boldsymbol{p}'s'_z} a^\dagger_{\boldsymbol{p}s_z})|\boldsymbol{p}s_z\rangle = 0,$$
$$(a^\dagger_{\boldsymbol{p}s_z} a_{\boldsymbol{p}'s'_z} + a_{\boldsymbol{p}'s'_z} a^\dagger_{\boldsymbol{p}s_z})|\boldsymbol{p}'s'_z\rangle = a^\dagger_{\boldsymbol{p}s_z}|\text{vac}\rangle + a_{\boldsymbol{p}'s'_z}|\boldsymbol{p}s_z, \boldsymbol{p}'s'_z\rangle = |\boldsymbol{p}s_z\rangle - |\boldsymbol{p}s_z\rangle = 0.$$

It is not difficult to show that the result remains zero when acting on any n-particle state and also on their linear combinations. Thus, we conclude that in the entire Fock space

$$\{a_{\boldsymbol{p}'s'_z}, a^\dagger_{\boldsymbol{p}s_z}\} = 0, \quad \text{if } (\boldsymbol{p}s_z) \neq (\boldsymbol{p}'s'_z).$$

In the case $(\boldsymbol{p}s_z) = (\boldsymbol{p}'s'_z)$ a similar calculation yields

$$\{a^\dagger_{\boldsymbol{p}s_z}, a_{\boldsymbol{p}s_z}\} = 1.$$

Therefore, for all values of $\boldsymbol{p}$, $\boldsymbol{p}'$, s_z and s'_z we can write

$$\{a^\dagger_{\boldsymbol{p}s_z}, a_{\boldsymbol{p}'s'_z}\} = \delta_{\boldsymbol{p},\boldsymbol{p}'}\delta_{s_z s'_z}. \tag{1.21}$$

Using similar arguments, one can show that

$$\{a^\dagger_{\boldsymbol{p}s_z}, a^\dagger_{\boldsymbol{p}'s'_z}\} = \{a_{\boldsymbol{p}s_z}, a_{\boldsymbol{p}'s'_z}\} = 0.$$

1.1.6 Creation and annihilation operators for photons

For photons that are bosons, the properties of creation and annihilation operators differ slightly from the fermion operators described above. Two or more photons can coexist in the same quantum state. Therefore, we determine the action of the photon creation operator $c^\dagger_{\boldsymbol{p}\tau}$[7] on a multi-photon state as

$$c^\dagger_{\boldsymbol{p}\tau}|\boldsymbol{p}_1\tau_1, \boldsymbol{p}_2\tau_2, \dots, \boldsymbol{p}_n\tau_n\rangle = |\boldsymbol{p}\tau, \boldsymbol{p}_1\tau_1, \boldsymbol{p}_2\tau_2, \dots, \boldsymbol{p}_n\tau_n\rangle,$$

regardless of whether there was a particle $(\boldsymbol{p}\tau)$ in the initial state or not. As in the case of fermions, boson annihilation operators $c_{\boldsymbol{p}\tau}$ are defined as Hermitian conjugates of the creation operators. The photon annihilation operator $c_{\boldsymbol{p}\tau}$ completely destroys a multi-photon state, so

$$c_{\boldsymbol{p}\tau}|\boldsymbol{p}_1\tau_1, \boldsymbol{p}_2\tau_2, \dots, \boldsymbol{p}_n\tau_n\rangle = 0$$

if the annihilated one-photon state $(\boldsymbol{p}\tau)$ was absent there. If the photon $(\boldsymbol{p}\tau)$ was present, then the annihilation operator $c_{\boldsymbol{p},\tau}$ simply removes this component, thus generating an $(n-1)$-photon state,

$$\begin{aligned} c_{\boldsymbol{p}_i\tau_i}&|\boldsymbol{p}_1\tau_1, \dots, \boldsymbol{p}_{i-1}\tau_{i-1}, \boldsymbol{p}_i\tau_i, \boldsymbol{p}_{i+1}\tau_{i+1}, \dots, \boldsymbol{p}_n\tau_n\rangle \\ &= |\boldsymbol{p}_1\tau_1, \dots, \boldsymbol{p}_{i-1}\tau_{i-1}, \boldsymbol{p}_{i+1}\tau_{i+1}, \dots, \boldsymbol{p}_n\tau_n\rangle. \end{aligned}$$

The above formulas can be extended without change to states where, in addition to photons, other particles are also present. Also, the action of operators extends by linearity to superpositions of basis vectors. From these rules, proceeding in analogy with Subsection 1.1.5, we obtain the following *commutation* relations for the photon annihilation and creation operators:

$$\begin{aligned} [c_{\boldsymbol{p}\tau}, c^\dagger_{\boldsymbol{p}'\tau'}] &= \delta_{\boldsymbol{p},\boldsymbol{p}'}\delta_{\tau\tau'}, \\ [c_{\boldsymbol{p}\tau}, c_{\boldsymbol{p}'\tau'}] &= [c^\dagger_{\boldsymbol{p}\tau}, c^\dagger_{\boldsymbol{p}'\tau'}] = 0. \end{aligned}$$

1.1.7 Particle number operators

With the help of creation and annihilation operators, we can build explicit expressions for various useful observables in the Fock space. Consider, for example, the product of two photon operators,

$$N_{\boldsymbol{p}\tau} = c^\dagger_{\boldsymbol{p}\tau}c_{\boldsymbol{p}\tau}. \tag{1.22}$$

---

7 The photon's momentum is $\boldsymbol{p}$ and τ is its helicity.

Acting on the state of two photons with quantum numbers $(\boldsymbol{p}\tau)$, this operator gives

$$\begin{aligned} N_{\boldsymbol{p}\tau}|\boldsymbol{p}\tau,\boldsymbol{p}\tau\rangle &= N_{\boldsymbol{p}\tau}c^{\dagger}_{\boldsymbol{p}\tau}c^{\dagger}_{\boldsymbol{p}\tau}|\text{vac}\rangle = c^{\dagger}_{\boldsymbol{p}\tau}c_{\boldsymbol{p}\tau}c^{\dagger}_{\boldsymbol{p}\tau}c^{\dagger}_{\boldsymbol{p}\tau}|\text{vac}\rangle \\ &= c^{\dagger}_{\boldsymbol{p}\tau}c^{\dagger}_{\boldsymbol{p}\tau}c_{\boldsymbol{p}\tau}c^{\dagger}_{\boldsymbol{p}\tau}|\text{vac}\rangle + c^{\dagger}_{\boldsymbol{p}\tau}c^{\dagger}_{\boldsymbol{p}\tau}|\text{vac}\rangle \\ &= c^{\dagger}_{\boldsymbol{p}\tau}c^{\dagger}_{\boldsymbol{p}\tau}c^{\dagger}_{\boldsymbol{p}\tau}c_{\boldsymbol{p}\tau}|\text{vac}\rangle + 2c^{\dagger}_{\boldsymbol{p}\tau}c^{\dagger}_{\boldsymbol{p}\tau}|\text{vac}\rangle \\ &= 2|\boldsymbol{p}\tau,\boldsymbol{p}\tau\rangle, \end{aligned}$$

but acting on the state $|\boldsymbol{p}\tau,\boldsymbol{p}'\tau'\rangle$, we get

$$\begin{aligned} N_{\boldsymbol{p}\tau}|\boldsymbol{p}\tau,\boldsymbol{p}'\tau'\rangle &= N_{\boldsymbol{p}\tau}c^{\dagger}_{\boldsymbol{p}\tau}c^{\dagger}_{\boldsymbol{p}'\tau'}|\text{vac}\rangle = c^{\dagger}_{\boldsymbol{p}\tau}c_{\boldsymbol{p}\tau}c^{\dagger}_{\boldsymbol{p}\tau}c^{\dagger}_{\boldsymbol{p}'\tau'}|\text{vac}\rangle \\ &= c^{\dagger}_{\boldsymbol{p}\tau}c^{\dagger}_{\boldsymbol{p}\tau}c_{\boldsymbol{p}\tau}c^{\dagger}_{\boldsymbol{p}'\tau'}|\text{vac}\rangle + c^{\dagger}_{\boldsymbol{p}\tau}c^{\dagger}_{\boldsymbol{p}'\tau'}|\text{vac}\rangle \\ &= c^{\dagger}_{\boldsymbol{p}\tau}c^{\dagger}_{\boldsymbol{p}\tau}c^{\dagger}_{\boldsymbol{p}'\tau'}c_{\boldsymbol{p}\tau}|\text{vac}\rangle + c^{\dagger}_{\boldsymbol{p}\tau}c^{\dagger}_{\boldsymbol{p}'\tau'}|\text{vac}\rangle \\ &= |\boldsymbol{p}\tau,\boldsymbol{p}'\tau'\rangle. \end{aligned}$$

These examples should convince us that the operator $N_{\boldsymbol{p}\tau}$ acts as a counter of photons with quantum numbers $(\boldsymbol{p}\tau)$.

1.1.8 Continuous spectrum of momentum

The properties of creation and annihilation operators presented in the previous subsections were derived for the case of discrete momentum spectrum. In reality, the momentum spectrum is continuous, and these results must be modified by taking the limit of a "very large box." It is not difficult to guess that in this limit equation (1.21) goes into

$$\{a_{\boldsymbol{p}'s'_z}, a^{\dagger}_{\boldsymbol{p}s_z}\} = \delta_{s_z s'_z}\delta(\boldsymbol{p}-\boldsymbol{p}'). \tag{1.23}$$

The sequence of formulas[8]

$$\begin{aligned} \delta_{s_z s'_z}\delta(\boldsymbol{p}-\boldsymbol{p}') &= \langle \boldsymbol{p}s_z|\boldsymbol{p}'s'_z\rangle = \langle\text{vac}|a_{\boldsymbol{p}s_z}a^{\dagger}_{\boldsymbol{p}'s'_z}|\text{vac}\rangle \\ &= -\langle\text{vac}|a^{\dagger}_{\boldsymbol{p}'s'_z}a_{\boldsymbol{p}s_z}|\text{vac}\rangle + \delta_{s_z s'_z}\delta(\boldsymbol{p}-\boldsymbol{p}') \\ &= \delta_{s_z s'_z}\delta(\boldsymbol{p}-\boldsymbol{p}') \end{aligned}$$

confirms the consistency of our choice (1.23).

8 The first equality is obtained from the normalization of momentum eigenvectors (**1**-5.21); the second equality follows from the definition of the creation operator (1.17); the third one from formula (1.23); and the fourth one from (1.20).

The same arguments can be applied to the operators of positrons ($b_{\boldsymbol{p}s_z}$ and $b^\dagger_{\boldsymbol{p}s_z}$), protons ($d_{\boldsymbol{p}s_z}$ and $d^\dagger_{\boldsymbol{p}s_z}$), antiprotons ($f_{\boldsymbol{p}s_z}$ and $f^\dagger_{\boldsymbol{p}s_z}$) and photons ($c_{\boldsymbol{p}\tau}$ and $c^\dagger_{\boldsymbol{p}\tau}$). So, finally, we get the following set of anticommutation and commutation relations relevant to QED:

$$\{a_{\boldsymbol{p}s_z}, a^\dagger_{\boldsymbol{p}'s'_z}\} = \{b_{\boldsymbol{p}s_z}, b^\dagger_{\boldsymbol{p}'s'_z}\} = \{d_{\boldsymbol{p}s_z}, d^\dagger_{\boldsymbol{p}'s'_z}\} = \{f_{\boldsymbol{p}s_z}, f^\dagger_{\boldsymbol{p}'s'_z}\} = \delta(\boldsymbol{p}-\boldsymbol{p}')\delta_{s_z s'_z}, \tag{1.24}$$

$$\begin{aligned}\{a_{\boldsymbol{p}s_z}, a_{\boldsymbol{p}'s'_z}\} &= \{b_{\boldsymbol{p}s_z}, b_{\boldsymbol{p}'s'_z}\} = \{d_{\boldsymbol{p}s_z}, d_{\boldsymbol{p}'s'_z}\} = \{f_{\boldsymbol{p}s_z}, f_{\boldsymbol{p}'s'_z}\}\\ &= \{a^\dagger_{\boldsymbol{p}s_z}, a^\dagger_{\boldsymbol{p}'s'_z}\} = \{b^\dagger_{\boldsymbol{p}s_z}, b^\dagger_{\boldsymbol{p}'s'_z}\} = \{d^\dagger_{\boldsymbol{p}s_z}, d^\dagger_{\boldsymbol{p}'s'_z}\}\\ &= \{f^\dagger_{\boldsymbol{p}s_z}, f^\dagger_{\boldsymbol{p}'s'_z}\} = 0,\end{aligned} \tag{1.25}$$

$$[c_{\boldsymbol{p}\tau}, c^\dagger_{\boldsymbol{p}'\tau'}] = \delta(\boldsymbol{p}-\boldsymbol{p}')\delta_{\tau\tau'}, \tag{1.26}$$

$$[c^\dagger_{\boldsymbol{p}\tau}, c^\dagger_{\boldsymbol{p}'\tau'}] = [c_{\boldsymbol{p}\tau}, c_{\boldsymbol{p}'\tau'}] = 0. \tag{1.27}$$

Commutators of operators related to different particles are always zero.

In the limit of continuous momentum, the counterpart of the particle counter (1.22) is the operator

$$\rho_{\boldsymbol{p}\tau} = c^\dagger_{\boldsymbol{p}\tau} c_{\boldsymbol{p}\tau}, \tag{1.28}$$

which can be interpreted as the *density* of photons with helicity τ and momentum $\boldsymbol{p}$. Having summed the density (1.28) by the photon polarizations and integrating it over the entire momentum space, we obtain the operator of the total number of photons in the system

$$N_{\text{ph}} = \sum_\tau \int d\boldsymbol{p} c^\dagger_{\boldsymbol{p}\tau} c_{\boldsymbol{p}\tau}. \tag{1.29}$$

We can also write similar expressions for the numbers of other particles. For example,

$$N_{\text{el}} = \sum_{s_z} \int d\boldsymbol{p} a^\dagger_{\boldsymbol{p}s_z} a_{\boldsymbol{p}s_z} \tag{1.30}$$

is the electron number operator. Then the operator

$$N = N_{\text{el}} + N_{\text{po}} + N_{\text{pr}} + N_{\text{an}} + N_{\text{ph}} \tag{1.31}$$

expresses the total number of particles in the system.

1.1.9 Normal ordering

It is necessary to note the important property of operators (1.29) and (1.30). Being expressed through particle creation and annihilation operators, they are applicable in the entire Fock space. We will follow this principle in our construction of other observables as well.

Thus, we intend to express operators in the Fock space in the form of polynomials in creation and annihilation operators. But for this, we need to overcome one notational problem related to the noncommutativity of particle operators: two different polynomials can represent the same operator. In order to have unified polynomial representatives, we will always agree to write the products of particle operators in the *normal order*, i. e., creation operators to the left and annihilation operators to the right. Using (anti)commutation relations (1.24)–(1.27), we can always convert any product of particle operators into a normally ordered form. Let us illustrate the above with one example. We have

$$\begin{aligned}
a_{\boldsymbol{p}'s_z'}c_{\boldsymbol{q}'\tau'}a^\dagger_{\boldsymbol{p}s_z}c^\dagger_{\boldsymbol{q}\tau} &= a_{\boldsymbol{p}'s_z'}a^\dagger_{\boldsymbol{p}s_z}c_{\boldsymbol{q}'\tau'}c^\dagger_{\boldsymbol{q}\tau} \\
&= (a^\dagger_{\boldsymbol{p}s_z}a_{\boldsymbol{p}'s_z'} + \delta(\boldsymbol{p}-\boldsymbol{p}')\delta_{s_zs_z'})(-c^\dagger_{\boldsymbol{q}\tau}c_{\boldsymbol{q}'\tau'} + \delta(\boldsymbol{q}-\boldsymbol{q}')\delta_{\tau\tau'}) \\
&= -a^\dagger_{\boldsymbol{p}s_z}c^\dagger_{\boldsymbol{q}\tau}a_{\boldsymbol{p}'s_z'}c_{\boldsymbol{q}'\tau'} + a^\dagger_{\boldsymbol{p}s_z}a_{\boldsymbol{p}'s_z'}\delta(\boldsymbol{q}-\boldsymbol{q}')\delta_{\tau\tau'} \\
&\quad - c^\dagger_{\boldsymbol{q}\tau}c_{\boldsymbol{q}'\tau'}\delta(\boldsymbol{p}-\boldsymbol{p}')\delta_{s_zs_z'} + \delta(\boldsymbol{p}-\boldsymbol{p}')\delta_{s_zs_z'}\delta(\boldsymbol{q}-\boldsymbol{q}')\delta_{\tau\tau'},
\end{aligned}$$

where the right-hand side is in a normally ordered form.

As can be seen from this example, the transition to the normal order is accomplished by moving all creation operators[9] $\alpha^\dagger_{\boldsymbol{p}}$ to the leftmost positions. Permutations of operators of different particles have no additional effect. When on its way to the left a creation operator $\alpha^\dagger_{\boldsymbol{p}}$ meets an annihilation operator of the same particle $\alpha_{\boldsymbol{q}}$, two terms appear[10] instead of one $(\alpha_{\boldsymbol{q}}\alpha^\dagger_{\boldsymbol{p}})$. In the first term, the creation operator simply "jumps over" the annihilation operator, leading to the product $\pm(\alpha^\dagger_{\boldsymbol{p}}\alpha_{\boldsymbol{q}})$. In the second term, the two operators *contract*, producing the delta function $\delta(\boldsymbol{p}-\boldsymbol{q})$.

The normal ordering in complex products of particle operators can be very laborious. Here, the celebrated *Wick theorem* comes to the rescue.

Theorem 1.1 (Wick). *When transformed to the normally ordered form, an arbitrary product $abc\cdots$ of particle operators becomes equal to the fully ordered term $:abc\cdots:$*[11] *plus the sum of terms with all possible contractions.*[12] *Each term in this sum includes the factor $(-1)^{\mathcal{P}}$, where $\mathcal{P}$ is the number of permutations of the fermionic operators needed in order to*

9 Here, for brevity, we drop the spin/polarization labels and use symbols $\alpha^\dagger$, α to denote generic particles operators (bosons and fermions).

10 They come from the (anti)commutation relation $\alpha_{\boldsymbol{q}}\alpha^\dagger_{\boldsymbol{p}} = \pm\alpha^\dagger_{\boldsymbol{p}}\alpha_{\boldsymbol{q}} + \delta(\boldsymbol{p}-\boldsymbol{q})$, where the minus (plus) sign refers to fermions (bosons).

11 The $:abc\cdots:$ symbol means that (i) particle operators are rearranged in the normal order and (ii) the resulting operator is multiplied by $(-1)^{\mathcal{P}}$, where $\mathcal{P}$ is the number of permutations of fermionic factors. For example, $:a_{\boldsymbol{p}'}c_{\boldsymbol{q}'}a^\dagger_{\boldsymbol{p}}c^\dagger_{\boldsymbol{q}}: = -a^\dagger_{\boldsymbol{p}}c^\dagger_{\boldsymbol{q}}a_{\boldsymbol{p}'}c_{\boldsymbol{q}'}$.

12 That is, contractions should be written for all pairs appearing in the "wrong" order $\alpha\ldots\alpha^\dagger$.

(i) *put the contracted operators next to each other (i. e., in the $aa^\dagger$ configuration) and*
(ii) *rearrange in the normal order the operators left after all contractions.*

The proof of this theorem can be found in many textbooks on quantum field theory, for instance in [1]. Here we simply illustrate this result by the example of the product of electron operators $a_{q'}a_{p'}a_p^\dagger a_q^\dagger$. According to Wick's theorem, in a normally ordered form, this operator is the sum of the fully ordered product and six contractions[13]:

$$
\begin{aligned}
:a_{q'}a_{p'}a_p^\dagger a_q^\dagger: &\equiv a_p^\dagger a_q^\dagger a_{q'}a_{p'},\\
a_{q'}\wick{\c1 a_{p'}\c1 a_p^\dagger} a_q^\dagger &\equiv -a_q^\dagger a_{q'}\delta(\mathbf{p}-\mathbf{p}'),\\
\wick{\c1 a_{q'}a_{p'}a_p^\dagger \c1 a_q^\dagger} &\equiv -a_p^\dagger a_{p'}\delta(\mathbf{q}-\mathbf{q}'),\\
\wick{\c1 a_{q'}a_{p'}\c1 a_p^\dagger} a_q^\dagger &\equiv a_q^\dagger a_{p'}\delta(\mathbf{p}-\mathbf{q}'),\\
a_{q'}\wick{\c1 a_{p'}a_p^\dagger \c1 a_q^\dagger} &\equiv a_p^\dagger a_{q'}\delta(\mathbf{q}-\mathbf{p}'),\\
\wick{\c2 a_{q'}\c1 a_{p'}\c1 a_p^\dagger \c2 a_q^\dagger} &\equiv \delta(\mathbf{p}-\mathbf{p}')\delta(\mathbf{q}-\mathbf{q}'),\\
\wick{\c1 a_{q'}\c2 a_{p'}\c1 a_p^\dagger \c2 a_q^\dagger} &\equiv -\delta(\mathbf{p}-\mathbf{q}')\delta(\mathbf{q}-\mathbf{p}').
\end{aligned}
$$

1.1.10 Noninteracting energy and momentum

Now we can fully appreciate the benefits of introducing creation and annihilation operators. In particular, with their help it is easy to obtain a compact expression for the noninteracting Hamiltonian H_0. It is obtained simply from the particle number operator (1.31), multiplying the integrands (particle densities in the momentum space) by the energies of free particles, i. e.,

$$H_0 = H_0^{\text{el+po}} + H_0^{\text{pr+an}} + H_0^{\text{ph}}, \tag{1.32}$$

$$
\begin{aligned}
H_0^{\text{el+po}} &= \int d\mathbf{p}\,\omega_{\mathbf{p}} \sum_{s_z=\pm 1/2} [a_{\mathbf{p}s_z}^\dagger a_{\mathbf{p}s_z} + b_{\mathbf{p}s_z}^\dagger b_{\mathbf{p}s_z}],\\
H_0^{\text{pr+an}} &= \int d\mathbf{p}\,\Omega_{\mathbf{p}} \sum_{s_z=\pm 1/2} [d_{\mathbf{p}s_z}^\dagger d_{\mathbf{p}s_z} + f_{\mathbf{p}s_z}^\dagger f_{\mathbf{p}s_z}],\\
H_0^{\text{ph}} &= c\int d\mathbf{p}\,p \sum_{\tau=\pm 1} c_{\mathbf{p}\tau}^\dagger c_{\mathbf{p}\tau}.
\end{aligned}
\tag{1.33}
$$

Here $\omega_{\mathbf{p}} = \sqrt{m_e^2c^4 + p^2c^2}$ are energies of free electrons and positrons, $\Omega_{\mathbf{p}} = \sqrt{m_p^2c^4 + p^2c^2}$ are energies of free protons and antiprotons and cp are energies of

13 Contracted pairs of operators are marked with overline signs.

free photons. It is not difficult to verify that H_0 in (1.32) acts on states from the sector $\mathcal{H}(1,0,0,0,0)$ in the same way as equation (1.12) and H_0 acts in the sector $\mathcal{H}(2,0,0,0,2)$ exactly as (1.13). So, we got a single expression for the energy that works equally well in all sectors of the Fock space.[14] Similar arguments show that the operator

$$\begin{aligned} \boldsymbol{P}_0 &= \boldsymbol{P}_0^{\text{el+po}} + \boldsymbol{P}_0^{\text{pr+an}} + \boldsymbol{P}_0^{\text{ph}}, \qquad (1.34)\\ \boldsymbol{P}_0^{\text{el+po}} &= \int d\boldsymbol{p}\boldsymbol{p} \sum_{s_z=\pm 1/2} [a^\dagger_{\boldsymbol{p}s_z} a_{\boldsymbol{p}s_z} + b^\dagger_{\boldsymbol{p}s_z} b_{\boldsymbol{p}s_z}],\\ \boldsymbol{P}_0^{\text{pr+an}} &= \int d\boldsymbol{p}\boldsymbol{p} \sum_{s_z=\pm 1/2} [d^\dagger_{\boldsymbol{p}s_z} d_{\boldsymbol{p}s_z} + f^\dagger_{\boldsymbol{p}s_z} f_{\boldsymbol{p}s_z}],\\ \boldsymbol{P}_0^{\text{ph}} &= \int d\boldsymbol{p}\boldsymbol{p} \sum_{\tau=\pm 1} c^\dagger_{\boldsymbol{p}\tau} c_{\boldsymbol{p}\tau} \qquad (1.35) \end{aligned}$$

has the meaning of the total momentum.

1.1.11 Noninteracting angular momentum and boost

Expressions for the generators $\boldsymbol{J}_0$ and $\boldsymbol{K}_0$ in the Fock space are more complicated, since they require the participation of derivatives of particle operators. For illustration, consider the example of a massive spinless particle with creation and annihilation operators $\alpha^\dagger_{\boldsymbol{p}}$ and $\alpha_{\boldsymbol{p}}$, respectively. The effect of the rotation $e^{-\frac{i}{\hbar}J_{0z}\varphi}$ on the one-particle state $|\boldsymbol{p}\rangle$ (see (**1**-5.10))

$$e^{-\frac{i}{\hbar}J_{0z}\varphi}|\boldsymbol{p}\rangle = |p_x \cos\varphi + p_y \sin\varphi, p_y \cos\varphi - p_x \sin\varphi, p_z\rangle \equiv |\boldsymbol{\varphi p}\rangle$$

can be imagined as annihilation of the initial state $|\boldsymbol{p}\rangle = |p_x, p_y, p_z\rangle$ followed by creation of the rotated state $|\boldsymbol{\varphi p}\rangle$, i. e.,

$$e^{-\frac{i}{\hbar}J_{0z}\varphi}|\boldsymbol{p}\rangle = \alpha^\dagger_{\boldsymbol{\varphi p}}\alpha_{\boldsymbol{p}}|p_x, p_y, p_z\rangle.$$

Therefore, for an arbitrary one-particle state the operator of finite rotation has the form

$$e^{-\frac{i}{\hbar}J_{0z}\varphi} = \int d\boldsymbol{p}\alpha^\dagger_{\boldsymbol{\varphi p}}\alpha_{\boldsymbol{p}}. \qquad (1.36)$$

It is not difficult to see that the same form is valid in the entire Fock space. Then the explicit expression for the generator J_{0z} is obtained by taking the derivative of (1.36)

14 Note that our expression for the energy does not contain the problematic infinite term (so-called vacuum energy) that is typical for approaches based on quantum fields; see, for example, formula (2.31) in [10].

with respect to φ,

$$J_{0z} = i\hbar \lim_{\varphi \to 0} \frac{d}{d\varphi} e^{-\frac{i}{\hbar} J_{0z}\varphi}$$
$$= i\hbar \lim_{\varphi \to 0} \frac{d}{d\varphi} \int d\boldsymbol{p} \alpha^\dagger_{p_x \cos\varphi + p_y \sin\varphi, p_y \cos\varphi - p_x \sin\varphi, p_z} \alpha_{\boldsymbol{p}}$$
$$= i\hbar \int d\boldsymbol{p} \left(p_y \frac{\partial \alpha^\dagger_{\boldsymbol{p}}}{\partial p_x} - p_x \frac{\partial \alpha^\dagger_{\boldsymbol{p}}}{\partial p_y} \right) \alpha_{\boldsymbol{p}}. \tag{1.37}$$

The action of a boost along the z-axis is obtained from (**1**-5.30) and (**1**-5.11). We have

$$e^{-\frac{ic}{\hbar} K_{0z}\theta} |\boldsymbol{p}\rangle = \sqrt{\frac{\omega_{\boldsymbol{\theta p}}}{\omega_{\boldsymbol{p}}}} |\boldsymbol{\theta p}\rangle, \tag{1.38}$$

where the rapidity vector is $\boldsymbol{\theta} = (0, 0, \theta)$. This transformation can be represented as annihilation of the state $|\boldsymbol{p}\rangle = |p_x, p_y, p_z\rangle$ and then creation of the state (1.38):

$$e^{-\frac{ic}{\hbar} K_{0z}\theta} |\boldsymbol{p}\rangle = \sqrt{\frac{\omega_{\boldsymbol{\theta p}}}{\omega_{\boldsymbol{p}}}} \alpha^\dagger_{\boldsymbol{\theta p}} \alpha_{\boldsymbol{p}} |\boldsymbol{p}\rangle.$$

Thus, for all states in the Fock space the finite boost operator is

$$e^{-\frac{ic}{\hbar} K_{0z}\theta} = \int d\boldsymbol{p} \sqrt{\frac{\omega_{\boldsymbol{\theta p}}}{\omega_{\boldsymbol{p}}}} \alpha^\dagger_{\boldsymbol{\theta p}} \alpha_{\boldsymbol{p}}.$$

The explicit formula for K_{0z} is obtained by taking the derivative of this expression with respect to θ,

$$K_{0z} = \frac{i\hbar}{c} \lim_{\theta \to 0} \frac{d}{d\theta} e^{-\frac{ic}{\hbar} K_{0z}\theta}$$
$$= \frac{i\hbar}{c} \lim_{\theta \to 0} \frac{d}{d\theta} \int d\boldsymbol{p} \sqrt{\frac{\omega_{\boldsymbol{p}} \cosh\theta + c p_z \sinh\theta}{\omega_{\boldsymbol{p}}}} \alpha^\dagger_{p_x, p_y, p_z \cosh\theta + \omega_{\boldsymbol{p}} \cosh\theta} \alpha_{\boldsymbol{p}}$$
$$= i\hbar \int d\boldsymbol{p} \left(\frac{p_z}{2\omega_{\boldsymbol{p}}} \alpha^\dagger_{\boldsymbol{p}} \alpha_{\boldsymbol{p}} + \frac{\omega_{\boldsymbol{p}}}{c^2} \frac{\partial \alpha^\dagger_{\boldsymbol{p}}}{\partial p_z} \alpha_{\boldsymbol{p}} \right). \tag{1.39}$$

Similar derivations can be done for other components of $\boldsymbol{J}_0$ and $\boldsymbol{K}_0$.

1.1.12 Poincaré transformations of particle operators

Having defined all ten generators $\{H_0, \boldsymbol{P}_0, \boldsymbol{J}_0, \boldsymbol{K}_0\}$ we secured the noninteracting representation

$$U_0(\boldsymbol{\theta}; \boldsymbol{\varphi}; \boldsymbol{r}; t) \equiv e^{-\frac{ic}{\hbar} \boldsymbol{K}_0 \cdot \boldsymbol{\theta}} e^{-\frac{i}{\hbar} \boldsymbol{J}_0 \cdot \boldsymbol{\varphi}} e^{-\frac{i}{\hbar} \boldsymbol{P}_0 \cdot \boldsymbol{r}} e^{\frac{i}{\hbar} H_0 t} \tag{1.40}$$

of the Poincaré group in the Fock space. By construction, this representation induces transformations (**1**-5.8)–(**1**-5.10), (**1**-5.30) of one-particle states. From this, it is not difficult to find out how creation and annihilation operators transform under the action of (1.40).

As an example, consider the boost transformation. For the electron creation operators, we get[15]

$$e^{-\frac{ic}{\hbar}\boldsymbol{K}_0\cdot\boldsymbol{\theta}} a^{\dagger}_{\boldsymbol{p}s_z} e^{\frac{ic}{\hbar}\boldsymbol{K}_0\cdot\boldsymbol{\theta}} |\mathrm{vac}\rangle = e^{-\frac{ic}{\hbar}\boldsymbol{K}_0\cdot\boldsymbol{\theta}} a^{\dagger}_{\boldsymbol{p}s_z} |\mathrm{vac}\rangle = e^{-\frac{ic}{\hbar}\boldsymbol{K}_0\cdot\boldsymbol{\theta}} |\boldsymbol{p}s_z\rangle$$
$$= \sqrt{\frac{\omega_{\boldsymbol{\theta p}}}{\omega_{\boldsymbol{p}}}} \sum_{s'_z} \mathscr{D}^{1/2}_{s'_z s_z}(\boldsymbol{\varphi}_W(\boldsymbol{p},\boldsymbol{\theta})) |(\boldsymbol{\theta p})s'_z\rangle$$
$$= \sqrt{\frac{\omega_{\boldsymbol{\theta p}}}{\omega_{\boldsymbol{p}}}} \sum_{s'_z} \mathscr{D}^{1/2}_{s'_z s_z}(\boldsymbol{\varphi}_W(\boldsymbol{p},\boldsymbol{\theta})) a^{\dagger}_{(\boldsymbol{\theta p})s'_z} |\mathrm{vac}\rangle.$$

Therefore[16]

$$e^{-\frac{ic}{\hbar}\boldsymbol{K}_0\cdot\boldsymbol{\theta}} a^{\dagger}_{\boldsymbol{p}s_z} e^{\frac{ic}{\hbar}\boldsymbol{K}_0\cdot\boldsymbol{\theta}} = \sqrt{\frac{\omega_{\boldsymbol{\theta p}}}{\omega_{\boldsymbol{p}}}} \sum_{s'_z} \mathscr{D}^{1/2}_{s'_z s_z}(\boldsymbol{\varphi}_W(\boldsymbol{p},\boldsymbol{\theta})) a^{\dagger}_{(\boldsymbol{\theta p})s'_z}$$
$$= \sqrt{\frac{\omega_{\boldsymbol{\theta p}}}{\omega_{\boldsymbol{p}}}} \sum_{s'_z} \mathscr{D}^{1/2*}_{s_z s'_z}(-\boldsymbol{\varphi}_W(\boldsymbol{p},\boldsymbol{\theta})) a^{\dagger}_{(\boldsymbol{\theta p})s'_z}. \tag{1.41}$$

The transformation law for annihilation operators is obtained by the Hermitian conjugation of (1.41),

$$e^{-\frac{ic}{\hbar}\boldsymbol{K}_0\cdot\boldsymbol{\theta}} a_{\boldsymbol{p}s_z} e^{\frac{ic}{\hbar}\boldsymbol{K}_0\cdot\boldsymbol{\theta}} = \sqrt{\frac{\omega_{\boldsymbol{\theta p}}}{\omega_{\boldsymbol{p}}}} \sum_{s'_z} \mathscr{D}^{1/2}_{s_z s'_z}(-\boldsymbol{\varphi}_W(\boldsymbol{p},\boldsymbol{\theta})) a_{(\boldsymbol{\theta p})s'_z}. \tag{1.42}$$

Actions of rotations and translations are derived in a similar way. We have

$$e^{-\frac{i}{\hbar}\boldsymbol{J}_0\cdot\boldsymbol{\varphi}} a^{\dagger}_{\boldsymbol{p}s_z} e^{\frac{i}{\hbar}\boldsymbol{J}_0\cdot\boldsymbol{\varphi}} = \sum_{s'_z} \mathscr{D}^{1/2*}_{s_z s'_z}(-\boldsymbol{\varphi}) a^{\dagger}_{(\boldsymbol{\varphi p})s'_z}, \tag{1.43}$$
$$e^{-\frac{i}{\hbar}\boldsymbol{J}_0\cdot\boldsymbol{\varphi}} a_{\boldsymbol{p}s_z} e^{\frac{i}{\hbar}\boldsymbol{J}_0\cdot\boldsymbol{\varphi}} = \sum_{s'_z} \mathscr{D}^{1/2}_{s_z s'_z}(-\boldsymbol{\varphi}) a_{(\boldsymbol{\varphi p})s'_z}, \tag{1.44}$$
$$e^{-\frac{i}{\hbar}\boldsymbol{P}_0\cdot\boldsymbol{r}} e^{\frac{i}{\hbar}H_0 t} a^{\dagger}_{\boldsymbol{p}s_z} e^{-\frac{i}{\hbar}H_0 t} e^{\frac{i}{\hbar}\boldsymbol{P}_0\cdot\boldsymbol{r}} = e^{-\frac{i}{\hbar}\boldsymbol{p}\cdot\boldsymbol{r}} e^{\frac{i}{\hbar}\omega_{\boldsymbol{p}} t} a^{\dagger}_{\boldsymbol{p}s_z}, \tag{1.45}$$
$$e^{-\frac{i}{\hbar}\boldsymbol{P}_0\cdot\boldsymbol{r}} e^{\frac{i}{\hbar}H_0 t} a_{\boldsymbol{p}s_z} e^{-\frac{i}{\hbar}H_0 t} e^{\frac{i}{\hbar}\boldsymbol{P}_0\cdot\boldsymbol{r}} = e^{\frac{i}{\hbar}\boldsymbol{p}\cdot\boldsymbol{r}} e^{-\frac{i}{\hbar}\omega_{\boldsymbol{p}} t} a_{\boldsymbol{p}s_z}. \tag{1.46}$$

---

15 Here we took into account the fact that the vacuum vector is invariant with respect to U_0 and used equation (**1**-5.30), where the Wigner angle $\boldsymbol{\varphi}_W(\boldsymbol{p},\boldsymbol{\theta})$ is defined by formula (**1**-5.18).

16 We took into account that for unitary representatives of rotations $\mathscr{D}^{1/2\,T*}(-\boldsymbol{\varphi}) \equiv \mathscr{D}^{1/2\dagger}(-\boldsymbol{\varphi}) = [\mathscr{D}^{1/2}(-\boldsymbol{\varphi})]^{-1} = \mathscr{D}^{1/2}(\boldsymbol{\varphi})$.

Transformations of photon operators are obtained from equation (**1**-5.69); we have

$$U_0(\Lambda;\boldsymbol{r};t)c^{\dagger}_{\boldsymbol{p}\tau}U_0^{-1}(\Lambda;\boldsymbol{r};t) = \sqrt{\frac{|\Lambda\boldsymbol{p}|}{p}}e^{-\frac{i}{\hbar}(\boldsymbol{p}\cdot\boldsymbol{r})+\frac{ic}{\hbar}pt}e^{i\tau\varphi_W(\boldsymbol{p},\Lambda)}c^{\dagger}_{(\Lambda\boldsymbol{p})\tau}, \tag{1.47}$$

$$U_0(\Lambda;\boldsymbol{r};t)c_{\boldsymbol{p}\tau}U_0^{-1}(\Lambda;\boldsymbol{r};t) = \sqrt{\frac{|\Lambda\boldsymbol{p}|}{p}}e^{\frac{i}{\hbar}(\boldsymbol{p}\cdot\boldsymbol{r})-\frac{ic}{\hbar}pt}e^{-i\tau\varphi_W(\boldsymbol{p},\Lambda)}c_{(\Lambda\boldsymbol{p})\tau}. \tag{1.48}$$

1.2 Interaction potentials

We would like to learn how to calculate the S-operator in QED, that is, the quantity most directly comparable with the experiment. Formulas derived in Section **1**-7.1 tell us that in order to achieve this goal, we need to know the interacting part V of the total Hamiltonian

$$H = H_0 + V.$$

The potential energy V in QED will be explicitly formulated only in Section 3.1. In the meantime, we will be interested in general properties of interactions and S-operators in the Fock space. In particular, we will try to find the limitations imposed on the choice of the operator V by a number of physical principles, such as conservation laws and cluster separability.

Note that in our approach we postulate that the interaction V has no effect on the structure of the state space (Fock space). All the properties of this space[17] defined in the noninteracting case remain true also in the presence of interactions. In Chapter 4 we will explain that even the necessity of renormalization will not force us to change the parameters (e. g., masses) of the particles from which the Fock space is constructed. In this respect, our approach differs from the axiomatic or constructive quantum field theory, where the Hilbert space of states has a non-Fock structure that depends on interactions. For more discussions see Volume 3.

1.2.1 Conservation laws

From the experiment, we know that electromagnetic interactions obey certain important constraints, which are called *conservation laws*. An observable F is referred to as *conserved* if it remains unchanged during the time evolution, i. e.,

$$F(t) \equiv e^{\frac{i}{\hbar}Ht}F(0)e^{-\frac{i}{\hbar}Ht} = F(0).$$

17 The inner product, the mutual orthogonality of n-particle sectors, the form of the particle number operators, etc.

It then follows that operators of conserved observables commute with the Hamiltonian $[F, H] = [F, H_0 + V] = 0$, which imposes some restrictions on the interaction operator V. For example, in the instant form of dynamics adopted in our book, the conservation of the total momentum and the total angular momentum means that[18]

$$[V, \boldsymbol{P}_0] = 0, \tag{1.49}$$

$$[V, \boldsymbol{J}_0] = 0. \tag{1.50}$$

It is also known that electromagnetic interactions conserve the *lepton charge*.[19] Therefore, $H = H_0 + V$ must commute with the lepton number operator

$$N_L = N_{\text{el}} - N_{\text{po}} = \sum_{s_z} \int d\boldsymbol{p}(a^\dagger_{\boldsymbol{p}s_z} a_{\boldsymbol{p}s_z} - b^\dagger_{\boldsymbol{p}s_z} b_{\boldsymbol{p}s_z}). \tag{1.51}$$

Since H_0 already commutes with N_L, we get

$$[V, N_L] = 0. \tag{1.52}$$

In addition, all known interactions preserve the *baryon charge*,[20] i. e.,

$$N_B = N_{\text{pr}} - N_{\text{an}} = \sum_{s_z} \int d\boldsymbol{p}(d^\dagger_{\boldsymbol{p}s_z} d_{\boldsymbol{p}s_z} - f^\dagger_{\boldsymbol{p}s_z} f_{\boldsymbol{p}s_z}). \tag{1.53}$$

Hence, V must commute with the baryon number operator, i. e.,

$$[V, N_B] = 0. \tag{1.54}$$

Taking into account that the electrons have a charge of $-e$, that the protons have a charge of $+e$ and that the charge of antiparticles is opposite to the charge of particles, we can introduce the *electric charge* operator

$$\begin{aligned} Q &= e(N_B - N_L) \\ &= e \sum_{s_z} \int d\boldsymbol{p}(b^\dagger_{\boldsymbol{p}s_z} b_{\boldsymbol{p}s_z} - a^\dagger_{\boldsymbol{p}s_z} a_{\boldsymbol{p}s_z} + d^\dagger_{\boldsymbol{p}s_z} d_{\boldsymbol{p}s_z} - f^\dagger_{\boldsymbol{p}s_z} f_{\boldsymbol{p}s_z}) \end{aligned} \tag{1.55}$$

and obtain the law of its conservation,

$$[H, Q] = [V, Q] = e[V, N_B - N_L] = 0, \tag{1.56}$$

from equations (1.52) and (1.54).

---

18 The conservation of energy is a consequence of the trivial equality $[H, H] = 0$.

19 In our case this is the number of electrons minus the number of positrons.

20 In our case this is the number of protons minus the number of antiprotons.

As we have just found out, in QED both operators H_0 and V commute with total momentum $\boldsymbol{P}_0$, total angular momentum $\boldsymbol{J}_0$, lepton charge N_L, baryon charge N_B and electric charge Q. Then, from the formulas in Section **1**-7.1 it follows that the scattering operators F, Σ and S also commute with $\boldsymbol{P}_0$, $\boldsymbol{J}_0$, N_L, N_B and Q. This means that the corresponding observables are conserved in collisions.

Although separately the numbers of particles of a certain type (for example, electrons or protons) may not be conserved, the conservation laws require that charged particles be born and destroyed only together with their antiparticles, i. e., in pairs. The pair production does not occur in low-energy reactions, because such processes require additional energy of $2m_ec^2 = 2\times 0.51\,\text{MeV} = 1.02\,\text{MeV}$ for an electron–positron pair and $2m_pc^2 = 1876.6\,\text{MeV}$ for a proton–antiproton pair. Such high-energy processes can be ignored in classical electrodynamics. However, even in the low-energy limit, it is necessary to take into account the emission of photons. Photons have zero mass, and the energy threshold for their creation is zero. Moreover, photons have zero charges (lepton, baryon and electric), so no conservation laws can limit their creation and destruction. Photons can be created (radiated) and annihilated (absorbed) in any quantities.

1.2.2 General form of interaction operators

The well-known theorem[21] claims that in the Fock space any operator V satisfying the conservation laws (1.49)–(1.50) can be written in the form of a polynomial in creation and annihilation operators,[22] i. e.,

$$V = \sum_{N=0}^{\infty}\sum_{M=0}^{\infty} V_{NM}, \tag{1.57}$$

$$\begin{aligned} V_{NM} = \sum_{\{\eta,\eta'\}} \int & d\boldsymbol{q}'_1 \cdots d\boldsymbol{q}'_N d\boldsymbol{q}_1 \cdots d\boldsymbol{q}_M \\ & \times D_{NM}(\boldsymbol{q}'_1\eta'_1,\ldots,\boldsymbol{q}'_N\eta'_N;\boldsymbol{q}_1\eta_1,\ldots,\boldsymbol{q}_M\eta_M) \\ & \times \delta\left(\sum_{i=1}^{N}\boldsymbol{q}'_i - \sum_{j=1}^{M}\boldsymbol{q}_j\right)\alpha^{\dagger}_{\boldsymbol{q}'_1\eta'_1}\cdots\alpha^{\dagger}_{\boldsymbol{q}'_N\eta'_N}\alpha_{\boldsymbol{q}_1\eta_1}\cdots\alpha_{\boldsymbol{q}_M\eta_M}, \end{aligned} \tag{1.58}$$

where the summation is over all spin/helicity indices η, η' and the integration is carried out over all particle momenta. The individual terms (monomials) V_{NM} in the expansion (1.57) will be called *potentials*. Each potential is a normally ordered product of

21 See p. 175 in [21].

22 Here symbols $\alpha^{\dagger}$, α refer to generic creation–annihilation operators without specifying the particle type.

N creation operators $\alpha^\dagger$ and M annihilation operators α. The pair of nonnegative integers $[N : M]$ will be called the *index* of the potential V_{NM}. A potential is called *bosonic* if it has an even number of fermion particle operators $N_f + M_f$. The conservation laws (1.52), (1.54) and (1.56),

$$[V, N_L] = [V, N_B] = [V, Q] = 0, \tag{1.59}$$

require that all interaction potentials in QED are bosonic. We are only interested in Hermitian operators V.

In (1.58) D_{NM} is a numerical *coefficient function*, which depends on the momenta and spin projections (or helicities) of all particles being created and destroyed. To satisfy the requirement $[V, \boldsymbol{J}_0] = 0$, this function must be rotationally invariant. The translational invariance ($[V, \boldsymbol{P}_0] = 0$) of (1.57)–(1.58) is guaranteed by the momentum delta function

$$\delta\left(\sum_{i=1}^{N} \boldsymbol{q}_i' - \sum_{j=1}^{M} \boldsymbol{q}_j\right).$$

This delta function also expresses the momentum conservation: the sum of the momenta of annihilated particles is equal to the sum of the momenta of created particles.

The potential energy operator V enters formulas for the S-operator (**1**-7.14), (**1**-7.17) and (**1**-7.18) in a t-dependent form, i. e.,

$$V(t) = e^{\frac{i}{\hbar}H_0 t} V e^{-\frac{i}{\hbar}H_0 t}. \tag{1.60}$$

We shall call *regular* those operators that satisfy conservation laws (1.49), (1.50) and (1.59) and whose t-dependence is determined by the free Hamiltonian H_0, as in equation (1.60). Equivalently, a t-dependent regular operator $V(t)$ satisfies the following differential equation:

$$\frac{d}{dt}V(t) = \frac{d}{dt}e^{\frac{i}{\hbar}H_0 t} V e^{-\frac{i}{\hbar}H_0 t} = \frac{i}{\hbar}e^{\frac{i}{\hbar}H_0 t}[H_0, V]e^{-\frac{i}{\hbar}H_0 t} = \frac{i}{\hbar}[H_0, V(t)]. \tag{1.61}$$

In our convention, if a regular operator V is written without its t-argument, then either this operator is t-independent (i. e., it commutes with H_0), or its value is taken at $t = 0$.

One final remark on notation. If the coefficient function of the potential V_{NM} is D_{NM}, then we will use the symbol $V_{NM} \circ \zeta$ for an operator whose coefficient function D_{NM}' is the product of D_{NM} and a numeric function ζ of the same arguments, i. e.,

$$\begin{aligned} &D_{NM}'(\boldsymbol{q}_1'\eta_1', \ldots, \boldsymbol{q}_N'\eta_N'; \boldsymbol{q}_1\eta_1, \ldots, \boldsymbol{q}_M\eta_M) \\ &\quad = D_{NM}(\boldsymbol{q}_1'\eta_1', \ldots, \boldsymbol{q}_N'\eta_N'; \boldsymbol{q}_1\eta_1, \ldots, \boldsymbol{q}_M\eta_M)\zeta(\boldsymbol{q}_1'\eta_1', \ldots, \boldsymbol{q}_N'\eta_N'; \boldsymbol{q}_1\eta_1, \ldots, \boldsymbol{q}_M\eta_M). \end{aligned}$$

Then, inserting (1.58) in (1.60) and using (1.45)–(1.48), we conclude that any regular potential $V_{NM}(t)$ takes the form

$$V_{NM}(t) = e^{\frac{i}{\hbar}H_0 t} V_{NM} e^{-\frac{i}{\hbar}H_0 t} = V_{NM} \circ e^{\frac{i}{\hbar}\mathcal{E}_{NM} t}, \tag{1.62}$$

where

$$\mathcal{E}_{NM}(\boldsymbol{q}'_1,\ldots,\boldsymbol{q}'_N,\boldsymbol{q}_1,\ldots,\boldsymbol{q}_M) \equiv \sum_{i=1}^{N}\sqrt{m_i^2c^4+q_i^2c^2} - \sum_{j=1}^{M}\sqrt{m_j^2c^4+q_j^2c^2} \tag{1.63}$$

is the difference between the energies of particles created and destroyed by the monomial V_{NM}. This difference is called the *energy function* of the potential V_{NM}. We can also extend this notation to general sums of potentials V_{NM} and write

$$V(t) = e^{\frac{i}{\hbar}H_0t}Ve^{-\frac{i}{\hbar}H_0t} = V\circ e^{\frac{i}{\hbar}\mathcal{E}_Vt}, \tag{1.64}$$

where $\mathcal{E}_V$ formally denotes energy functions of the monomials in V. In this economical notation we obtain[23]

$$\frac{d}{dt}V(t) = V(t)\circ\left(\frac{i}{\hbar}\mathcal{E}_V\right),$$

$$\underline{V(t)} = -\frac{i}{\hbar}\int_{-\infty}^{t}V(t')dt' = V(t)\circ\left(\frac{-1}{\mathcal{E}_V}\right), \tag{1.65}$$

$$\underbrace{V} \equiv -\frac{i}{\hbar}\int_{-\infty}^{\infty}V(t)dt = -2\pi iV\circ\delta(\mathcal{E}_V). \tag{1.66}$$

For example, formula (1.66) means that each monomial in $\underbrace{V}$ is different from zero only on the surface that is a solution of the equation

$$\mathcal{E}_{NM}(\boldsymbol{q}'_1,\ldots,\boldsymbol{q}'_N,\boldsymbol{q}_1,\ldots,\boldsymbol{q}_M) = 0 \tag{1.67}$$

(if such a solution exists). This surface in the momentum space is called the *energy surface* or the *energy shell* of the potential V_{NM}. We will also say that the operator $\underbrace{V}$ in equation (1.66) is zero outside its energy shell $\mathcal{E}_V = 0$. Note that the scattering operator (**1**-7.14) $S = 1 + \underbrace{\Sigma}$ is different from 1 only on the energy shell, i. e., where the energy conservation condition (1.67) is fulfilled.

It is easy to verify that the energy function of the product of two regular operators is equal to the sum of their energy functions, i. e., $\mathcal{E}_{AB} = \mathcal{E}_A + \mathcal{E}_B$. This implies the following equality:

$$\begin{aligned}\underbrace{\underline{AB}} &= -\underbrace{AB\circ(\mathcal{E}_B)^{-1}} = -AB\circ(\mathcal{E}_B)^{-1}\delta(\mathcal{E}_A+\mathcal{E}_B)\\ &= AB\circ(\mathcal{E}_A)^{-1}\delta(\mathcal{E}_A+\mathcal{E}_B) = -\underbrace{\underline{AB}},\end{aligned} \tag{1.68}$$

which we will find useful in the third volume.

23 Here we tacitly assume the adiabatic switching of the interaction (**1**-7.26) and use formulas (**1**-7.12), (**1**-7.13) and (**1**-7.27).

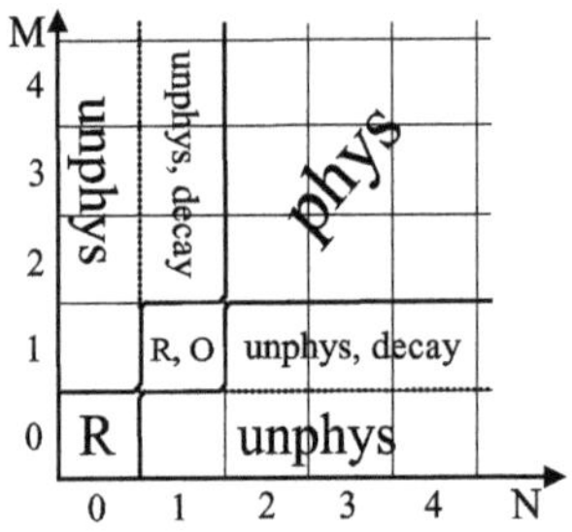

Figure 1.1: Locations of different types of operators in the "index space" $[N : M]$. R = *renorm*, O = *oscillation*.

1.2.3 Five types of regular potentials

In this subsection we are going to introduce a classification of regular potentials (1.58), by dividing them into five groups depending on the index $[N : M]$. We call these types of operators *renorm*, *oscillation*, *decay*, *phys* and *unphys*.[24] This classification will help in our study of renormalization in Chapter 4 and also in Volume 3, where we will formulate the "dressing" approach to QFT.

Renorm potentials have either index $[0 : 0]$ or index $[1 : 1]$. In the former case, the operator simply multiplies states by a numerical constant C. In the latter case, it is assumed that the particles that are produced and destroyed have the same type. In QED, the most general form of a *renorm potential* is[25]

$$R \propto a^\dagger a + b^\dagger b + d^\dagger d + f^\dagger f + c^\dagger c + C. \tag{1.69}$$

Renorm potentials are characterized by the property that their energy functions (1.63) are identically zero. This means that such potentials always have an energy shell, where they do not vanish.

Lemma 1.2. *Any two* renorm *operators commute with each other.*

Proof. A general *renorm* operator is the sum (1.69). The summands referring to different particles commute, because particle operators of different particles always commute. It is not difficult to verify that two *renorm* operators, corresponding to the same particle, commute as well:

$$\left[\int d\boldsymbol{p} f(\boldsymbol{p}) \alpha_{\boldsymbol{p}}^\dagger \alpha_{\boldsymbol{p}}, \int d\boldsymbol{q} g(\boldsymbol{q}) \alpha_{\boldsymbol{q}}^\dagger \alpha_{\boldsymbol{q}}\right] = 0. \qquad \square$$

The free Hamiltonian (1.32) and the total momentum (1.34) are examples of *renorm* operators. In particular, this implies that renorm potentials commute with H_0, so regular *renorm* operators are independent of t.

24 The correlation between potential's index $[N : M]$ and its type is shown in Figure 1.1.

25 For brevity, here we write only the operator structure of R, omitting numerical multipliers, indices, summation and integration signs.

Oscillation potentials have index $[1:1]$. In contrast to *renorm* potentials with the same index, *oscillation* potentials create and destroy different types of particles having different masses. For this reason, their energy functions (1.63) never turn to zero, so they do not have energy shells. In nature, *oscillation* potentials act on particles such as kaons and neutrinos. A vivid experimental manifestation of such interactions are time-dependent oscillations between different types of particles [6].

In QED there cannot be *oscillation* interactions, because they would violate the lepton and/or baryon conservation laws.

Decay potentials satisfy two conditions:

(1) their indices are either $[1:N]$ or $[N:1]$ with $N \geq 2$;
(2) they have a nonempty energy shell, where their coefficient functions do not vanish.

These potentials describe decay processes $1 \to N$,[26] in which one particle decays into N products. Moreover, we require that the laws of conservation of energy and momentum are fulfilled in the decay, i. e., there is a nontrivial energy shell, where the coefficient function does not vanish. *Decay* terms are not present in the QED Hamiltonian and in the corresponding S-matrix, because decays of electrons, protons or photons would be against conservation laws.[27] Nevertheless, decays of elementary particles play a huge role in other branches of high-energy physics, and we will discuss them in the third volume.

Phys potentials have at least two creation operators and at least two annihilation operators (they have indices $[N : M]$ where $N \geq 2$ and $M \geq 2$). For *phys* potentials, the energy shell always exists. For example, in the case of the *phys* operator[28] $d^\dagger_{(\boldsymbol{p}+\boldsymbol{k})\rho} f^\dagger_{(\boldsymbol{q}-\boldsymbol{k})\sigma} a_{\boldsymbol{p}\tau} b_{\boldsymbol{q}\eta}$ the energy shell is the set of solutions of the equation $\Omega_{\boldsymbol{p}+\boldsymbol{k}} + \Omega_{\boldsymbol{q}-\boldsymbol{k}} = \omega_{\boldsymbol{p}} + \omega_{\boldsymbol{q}}$ in the nine-dimensional momentum space $\{\boldsymbol{p}, \boldsymbol{q}, \boldsymbol{k}\}$. This equation has nontrivial solutions, so the energy shell is not empty.

All regular operators that do not belong to any of the four above classes will be called **unphys potentials.** They can be divided into two subclasses with the following indices:

(1) $[0 : N]$ or $[N : 0]$, where $N \geq 1$. Obviously, in this case the energy shell is absent.

---

26 And also inverse processes $N \to 1$.

27 In principle, one photon can decay into an odd number of other photons without violating the conservation laws. For example, such a process could be described by the potential $c^\dagger_{\boldsymbol{k}_1\tau_1} c^\dagger_{\boldsymbol{k}_2\tau_2} c^\dagger_{\boldsymbol{k}_3\tau_3} c_{(\boldsymbol{k}_1+\boldsymbol{k}_2+\boldsymbol{k}_3)\tau_4}$, which formally satisfies all conservation laws if the momenta of all involved photons are collinear and $k_1 + k_2 + k_3 - |\boldsymbol{k}_1 + \boldsymbol{k}_2 + \boldsymbol{k}_3| = 0$. However, as shown in [5], such terms in the S-operator are zero on the energy shell, so photon decays are forbidden in QED.

28 This operator describes the conversion reaction electron + positron $\to$ proton + antiproton. In the arguments of particle operators, we have already taken into account the momentum conservation law.

(2) [1 : N] or [N : 1], where $N \geq 2$. These are the same indices as for *decay* potentials, but here we demand that either the energy shell does not exist or that the coefficient function vanishes on the energy shell.

Here is an example of an *unphys* potential with condition (2):

$$a^{\dagger}_{(\boldsymbol{p}-\boldsymbol{k})\sigma} c^{\dagger}_{\boldsymbol{k}\tau} a_{\boldsymbol{p}\rho}. \tag{1.70}$$

The energy shell equation $\omega_{\boldsymbol{p}-\boldsymbol{k}} + ck = \omega_{\boldsymbol{p}}$ has only one solution, $\boldsymbol{k} = \boldsymbol{0}$. However, the zero vector is excluded from the photon's momentum spectrum (see Subsection **1**-5.4.1), so the potential (1.70) has an empty energy shell. This means that a free electron cannot emit a photon without violating the energy–momentum conservation law.

Table 1.1: Types of regular potentials in the Fock space.

Potential	Index [N : M]	Energy shell	Example
renorm	[0 : 0],[1 : 1]	yes	$a^{\dagger}_{\boldsymbol{p}} a_{\boldsymbol{p}}$
oscillation	[1 : 1]	no	forbidden in QED
unphys	[0 : M ≥ 1],[N ≥ 1 : 0]	no	$a^{\dagger}_{\boldsymbol{p}} b^{\dagger}_{-\boldsymbol{p}-\boldsymbol{k}} c^{\dagger}_{\boldsymbol{k}}$
unphys	[1 : M ≥ 2],[N ≥ 2 : 1]	no	$a^{\dagger}_{\boldsymbol{p}} a_{\boldsymbol{p}-\boldsymbol{k}} c_{\boldsymbol{k}}$
decay	[1 : M ≥ 2],[N ≥ 2 : 1]	yes	forbidden in QED
phys	[N ≥ 2 : M ≥ 2]	yes	$d^{\dagger}_{\boldsymbol{q}+\boldsymbol{k}} a^{\dagger}_{\boldsymbol{p}-\boldsymbol{k}} d_{\boldsymbol{q}} a_{\boldsymbol{p}}$

The properties of potentials considered above are summarized in Table 1.1. These five types of interactions exhaust all possibilities; therefore any regular operator V must have a unique expansion

$$V = V^{\text{ren}} + V^{\text{unp}} + V^{\text{dec}} + V^{\text{phys}} + V^{\text{osc}}.$$

As mentioned above, there are no *oscillation* and *decay* interactions in QED, so everywhere in this volume we will assume that the most general potential is equal to the sum of *renorm*, *unphys* and *phys* parts:

$$V^{\text{QED}} = V^{\text{ren}} + V^{\text{unp}} + V^{\text{phys}}.$$

Now we need to figure out how to perform various manipulations with these three classes of potentials. In particular, we want to learn how to calculate products, commutators and t-integrals that are necessary for computing scattering operators from Section **1**-7.1.

1.2.4 Products and commutators of regular potentials

Let us first prove a few simple results.

Lemma 1.3. *The product of two or more regular operators is regular.*

Proof. By definition, if operators $A(t)$ and $B(t)$ are regular, then

$$A(t) = e^{\frac{i}{\hbar}H_0 t} A e^{-\frac{i}{\hbar}H_0 t},$$
$$B(t) = e^{\frac{i}{\hbar}H_0 t} B e^{-\frac{i}{\hbar}H_0 t}.$$

Hence, their product $C(t) = A(t)B(t)$ has the t-dependence

$$C(t) = e^{\frac{i}{\hbar}H_0 t} A e^{-\frac{i}{\hbar}H_0 t} e^{\frac{i}{\hbar}H_0 t} B e^{-\frac{i}{\hbar}H_0 t} = e^{\frac{i}{\hbar}H_0 t} AB e^{-\frac{i}{\hbar}H_0 t}$$

characteristic of regular operators. The conservation laws (1.49), (1.50) and (1.59) are valid for the product AB, just as they are valid for A and B separately. Therefore, $C(t)$ is regular. □

Theorem 1.4. *A Hermitian operator P is* phys *if and only if it yields zero when acting on both the vacuum vector $|\text{vac}\rangle$ and one-particle states $|1\rangle \equiv \alpha^\dagger|\text{vac}\rangle$*[29]*:*

$$P|\text{vac}\rangle = 0, \tag{1.71}$$
$$P|1\rangle \equiv P\alpha^\dagger|\text{vac}\rangle = 0. \tag{1.72}$$

Proof. By definition, normally ordered *phys* potentials have (at least) two annihilation operators on the right. So, they yield zero when applied to the vacuum or any one-particle state. Therefore, equations (1.71) and (1.72) are satisfied for any *phys* operator P.

Let us now prove the converse. *Renorm* operators cannot satisfy requirements (1.71) and (1.72), because they preserve the number of particles. *Unphys* operators $[1 : M]$ can satisfy these requirements. For example,

$$\alpha_1^\dagger \alpha_2 \alpha_3 |\text{vac}\rangle = 0,$$
$$\alpha_1^\dagger \alpha_2 \alpha_3 |1\rangle = 0.$$

However, in order to be Hermitian, such operators must always be present in pairs with $[M : 1]$ operators, like $\alpha_2^\dagger \alpha_3^\dagger \alpha_1$. Then there exists at least one single-particle state $|1\rangle$ for which equation (1.72) is not valid, that is,

$$\alpha_3^\dagger \alpha_2^\dagger \alpha_1 |1\rangle = \alpha_3^\dagger \alpha_2^\dagger |\text{vac}\rangle \neq 0.$$

29 Here α means any of the five particle operators (a, b, d, f, c) related to QED. Momentum and spin labels are omitted for brevity.

Similar arguments apply to *unphys* operators with indices $[0 : M]$ and $[N : 0]$. Hence, the only remaining possibility for the potential P is to be *phys*. □

Lemma 1.5. *The product and commutator of any* phys *operators A and B are also* phys.

Proof. By Theorem 1.4, if A and B are *phys*, then

$$A|\text{vac}\rangle = B|\text{vac}\rangle = A|1\rangle = B|1\rangle = 0.$$

The same properties are valid for the Hermitian combinations $i(AB - BA)$ and $AB + BA$. Hence, both the commutator $[A, B]$ and the anticommutator $\{A, B\}$ are *phys*. The same conclusion is true for the product, which can be expressed as the sum

$$AB = \frac{1}{2}\{A, B\} + \frac{1}{2}[A, B].$$ □

Lemma 1.6. *If R is a* renorm *operator, P is a* phys *operator and* $[P, R] \neq 0$, *then operator* $[P, R]$ *is of the* phys *type.*

Proof. Let us first check how this commutator acts on the vacuum and single-particle states.[30] We have

$$\begin{aligned} i(PR - RP)|\text{vac}\rangle &= iPR|\text{vac}\rangle = iPC_0|\text{vac}\rangle = 0, \\ i(PR - RP)|1\rangle &= iPR|1\rangle = iPC_1|1'\rangle = 0. \end{aligned}$$

This means that the Hermitian commutator $i[P, R]$ turns vectors $|\text{vac}\rangle$ and $|1\rangle$ to zero. By Lemma 1.4 this operator is *phys*. □

Lemma 1.7. *If R is a* renorm *operator, U is an* unphys *operator and* $[U, R] \neq 0$, *then operator* $[U, R]$ *has the* unphys *type.*

Idea of the proof. Let us first calculate the commutator of the *renorm* operator $R = \int d\boldsymbol{p} f(\boldsymbol{p})\alpha^\dagger_{\boldsymbol{p}}\alpha_{\boldsymbol{p}}$ with a particle creation operator[31] We have

$$\begin{aligned} [\alpha^\dagger_{\boldsymbol{q}}, R] &= \alpha^\dagger_{\boldsymbol{q}}\left(\int d\boldsymbol{p} f(\boldsymbol{p})\alpha^\dagger_{\boldsymbol{p}}\alpha_{\boldsymbol{p}}\right) - \left(\int d\boldsymbol{p} f(\boldsymbol{p})\alpha^\dagger_{\boldsymbol{p}}\alpha_{\boldsymbol{p}}\right)\alpha^\dagger_{\boldsymbol{q}} \\ &= \pm\int d\boldsymbol{p} f(\boldsymbol{p})\alpha^\dagger_{\boldsymbol{p}}\alpha^\dagger_{\boldsymbol{q}}\alpha_{\boldsymbol{p}} - \int d\boldsymbol{p} f(\boldsymbol{p})\alpha^\dagger_{\boldsymbol{p}}\alpha_{\boldsymbol{p}}\alpha^\dagger_{\boldsymbol{q}} \\ &= \int d\boldsymbol{p} f(\boldsymbol{p})\alpha^\dagger_{\boldsymbol{p}}\alpha_{\boldsymbol{p}}\alpha^\dagger_{\boldsymbol{q}} - \int d\boldsymbol{p} f(\boldsymbol{p})\alpha^\dagger_{\boldsymbol{p}}\delta(\boldsymbol{p} - \boldsymbol{q}) - \int d\boldsymbol{p} f(\boldsymbol{p})\alpha^\dagger_{\boldsymbol{p}}\alpha_{\boldsymbol{p}}\alpha^\dagger_{\boldsymbol{q}} \\ &= -f(\boldsymbol{q})\alpha^\dagger_{\boldsymbol{q}}. \end{aligned}$$

30 Here we took into account that *renorm* operators preserve the number of particles: $R|\text{vac}\rangle = \text{const}\times|\text{vac}\rangle$, $R|1\rangle = |1'\rangle$ and *phys* operators turn the states $|\text{vac}\rangle$ and $|1\rangle$ to zero.

31 The upper sign is for bosons, the lower sign is for fermions.

Similarly, we obtain the commutator with an annihilation operator:

$$[\alpha_{\boldsymbol{q}}, R] = f(\boldsymbol{q})\alpha_{\boldsymbol{q}}.$$

Now, as an example of an *unphys* operator, we take a potential with index [2 : 1],

$$U = \int d\boldsymbol{q}_1 d\boldsymbol{q}_2 d\boldsymbol{p} D(\boldsymbol{q}_1, \boldsymbol{q}_2; \boldsymbol{p})\delta(\boldsymbol{q}_1 + \boldsymbol{q}_2 - \boldsymbol{p})\alpha^\dagger_{\boldsymbol{q}_1}\alpha^\dagger_{\boldsymbol{q}_2}\alpha_{\boldsymbol{p}}.$$

The index of the commutator is also [2 : 1]. We have

$$\begin{aligned}
[U, R] &= \int d\boldsymbol{q}_1 d\boldsymbol{q}_2 d\boldsymbol{p} D(\boldsymbol{q}_1, \boldsymbol{q}_2; \boldsymbol{p})\delta(\boldsymbol{q}_1 + \boldsymbol{q}_2 - \boldsymbol{p})\alpha^\dagger_{\boldsymbol{q}_1}\alpha^\dagger_{\boldsymbol{q}_2}[\alpha_{\boldsymbol{p}}, R] \\
&+ \int d\boldsymbol{q}_1 d\boldsymbol{q}_2 d\boldsymbol{p} D(\boldsymbol{q}_1, \boldsymbol{q}_2; \boldsymbol{p})\delta(\boldsymbol{q}_1 + \boldsymbol{q}_2 - \boldsymbol{p})\alpha^\dagger_{\boldsymbol{q}_1}[\alpha^\dagger_{\boldsymbol{q}_2}, R]\alpha_{\boldsymbol{p}} \\
&+ \int d\boldsymbol{q}_1 d\boldsymbol{q}_2 d\boldsymbol{p} D(\boldsymbol{q}_1, \boldsymbol{q}_2; \boldsymbol{p})\delta(\boldsymbol{q}_1 + \boldsymbol{q}_2 - \boldsymbol{p})[\alpha^\dagger_{\boldsymbol{q}_1}, R]\alpha^\dagger_{\boldsymbol{q}_2}\alpha_{\boldsymbol{p}} \\
&= \int d\boldsymbol{q}_1 d\boldsymbol{q}_2 d\boldsymbol{p} D(\boldsymbol{q}_1, \boldsymbol{q}_2; \boldsymbol{p}) f(\boldsymbol{p})\delta(\boldsymbol{q}_1 + \boldsymbol{q}_2 - \boldsymbol{p})\alpha^\dagger_{\boldsymbol{q}_1}\alpha^\dagger_{\boldsymbol{q}_2}\alpha_{\boldsymbol{p}} \\
&- \int d\boldsymbol{q}_1 d\boldsymbol{q}_2 d\boldsymbol{p} D(\boldsymbol{q}_1, \boldsymbol{q}_2; \boldsymbol{p}) f(\boldsymbol{q}_2)\delta(\boldsymbol{q}_1 + \boldsymbol{q}_2 - \boldsymbol{p})\alpha^\dagger_{\boldsymbol{q}_1}\alpha^\dagger_{\boldsymbol{q}_2}\alpha_{\boldsymbol{p}} \\
&- \int d\boldsymbol{q}_1 d\boldsymbol{q}_2 d\boldsymbol{p} D(\boldsymbol{q}_1, \boldsymbol{q}_2; \boldsymbol{p}) f(\boldsymbol{q}_1)\delta(\boldsymbol{q}_1 + \boldsymbol{q}_2 - \boldsymbol{p})\alpha^\dagger_{\boldsymbol{q}_1}\alpha^\dagger_{\boldsymbol{q}_2}\alpha_{\boldsymbol{p}}.
\end{aligned}$$

Moreover, if the operator U does not have an energy shell, then $[U, R]$ also does not have it, i. e., its type is *unphys*. If U has an energy shell where the coefficient function $D(\boldsymbol{q}_1, \boldsymbol{q}_2; \boldsymbol{p})$ is zero, then $[U, R]$ also has this property. □

Lemma 1.8. *The commutator $[P, U]$ of an Hermitian* phys *operator P and an Hermitian* unphys *operator U cannot contain* renorm *terms.*

Proof. Applying the operator $[P, U]$ to a single-particle state $|1\rangle$ and using (1.72), we obtain

$$[P, U]|1\rangle = (PU - UP)|1\rangle = PU|1\rangle. \tag{1.73}$$

If the commutator $[P, U]$ contained *renorm* terms, then the right-hand side of (1.73) would have a nonzero one-particle component. However, the range of any *phys P* does not include the one-particle sector. This implies $[P, U]^{\text{ren}} = 0$. □

Finally, it is easy to verify that there are no restrictions on the type of the commutator of two *unphys* operators $[U, U']$. It can contain *unphys*, *phys* and *renorm* parts.

The above results are summarized in Table 1.2.

1.2.5 More about *t*-integrals

Lemma 1.9. *The t-derivative of a regular operator $A(t)$ is regular, and its* renorm *part vanishes.*

Table 1.2: Commutators, t-derivatives and t-integrals with regular operator A in the Fock space. (Notation: P = *phys*, U = *unphys*, R = *renorm*, NR = nonregular.)

Type of A	$[A,P]$	$[A,U]$	$[A,R]$	$\frac{dA}{dt}$	$\underline{A}$	$\underbrace{A}$
P	P	P+U	P	P	P	P
U	P+U	P+U+R	U	U	U	0
R	P	U	0	0	NR	∞

Proof. According to (1.61), the derivative of $A(t)$ is equal to the commutator with regular H_0. Then by Lemma 1.3 this derivative is regular.

Suppose, by contradiction, that $\frac{d}{dt}A(t)$ has a nonzero *renorm* part R. Then R does not depend on t, because it is regular. It follows that the most general form of $A(t)$ is $A(t) = Rt + S$, where S is any operator independent of t. From the condition that the *renorm* part of the regular operator $A(t)$ cannot depend on t, we obtain $R = 0$. □

From equation (1.65) we conclude that t-integrals of regular *phys* and *unphys* operators are regular. However, this property does not hold for t-integrals of *renorm* operators. As we know, *renorm* operators are independent of t. Hence, when the interaction is adiabatically switched on, as in (**1**-7.26), we obtain

$$\begin{aligned}
\underline{V^{\text{ren}}(t)} &= \lim_{\epsilon\to+0}\left(-\frac{i}{\hbar}\int\limits_{-\infty}^{0} V^{\text{ren}}e^{\epsilon t'}dt' - \frac{i}{\hbar}\int\limits_{0}^{t} V^{\text{ren}}e^{-\epsilon t'}dt'\right) \\
&= -\left(\frac{i}{\hbar}\right)V^{\text{ren}} \circ \lim_{\epsilon\to+0}\left(\frac{e^{\epsilon t}}{\epsilon}\bigg|_{t=-\infty}^{t=0} - \frac{e^{-\epsilon t}}{\epsilon}\bigg|_{t=0}^{t=t}\right) \\
&= -\left(\frac{i}{\hbar}\right)V^{\text{ren}} \circ \lim_{\epsilon\to+0}\left(\frac{1}{\epsilon} - \frac{e^{-\epsilon t}}{\epsilon} + \frac{1}{\epsilon}\right) \\
&= -\left(\frac{i}{\hbar}\right)V^{\text{ren}} \circ \lim_{\epsilon\to+0}\left(\frac{1}{\epsilon} + t + \cdots\right),
\end{aligned} \tag{1.74}$$

$$\underbrace{V^{\text{ren}}} = \lim_{t\to\infty}\underline{V^{\text{ren}}(t)} = \infty. \tag{1.75}$$

Hence, *renorm* operators differ from all others in that their t-integrals (1.74)–(1.75) are infinite and nonregular.[32]

By definition, an *unphys* operator V^{unp} either does not have an energy shell, or its coefficient function vanishes on the energy shell. Then, from equation (1.66) it follows that for any *unphys* operator

$$\underbrace{V^{\text{unp}}} = 0. \tag{1.76}$$

Results obtained in this subsection are shown in the last three columns of Table 1.2.

---

32 As we shall see in Subsection 4.1.1, correctly renormalized expressions for scattering operators should not contain *renorm* terms and pathological constructs like (1.74)–(1.75).

1.2.6 Solution of one commutator equation

Quite often we need solutions of equations of the type[33]

$$i[H_0, A] = V, \tag{1.77}$$

where H_0 is the free Hamiltonian, V is a given regular Hermitian operator, having zero *renorm* part, and A is the desired solution (yet unknown regular Hermitian operator). What can we say about this solution? Let us first multiply both sides of equation (1.77) by the usual t-exponents $e^{\frac{i}{\hbar}H_0 t} \ldots e^{-\frac{i}{\hbar}H_0 t}$. Then we obtain

$$i[H_0, A(t)] = V(t).$$

Using (1.61), we can rewrite this equation in the form[34]

$$\hbar \frac{d}{dt} A(t) = V(t). \tag{1.78}$$

Next, we assume that the usual "adiabatic switching" (**1**-7.26) is enforced, such that $V(-\infty) = 0$. As our initial condition, we will assume that the solution $A(t)$ has the same property, i. e.,

$$A(-\infty) = 0. \tag{1.79}$$

Then equation (1.78) has a simple solution, i. e.,

$$A(t) = \frac{1}{\hbar} \int_{-\infty}^{t} V(t')dt' = i\underline{V(t)}.$$

In order to get the t-independent solution of our original equation (1.77), we simply set $t = 0$ and obtain

$$A \equiv A(0) = i\underline{V} \equiv V \circ \frac{-i}{\mathcal{E}_V} \tag{1.80}$$

or

$$[\underline{V}, H_0] = V. \tag{1.81}$$

33 A solution of this equation in the case of *phys* V can be found in § 34 of [15].

34 Here it becomes clear that if we chose $V^{\text{ren}} \neq 0$, we would come into contradiction with Lemma 1.9, which forbids *renorm* terms on the left-hand side of (1.78).

1.2.7 Two-particle potentials

We already know that quantum-mechanical cluster-separable interactions can be written as sums of smooth potentials (**1**-6.55), depending on particle observables (positions, momenta and spins). However, this notation is very inconvenient, because such sums have very different forms in different Fock sectors. For example, the Coulomb potential looks like (**1**-6.52) in the two-particle sector and like (**1**-6.53) in the three-particle sector. It would be preferable to have a single formula that is equally valid in all n-particle sectors.

This is achieved by writing n-particle potentials ($n \geq 2$) in the formalism of particle operators. These potentials preserve the number of particles and their types, so they should have equal numbers of annihilation and creation operators ($N = M$, $N \geq 2$, $M \geq 2$). Therefore, here we are interested in *phys* potentials.

As an example, consider the two-electron subspace $\mathscr{H}(2,0,0,0,0)$ of the Fock space. *Phys* operators that act nontrivially in this subspace, and at the same time leave it invariant, should have index $[2:2]$. In accordance with equation (1.58), we write them in the form[35]

$$\begin{aligned} V &= \int d\boldsymbol{p}d\boldsymbol{q}d\boldsymbol{p}'d\boldsymbol{q}'D_{22}(\boldsymbol{p},\boldsymbol{q};\boldsymbol{p}',\boldsymbol{q}')\delta(\boldsymbol{p}+\boldsymbol{q}-\boldsymbol{p}'-\boldsymbol{q}')a^{\dagger}_{\boldsymbol{q}}a^{\dagger}_{\boldsymbol{p}}a_{\boldsymbol{p}'}a_{\boldsymbol{q}'} \\ &= \int d\boldsymbol{p}d\boldsymbol{q}d\boldsymbol{p}'D_{22}(\boldsymbol{p},\boldsymbol{q};\boldsymbol{p}',\boldsymbol{p}+\boldsymbol{q}-\boldsymbol{p}')a^{\dagger}_{\boldsymbol{q}}a^{\dagger}_{\boldsymbol{p}}a_{\boldsymbol{p}'}a_{\boldsymbol{p}+\boldsymbol{q}-\boldsymbol{p}'} \\ &= \int d\boldsymbol{p}d\boldsymbol{p}'d\boldsymbol{k}D(\boldsymbol{p},\boldsymbol{p}',\boldsymbol{k})a^{\dagger}_{\boldsymbol{p}'-\boldsymbol{k}}a^{\dagger}_{\boldsymbol{p}+\boldsymbol{k}}a_{\boldsymbol{p}}a_{\boldsymbol{p}'}, \end{aligned} \tag{1.82}$$

where we denoted $\boldsymbol{k} = \boldsymbol{q}' - \boldsymbol{q}$ the "transferred momentum" and

$$D(\boldsymbol{p},\boldsymbol{p}',\boldsymbol{k}) \equiv D_{22}(\boldsymbol{p}+\boldsymbol{k},\boldsymbol{p}'-\boldsymbol{k};\boldsymbol{p},\boldsymbol{p}').$$

The next step is to find out how the potential (1.82) acts on two-electron states. Applying this operator to an arbitrary two-electron state

$$|\Psi\rangle = \int d\boldsymbol{q}d\boldsymbol{q}'\Psi(\boldsymbol{q},\boldsymbol{q}')a^{\dagger}_{\boldsymbol{q}}a^{\dagger}_{\boldsymbol{q}'}|\text{vac}\rangle, \tag{1.83}$$

we obtain

$$V|\Psi\rangle = \int d\boldsymbol{p}d\boldsymbol{p}'d\boldsymbol{k}D(\boldsymbol{p},\boldsymbol{p}',\boldsymbol{k})a^{\dagger}_{\boldsymbol{p}'-\boldsymbol{k}}a^{\dagger}_{\boldsymbol{p}+\boldsymbol{k}}a_{\boldsymbol{p}}a_{\boldsymbol{p}'}\int d\boldsymbol{q}d\boldsymbol{q}'\Psi(\boldsymbol{q},\boldsymbol{q}')a^{\dagger}_{\boldsymbol{q}}a^{\dagger}_{\boldsymbol{q}'}|\text{vac}\rangle.$$

To understand how the above product of particle operators acts on the vacuum vector

$$a^{\dagger}_{\boldsymbol{p}'-\boldsymbol{k}}a^{\dagger}_{\boldsymbol{p}+\boldsymbol{k}}a_{\boldsymbol{p}}a_{\boldsymbol{p}'}a^{\dagger}_{\boldsymbol{q}}a^{\dagger}_{\boldsymbol{q}'}|\text{vac}\rangle,$$

35 In this subsection, for brevity, we omit spin indices.

we apply the Wick theorem, Theorem 1.1, and convert this product to the normal order. Since the resulting polynomial in $a^\dagger$ and a acts on the vacuum, all terms that have annihilation operators (on the right) are irrelevant. Therefore, we are only interested in terms where two contractions have removed all annihilation operators, i. e.,

$$
\begin{aligned}
& a^\dagger_{\boldsymbol{p}'-\boldsymbol{k}} a^\dagger_{\boldsymbol{p}+\boldsymbol{k}} a_{\boldsymbol{p}} a_{\boldsymbol{p}'} a^\dagger_{\boldsymbol{q}} a^\dagger_{\boldsymbol{q}'} |\text{vac}\rangle \\
&= a^\dagger_{\boldsymbol{p}'-\boldsymbol{k}} a^\dagger_{\boldsymbol{p}+\boldsymbol{k}} a_{\boldsymbol{p}} a_{\boldsymbol{p}'} a^\dagger_{\boldsymbol{q}} a^\dagger_{\boldsymbol{q}'} |\text{vac}\rangle + a^\dagger_{\boldsymbol{p}'-\boldsymbol{k}} a^\dagger_{\boldsymbol{p}+\boldsymbol{k}} a_{\boldsymbol{p}} a_{\boldsymbol{p}'} a^\dagger_{\boldsymbol{q}} a^\dagger_{\boldsymbol{q}'} |\text{vac}\rangle + \cdots \\
&= -a^\dagger_{\boldsymbol{p}'-\boldsymbol{k}} a^\dagger_{\boldsymbol{p}+\boldsymbol{k}} \delta(\boldsymbol{q}' - \boldsymbol{p}')\delta(\boldsymbol{q} - \boldsymbol{p})|\text{vac}\rangle + a^\dagger_{\boldsymbol{p}'-\boldsymbol{k}} a^\dagger_{\boldsymbol{p}+\boldsymbol{k}} \delta(\boldsymbol{q}' - \boldsymbol{p})\delta(\boldsymbol{q} - \boldsymbol{p}')|\text{vac}\rangle + \cdots .
\end{aligned}
$$

Hence[36]

$$
\begin{aligned}
V|\Psi\rangle &= -\int d\boldsymbol{p}d\boldsymbol{p}'d\boldsymbol{k} \int d\boldsymbol{q}d\boldsymbol{q}' D(\boldsymbol{p},\boldsymbol{p}',\boldsymbol{k})\Psi(\boldsymbol{q},\boldsymbol{q}')\delta(\boldsymbol{q}' - \boldsymbol{p}')\delta(\boldsymbol{q} - \boldsymbol{p}) a^\dagger_{\boldsymbol{p}'-\boldsymbol{k}} a^\dagger_{\boldsymbol{p}+\boldsymbol{k}} |\text{vac}\rangle \\
&\quad + \int d\boldsymbol{p}d\boldsymbol{p}'d\boldsymbol{k} \int d\boldsymbol{q}d\boldsymbol{q}' D(\boldsymbol{p},\boldsymbol{p}',\boldsymbol{k})\Psi(\boldsymbol{q},\boldsymbol{q}')\delta(\boldsymbol{q}' - \boldsymbol{p})\delta(\boldsymbol{q} - \boldsymbol{p}') a^\dagger_{\boldsymbol{p}'-\boldsymbol{k}} a^\dagger_{\boldsymbol{p}+\boldsymbol{k}} |\text{vac}\rangle \\
&= -\int d\boldsymbol{p}d\boldsymbol{p}'d\boldsymbol{k} D(\boldsymbol{p},\boldsymbol{p}',\boldsymbol{k})\Psi(\boldsymbol{p},\boldsymbol{p}') a^\dagger_{\boldsymbol{p}'-\boldsymbol{k}} a^\dagger_{\boldsymbol{p}+\boldsymbol{k}} |\text{vac}\rangle \\
&\quad + \int d\boldsymbol{p}d\boldsymbol{p}'d\boldsymbol{k} D(\boldsymbol{p},\boldsymbol{p}',\boldsymbol{k})\Psi(\boldsymbol{p}',\boldsymbol{p}) a^\dagger_{\boldsymbol{p}'-\boldsymbol{k}} a^\dagger_{\boldsymbol{p}+\boldsymbol{k}} |\text{vac}\rangle \\
&= 2\int d\boldsymbol{p}d\boldsymbol{p}'d\boldsymbol{k} D(\boldsymbol{p},\boldsymbol{p}',\boldsymbol{k})\Psi(\boldsymbol{p}',\boldsymbol{p}) a^\dagger_{\boldsymbol{p}'-\boldsymbol{k}} a^\dagger_{\boldsymbol{p}+\boldsymbol{k}} |\text{vac}\rangle \\
&= \int d\boldsymbol{q}d\boldsymbol{q}' \left(2\int d\boldsymbol{k} D(\boldsymbol{q}' - \boldsymbol{k}, \boldsymbol{q} + \boldsymbol{k}, \boldsymbol{k})\Psi(\boldsymbol{q} + \boldsymbol{k}, \boldsymbol{q}' - \boldsymbol{k}) \right) a^\dagger_{\boldsymbol{q}} a^\dagger_{\boldsymbol{q}'} |\text{vac}\rangle.
\end{aligned}
$$

Comparing this result with (1.83), we see that under the action of the operator V the wave function $\Psi(\boldsymbol{q},\boldsymbol{q}')$ has been transformed into a new wave function, namely

$$
\Psi'(\boldsymbol{q},\boldsymbol{q}') \equiv \hat{V}\Psi(\boldsymbol{q},\boldsymbol{q}') = 2\int d\boldsymbol{k} D(\boldsymbol{q}' - \boldsymbol{k}, \boldsymbol{q} + \boldsymbol{k}, \boldsymbol{k})\Psi(\boldsymbol{q} + \boldsymbol{k}, \boldsymbol{q}' - \boldsymbol{k}). \tag{1.84}
$$

This is the most general linear transformation of a two-particle momentum–space wave function that preserves the total momentum. For comparison with traditional interparticle potentials, it will be more convenient to express the operator V in the position representation. This can be achieved by means of the Fourier transform (**1**-5.49). Then we have

$$
\begin{aligned}
\Psi'(\boldsymbol{x},\boldsymbol{y}) &= \frac{1}{(2\pi\hbar)^3} \int d\boldsymbol{q}d\boldsymbol{q}' e^{\frac{i}{\hbar}\boldsymbol{q}\cdot\boldsymbol{x} + \frac{i}{\hbar}\boldsymbol{q}'\cdot\boldsymbol{y}} \Psi'(\boldsymbol{q},\boldsymbol{q}') \\
&= \frac{2}{(2\pi\hbar)^3} \int d\boldsymbol{q}d\boldsymbol{q}' e^{\frac{i}{\hbar}\boldsymbol{q}\cdot\boldsymbol{x} + \frac{i}{\hbar}\boldsymbol{q}'\cdot\boldsymbol{y}} \int d\boldsymbol{k} D(\boldsymbol{q}' - \boldsymbol{k}, \boldsymbol{q} + \boldsymbol{k}, \boldsymbol{k})\Psi(\boldsymbol{q} + \boldsymbol{k}, \boldsymbol{q}' - \boldsymbol{k})
\end{aligned}
$$

---

36 Here we use the antisymmetric character of the two-fermion wave function: $\Psi(\boldsymbol{q},\boldsymbol{q}') = -\Psi(\boldsymbol{q}',\boldsymbol{q})$.

$$= \frac{2}{(2\pi\hbar)^3} \int d\boldsymbol{p}d\boldsymbol{p}' e^{\frac{i}{\hbar}(\boldsymbol{p}-\boldsymbol{k})\cdot\boldsymbol{x}+\frac{i}{\hbar}(\boldsymbol{p}'+\boldsymbol{k})\cdot\boldsymbol{y}} \int d\boldsymbol{k} D(\boldsymbol{p}',\boldsymbol{p},\boldsymbol{k})\Psi(\boldsymbol{p},\boldsymbol{p}')$$

$$= \left(2\int d\boldsymbol{k} e^{\frac{i}{\hbar}\boldsymbol{k}\cdot(\boldsymbol{y}-\boldsymbol{x})} D(\boldsymbol{p}',\boldsymbol{p},\boldsymbol{k})\right)\left[\frac{1}{(2\pi\hbar)^3}\int d\boldsymbol{p}d\boldsymbol{p}' e^{\frac{i}{\hbar}\boldsymbol{p}\cdot\boldsymbol{x}+\frac{i}{\hbar}\boldsymbol{p}'\cdot\boldsymbol{y}}\Psi(\boldsymbol{p},\boldsymbol{p}')\right], \quad (1.85)$$

where in square brackets we recognize the original position–space wave function

$$\Psi(\boldsymbol{x},\boldsymbol{y}) = \frac{1}{(2\pi\hbar)^3}\int d\boldsymbol{p}d\boldsymbol{p}' e^{\frac{i}{\hbar}\boldsymbol{p}\cdot\boldsymbol{x}+\frac{i}{\hbar}\boldsymbol{p}'\cdot\boldsymbol{y}}\Psi(\boldsymbol{p},\boldsymbol{p}') \quad (1.86)$$

and the expression in the parenthesis is the operator $\hat{V}$ acting on it. This operator takes on a particularly simple form if we assume that the coefficient function $D(\boldsymbol{p}',\boldsymbol{p},\boldsymbol{k})$ does not depend on $\boldsymbol{p}$ and $\boldsymbol{p}'$. Then, introducing notation

$$D(\boldsymbol{p}',\boldsymbol{p},\boldsymbol{k}) \equiv D(\boldsymbol{k}).$$

we obtain

$$\hat{V}\Psi(\boldsymbol{x},\boldsymbol{y}) = 2\int d\boldsymbol{k} e^{\frac{i}{\hbar}\boldsymbol{k}\cdot(\boldsymbol{y}-\boldsymbol{x})} D(\boldsymbol{k})\Psi(\boldsymbol{x},\boldsymbol{y}) = w(\boldsymbol{y}-\boldsymbol{x})\Psi(\boldsymbol{x},\boldsymbol{y}), \quad (1.87)$$

where

$$w(\boldsymbol{r}) = 2\int d\boldsymbol{k} e^{\frac{i}{\hbar}\boldsymbol{k}\cdot\boldsymbol{r}} D(\boldsymbol{k})$$

is the Fourier transform of the function $D(\boldsymbol{k})$. We see that $\hat{V}$ acts on wave functions in the position representation simply by multiplying them on the function $w(\boldsymbol{r})$. Hence, this is the usual position–space potential. Note that conservation of the total momentum means that the potential $w(\boldsymbol{r})$ depends only on the relative position $\boldsymbol{r} \equiv \boldsymbol{y} - \boldsymbol{x}$ of the two particles. Conservation of the angular momentum (= rotational invariance) imposes the additional restriction that the potential may depend only on the interparticle distance $w(\boldsymbol{r}) = w(r)$.

As an example, consider an interaction operator of the form (1.82)

$$V = \frac{q_1 q_2}{2(2\pi)^3\hbar}\int \frac{d\boldsymbol{p}d\boldsymbol{p}'d\boldsymbol{k}}{k^2+\lambda^2c^2} a^\dagger_{\boldsymbol{p}'-\boldsymbol{k}} a^\dagger_{\boldsymbol{p}+\boldsymbol{k}} a_{\boldsymbol{p}} a_{\boldsymbol{p}'}, \quad (1.88)$$

where the constants q_1 and q_2 are interpreted as charges of the two particles and $D(\boldsymbol{k}) = q_1q_2/(16\pi^3\hbar(k^2+\lambda^2c^2))$. Then in the position representation this interaction turns into the *Yukawa potential*

$$w(\boldsymbol{r}) = \frac{q_1 q_2}{(2\pi)^3\hbar}\int \frac{d\boldsymbol{k}}{k^2+\lambda^2c^2} e^{\frac{i}{\hbar}\boldsymbol{k}\cdot\boldsymbol{r}} = \frac{q_1q_2}{4\pi r} e^{-\lambda cr/\hbar}. \quad (1.89)$$

It becomes the familiar Coulomb potential $w(\boldsymbol{r}) \to q_1q_2/(4\pi r)$ in the limit $\lambda \to 0$. In many-electron sectors, the operator (1.88) is a sum of two-particle potentials over all particle pairs, as in (**1**-6.53).

1.2.8 Momentum-dependent potentials

Let us now consider the general case (1.85). Without loss of generality, we can represent the function $D(\boldsymbol{p}', \boldsymbol{p}, \boldsymbol{k})$ as a series,[37] i. e.,

$$D(\boldsymbol{p}', \boldsymbol{p}, \boldsymbol{k}) = \sum_j \chi_j(\boldsymbol{p}, \boldsymbol{p}') D_j(\boldsymbol{k}).$$

Then we obtain

$$\begin{aligned}
V\Psi(\boldsymbol{x}, \boldsymbol{y}) &= 2\sum_j \int d\boldsymbol{k} e^{\frac{i}{\hbar}\boldsymbol{k}\cdot(\boldsymbol{y}-\boldsymbol{x})} D_j(\boldsymbol{k}) \left[\frac{1}{(2\pi\hbar)^3} \int d\boldsymbol{p} d\boldsymbol{q} \chi_j(\boldsymbol{p}, \boldsymbol{q}) e^{\frac{i}{\hbar}\boldsymbol{p}\cdot\boldsymbol{x} + \frac{i}{\hbar}\boldsymbol{q}\cdot\boldsymbol{y}} \Psi(\boldsymbol{p}, \boldsymbol{q}) \right] \\
&= \sum_j w_j(\boldsymbol{y} - \boldsymbol{x}) \chi_j(\hat{\boldsymbol{p}}, \hat{\boldsymbol{q}}) \left[\frac{1}{(2\pi\hbar)^3} \int d\boldsymbol{p} d\boldsymbol{q} e^{\frac{i}{\hbar}\boldsymbol{p}\cdot\boldsymbol{x} + \frac{i}{\hbar}\boldsymbol{q}\cdot\boldsymbol{y}} \Psi(\boldsymbol{p}, \boldsymbol{q}) \right] \\
&= \sum_j w_j(\boldsymbol{y} - \boldsymbol{x}) \chi_j(\hat{\boldsymbol{p}}, \hat{\boldsymbol{q}}) \Psi(\boldsymbol{x}, \boldsymbol{y}),
\end{aligned} \tag{1.90}$$

where $\hat{\boldsymbol{p}} = -i\hbar(d/d\boldsymbol{x})$ and $\hat{\boldsymbol{q}} = -i\hbar(d/d\boldsymbol{y})$ are position space representatives of the momentum operators of the two particles and

$$w_j(\boldsymbol{r}) \equiv 2 \int d\boldsymbol{k} e^{\frac{i}{\hbar}\boldsymbol{k}\cdot\boldsymbol{r}} D_j(\boldsymbol{k}).$$

Thus, it follows from formula (1.90) that interaction of the type $a^\dagger a^\dagger aa$ can always be represented as a two-particle potential, which depends on the distance between the particles and on their momenta. This knowledge will be helpful to us in the third volume of the book.

37 For example, a series of this kind is obtained in a Taylor expansion with respect to the variable $\boldsymbol{k}$, where χ_j are coefficients that depend on $\boldsymbol{p}$ and $\boldsymbol{p}'$.

2 Scattering in Fock space

There are many theorists, myself included, who feel that we're in a totally unique time, where the questions on the table are the really huge, structural ones, not the details of the next particle. We're very lucky to get to live in a period like this – even if there may not be major, verified progress in our lifetimes.
Nima Arkani-Hamed

2.1 Toy model theory

Before turning to the full-fledged quantum electrodynamics, in this section we are going to practice on a simple – but at the same time quite realistic – model theory with a variable number of particles. In this model, the perturbation theory series for the S-operator (**1**-7.14)–(**1**-7.15) can be calculated with minimal effort using a convenient *diagram technique*.

2.1.1 Fock space and Hamiltonian

The toy model, which we are going to study, is a distant relative of QED. In this model, only two types of particles are present: electrons and photons. So, the part of the full Fock space that interests us here is the direct sum of electron–photon sectors with arbitrary particle numbers, as in (1.2)–(1.10). For simplicity, we also assume that the interaction does not affect the electron's spin and the photon's polarization. So, we will omit the corresponding state labels. For reasons that will become clear in Chapter 4, it is assumed that the photons have a small nonzero mass $\lambda > 0$ and the energy $\varepsilon_{\boldsymbol{p}} = \sqrt{\lambda^2 c^4 + p^2 c^2}$. One can always go to physical massless photons by taking the $\lambda \to 0$ limit.

The necessary (anti)commutation relations of the particle operators are taken from (1.24)–(1.27):

$$\{a_{\boldsymbol{p}}, a^\dagger_{\boldsymbol{p}'}\} = \delta(\boldsymbol{p} - \boldsymbol{p}'), \tag{2.1}$$

$$[c_{\boldsymbol{p}}, c^\dagger_{\boldsymbol{p}'}] = \delta(\boldsymbol{p} - \boldsymbol{p}'), \tag{2.2}$$

$$\{a_{\boldsymbol{p}}, a_{\boldsymbol{p}'}\} = \{a^\dagger_{\boldsymbol{p}}, a^\dagger_{\boldsymbol{p}'}\} = 0, \tag{2.3}$$

$$[c_{\boldsymbol{p}}, c_{\boldsymbol{p}'}] = [c^\dagger_{\boldsymbol{p}}, c^\dagger_{\boldsymbol{p}'}] = 0, \tag{2.4}$$

$$[a^\dagger_{\boldsymbol{p}}, c^\dagger_{\boldsymbol{p}'}] = [a^\dagger_{\boldsymbol{p}}, c_{\boldsymbol{p}'}] = [a_{\boldsymbol{p}}, c^\dagger_{\boldsymbol{p}'}] = [a_{\boldsymbol{p}}, c_{\boldsymbol{p}'}] = 0. \tag{2.5}$$

https://doi.org/10.1515/9783110493207-002

The full Hamiltonian[1]

$$H^n = H_0 + V_1, \tag{2.6}$$

as usual, is the sum of the free Hamiltonian

$$H_0 = \int d\boldsymbol{p}\omega_{\boldsymbol{p}} a^\dagger_{\boldsymbol{p}} a_{\boldsymbol{p}} + \int d\boldsymbol{k}\varepsilon_{\boldsymbol{k}} c^\dagger_{\boldsymbol{k}} c_{\boldsymbol{k}}$$

and the interaction, which we will choose in the following (*unphys*) form[2]:

$$V_1 = \frac{e\hbar c}{(2\pi\hbar)^{3/2}} \int \frac{d\boldsymbol{p}d\boldsymbol{k}}{\sqrt{2\varepsilon_{\boldsymbol{k}}}} a^\dagger_{\boldsymbol{p}} c^\dagger_{\boldsymbol{k}} a_{\boldsymbol{p}+\boldsymbol{k}} + \frac{e\hbar c}{(2\pi\hbar)^{3/2}} \int \frac{d\boldsymbol{p}d\boldsymbol{k}}{\sqrt{2\varepsilon_{\boldsymbol{k}}}} a^\dagger_{\boldsymbol{p}} a_{\boldsymbol{p}-\boldsymbol{k}} c_{\boldsymbol{k}}. \tag{2.7}$$

The *coupling constant* e is equal to the absolute value of the electron charge.

In the theory constructed above, the conservation laws are valid, in particular,

$$[H^n, \boldsymbol{P}_0] = [H^n, \boldsymbol{J}_0] = [H^n, Q] = 0,$$

where operators $\boldsymbol{P}_0, \boldsymbol{J}_0$ and Q refer to the total momentum

$$\boldsymbol{P}_0 = \int d\boldsymbol{p}\boldsymbol{p}(a^\dagger_{\boldsymbol{p}} a_{\boldsymbol{p}} + c^\dagger_{\boldsymbol{p}} c_{\boldsymbol{p}}),$$

the total angular momentum $\boldsymbol{J}_0$ and the electric charge

$$Q = -e \int d\boldsymbol{p} a^\dagger_{\boldsymbol{p}} a_{\boldsymbol{p}},$$

respectively. The number of electrons is conserved due to the charge conservation law, but the number of photons can vary without restrictions. Therefore, our model is capable of describing important processes of photon emission and absorption.

Our toy model has one important drawback: it is not relativistically invariant. In other words, it is not possible to construct an interacting boost operator $\boldsymbol{K}$ having the required Poincaré commutation relations with seven other generators $\{H^n, \boldsymbol{P}_0, \boldsymbol{J}_0\}$. Here we will close our eyes to this flaw, but in Chapter 3 we shall show how relativistic invariance can be ensured in a more satisfactory theory – QED.

Usually, in nonrelativistic quantum mechanics, investigation of a physical system begins with the diagonalization of its Hamiltonian, determination of the energy spectrum, stationary wave functions, etc. However, in quantum field theories[3] such studies are very difficult, if not impossible. Therefore, as a rule, field theoreticians are concerned with calculations of the scattering operator, which is a simpler task.

---

1 Here we label this Hamiltonian with the superscript "n" from the word "naïve." In Section 2.2 and Chapter 4 we will see that Hamiltonians of the type H^n are not suitable for describing scattering in higher orders of perturbation theory. Renormalization counterterms should be added to such Hamiltonians.

2 Such a trilinear interaction is widespread in various models of quantum field theory. Compare, for example, with the first two terms in the QED interaction (D.9).

3 The model considered here is also a simple example of quantum field theory, although we will introduce the concept of the quantum field only in Chapter 3.

2.1.2 *S*-operator in second order

Our plan is to calculate the S-operator for the interaction (2.7) using perturbation theory formulas (**1**-7.14)–(**1**-7.15), namely

$$\begin{aligned} S &= 1 + \underbrace{\Sigma_1} + \underbrace{\Sigma_2} + \underbrace{\Sigma_3} + \cdots \\ &= 1 + \underbrace{V_1} + \underbrace{V_1\underline{V_1}} + \underbrace{V_1\underline{V_1\underline{V_1}}} + \cdots . \end{aligned} \tag{2.8}$$

This formula contains terms (operators) proportional to the powers of the interaction operator and, consequently, to the powers of the coupling constant *e*. We will indicate these powers with a subscript and call them *perturbation orders*. For example, the free Hamiltonian H_0 does not depend on e, hence its order is zero; the order of the operator V_1 is equal to one, etc.

Since the coupling constant is relatively small, we will assume that terms of higher orders are negligible, so that series of the type (2.8) converge, and the first few terms give a good approximation for the whole series. In fact, these assumptions require additional justification, but we will not do this in our book.

Now let us proceed to the direct calculation of the S-operator (2.8). In the lowest perturbation orders, we have

$$\Sigma_1 = V_1, \tag{2.9}$$

$$\Sigma_2 = (V_1\underline{V_1})^{\text{unp}} + (V_1\underline{V_1})^{\text{phys}} + (V_1\underline{V_1})^{\text{ren}}. \tag{2.10}$$

As we already know, in order to obtain the corresponding contributions to the S-operator, we have to take t-integrals, i. e.,

$$S = 1 + \underbrace{\Sigma_1} + \underbrace{\Sigma_2} + \cdots . \tag{2.11}$$

First, we notice that the operator V_1 is *unphys*. Therefore, according to (1.76), the second term on the right-hand side of (2.11) vanishes. In general, *unphys* contributions can be omitted when calculating S and Σ, so

$$S = 1 + \underbrace{\Sigma_2^{\text{phys}}} + \underbrace{\Sigma_2^{\text{ren}}} + \underbrace{\Sigma_3^{\text{phys}}} + \cdots . \tag{2.12}$$

At this stage we will also drop the *renorm* contribution in (2.12).[4] Then, in the second order of perturbation theory, we get only one relevant term,

$$S_2 = \underbrace{(V_1\underline{V_1})^{\text{phys}}} + \cdots . \tag{2.13}$$

4 In fact, in a consistent theory, one should add to the Hamiltonian (2.6) a renormalization counterterm which exactly compensates the *renorm* term $\underbrace{\Sigma_2^{\text{ren}}}$. We will discuss renormalization in more detail in Section 2.2 and in Chapter 4.

The problem has been reduced to considering the product $V_1\underline{V_1}$. This calculation is not difficult. The t-integral (1.65),

$$\begin{aligned}\underline{V_1} = &-\frac{e\hbar c}{(2\pi\hbar)^{3/2}}\int \frac{d\boldsymbol{p}d\boldsymbol{k}}{\sqrt{2\varepsilon_{\boldsymbol{k}}}}\cdot\frac{a^\dagger_{\boldsymbol{p}}c^\dagger_{\boldsymbol{k}}a_{\boldsymbol{p}+\boldsymbol{k}}}{\omega_{\boldsymbol{p}}+\varepsilon_{\boldsymbol{k}}-\omega_{\boldsymbol{p}+\boldsymbol{k}}}\\ &-\frac{e\hbar c}{(2\pi\hbar)^{3/2}}\int \frac{d\boldsymbol{p}d\boldsymbol{k}}{\sqrt{2\varepsilon_{\boldsymbol{k}}}}\cdot\frac{a^\dagger_{\boldsymbol{p}}a_{\boldsymbol{p}-\boldsymbol{k}}c_{\boldsymbol{k}}}{\omega_{\boldsymbol{p}}-\varepsilon_{\boldsymbol{k}}-\omega_{\boldsymbol{p}-\boldsymbol{k}}},\end{aligned} \tag{2.14}$$

differs from V_1 only by the denominator $-\mathcal{E}^{-1}_{V_1}$. Omitting numerical factors, we see that the product in (2.13) has the following operator form:

$$\Sigma_2 = (V_1\underline{V_1})^{\text{phys}} = (a^\dagger c^\dagger a + a^\dagger ac)\underline{(a^\dagger c^\dagger a + a^\dagger ac)}. \tag{2.15}$$

To advance, it is necessary to do three things:
(1) Expand the parentheses in this product.
(2) Bring the resulting terms to the normal order.
(3) Here we are interested in the $e^- + e^-$ scattering. So, we should focus on terms with the corresponding operator structure $a^\dagger a^\dagger aa$.

After the step (1) we obtain a sum of four "primary" terms, each of which is a product of $3 \times 2 = 6$ particle operators in mixed order.

The transition to the normal order (step (2)) is carried out using contractions according to Wick's theorem 1.1. We are only interested in terms that have the structure $a^\dagger a^\dagger aa$, so not all primary terms and not all Wick contractions will be relevant. For example, the last term in the second parentheses has a photon annihilation operator on the right, and this operator will remain there after all reorderings. So, this term is not useful. Similarly, the first term in the first parenthesis has a photon creation operator in the leftmost position. This term can also be ignored.

It is not difficult to understand that there is only one contraction that yields the desired electron–electron scattering term $a^\dagger a^\dagger aa$, i. e.,

$$(\cdots + a^\dagger a\overbrace{c)\underline{(a^\dagger c^\dagger}}\underline{a + \cdots)}. \tag{2.16}$$

2.1.3 Drawing diagrams in toy model

Along with Wick's theorem, the diagram technique is another convenient method for calculating and analyzing contributions to the S-operator. The diagrams in this subsection are prototypes of Feynman diagrams in QED, which will be introduced in Subsection 3.2.4.

Let us first represent graphically the two terms in the interaction potential (2.7) as *vertices* in Figures 2.1 (a) and (b). Annihilation/creation operators in V_1 are shown by

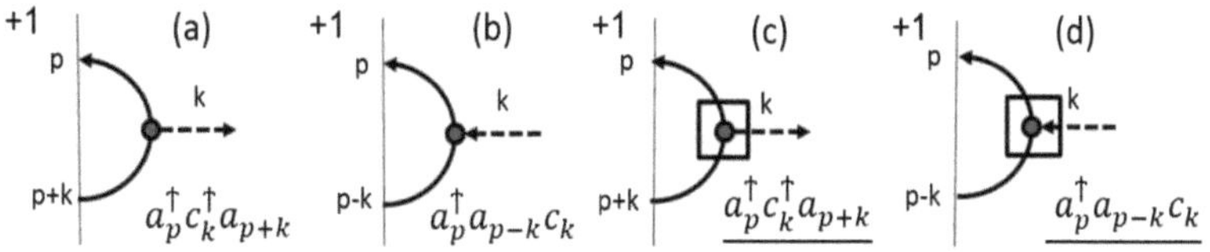

Figure 2.1: Diagram representation of the operators V_1 = (a) + (b) and $\underline{V_1}$ = (c) + (d).

directed *lines* (arrows). The annihilation line *enters* the vertex, and the creation line *leaves* the vertex. Electron lines are shown by solid arcs and photon lines by straight dashed arrows. Hence, each vertex has three lines touching it: two electron lines (one incoming and one outgoing) and one photon line. Each line is labeled with the momentum of the corresponding particle operator. The free ends of electronic lines touch the vertical "order axis" on the left of each graph. The points where these *external lines* touch the axis (from the bottom of the axis upwards) correspond to the order in which the electron operators are placed in the potential (from right to left). For example, in the potential shown in Figure 2.1 (a), the annihilator $a_{\boldsymbol{p}+\boldsymbol{k}}$ is to the right of the creator $a_{\boldsymbol{p}}^\dagger$, so the line of the former operator touches the order axis at a lower point. The numerical multiplier associated with each diagram is indicated in the upper left corner.

In Figures 2.1 (c) and (d) we showed diagrams for the two terms in $\underline{V_1}$ (2.14). The energy denominators are represented by rectangles surrounding the vertices. The rectangles are drawn in such a way that the energies of the crossing lines give correct contributions to the energy denominator. Lines leaving the rectangle (= created particles) contribute their energy with a positive sign to the total energy function $\mathcal{E}_{V_1}$. Lines entering the rectangle (= annihilated particles) give negative contributions to $\mathcal{E}_{V_1}$. In addition, a common factor (−1) is associated with each rectangle. In more complex diagrams, there can be several such rectangles (and denominators), and rectangles may enclose several vertices.

The diagram corresponding to the product of two operators (diagrams) *AB* is obtained simply by placing the diagram *B* under the diagram *A* in one graph. For example, the product of the second term in (2.7) (Figure 2.1 (b)) and the first term in (2.14) (Figure 2.1 (c)),

$$V_1 \underline{V_1} \propto (a_{\boldsymbol{p}}^\dagger a_{\boldsymbol{p}-\boldsymbol{k}} c_{\boldsymbol{k}}) \underline{(a_{\boldsymbol{q}}^\dagger c_{\boldsymbol{k}'}^\dagger a_{\boldsymbol{q}+\boldsymbol{k}'})} + \cdots, \tag{2.17}$$

is shown in Figure 2.2 (a).[5] This product should be converted to a normally ordered form, i. e., all creation operators moved to the left, which corresponds to moving all

5 We agree to place free ends of the external photon lines on the right side of the diagram. The order of these free ends (from bottom to top) corresponds to the order of the photon operators in the algebraic expression (from right to left). For example, in Figure 2.2 (a) the incoming photon line is above the outgoing line, which corresponds to the order $c \cdots c^\dagger$ of the photon operators in (2.17).

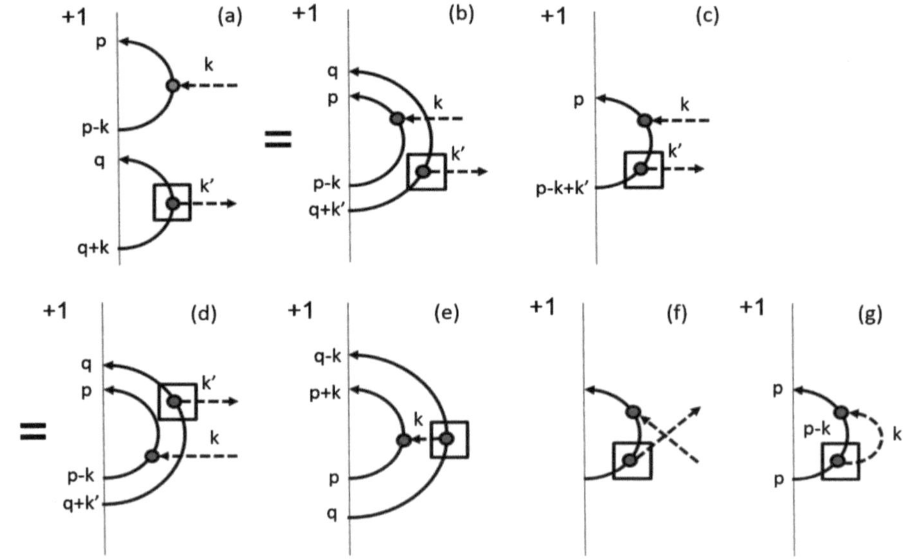

Figure 2.2: Normal ordering of the product 2.1 (b) × 2.1 (c).

outgoing lines to the top of the diagram. During these transformations, we will need to perform permutations of operators/lines (outgoing)(outgoing') → (outgoing')(outgoing) and (incoming)(outgoing) → (outgoing)(incoming). Due to the (anti)commutation relations (2.3), each permutation of two electron creation operators (two outgoing solid lines) changes the sign of the diagram. From (2.4) it follows that the permutation of two outgoing photon lines has no additional effect. The permutation of incoming and outgoing lines of identical particles has more significant consequences. According to (2.1) and (2.2), such a rearrangement leads to the formation of a new (secondary) operator expression and the corresponding new diagram. In this expression, the two swapped operators are replaced by a delta function. In the diagram, the contracted external lines are combined into one (internal) line, which connects the two vertices directly. Permuting operators (lines) of different particles has no effect due to (2.5).

Let us apply these rules to (2.17), i. e., the diagram 2.2 (a). First we move the electron creation operator $a^\dagger_{\boldsymbol{q}}$ to the leftmost position and add an extra term (diagram) due to the anticommutator $a_{\boldsymbol{p}-\boldsymbol{k}}a^\dagger_{\boldsymbol{q}} = -a^\dagger_{\boldsymbol{q}}a_{\boldsymbol{p}-\boldsymbol{k}} + \delta(\boldsymbol{q}-\boldsymbol{p}+\boldsymbol{k})$. Then we have

$$\begin{aligned}\text{Figure 2.2 (a)} &\propto a^\dagger_{\boldsymbol{q}}a^\dagger_{\boldsymbol{p}}a_{\boldsymbol{p}-\boldsymbol{k}}a_{\boldsymbol{q}+\boldsymbol{k}'}c_{\boldsymbol{k}}c^\dagger_{\boldsymbol{k}'} + \delta(\boldsymbol{q}-\boldsymbol{p}+\boldsymbol{k})a^\dagger_{\boldsymbol{p}}a_{\boldsymbol{q}+\boldsymbol{k}'}c_{\boldsymbol{k}}c^\dagger_{\boldsymbol{k}'} \\ &= a^\dagger_{\boldsymbol{q}}a^\dagger_{\boldsymbol{p}}a_{\boldsymbol{p}-\boldsymbol{k}}a_{\boldsymbol{q}+\boldsymbol{k}'}c_{\boldsymbol{k}}c^\dagger_{\boldsymbol{k}'} + a^\dagger_{\boldsymbol{p}}a_{\boldsymbol{p}-\boldsymbol{k}+\boldsymbol{k}'}c_{\boldsymbol{k}}c^\dagger_{\boldsymbol{k}'}.\end{aligned} \tag{2.18}$$

Expression (2.18) is represented by two diagrams in Figures 2.2 (b) and 2.2 (c). The diagram 2.2 (b) was obtained by moving the electron line labeled $\boldsymbol{q}$ to the top of the order axis. During this move, the line $\boldsymbol{q}$ has crossed other electron lines twice, so the sign of the diagram remained unchanged. On the diagram 2.2 (c), we contracted the incoming electron line $\boldsymbol{p}-\boldsymbol{k}$ with the outgoing line $\boldsymbol{q}$, thus obtaining an internal line carrying the momentum $\boldsymbol{p}-\boldsymbol{k}$ between two vertices.

In the expression (2.18), the electron operators have settled in the normal order – but the photon operators not yet. In the following step we reorder the photon operators:

$$\begin{aligned}\text{Figure 2.2 (a)} \propto\; & a^\dagger_{\boldsymbol{q}} a^\dagger_{\boldsymbol{p}} c^\dagger_{\boldsymbol{k}'} a_{\boldsymbol{p}-\boldsymbol{k}} a_{\boldsymbol{q}+\boldsymbol{k}'} c_{\boldsymbol{k}} + a^\dagger_{\boldsymbol{q}-\boldsymbol{k}} a^\dagger_{\boldsymbol{p}+\boldsymbol{k}'} a_{\boldsymbol{p}} a_{\boldsymbol{q}} \delta(\boldsymbol{k}'-\boldsymbol{k}) \\ & + a^\dagger_{\boldsymbol{p}} c^\dagger_{\boldsymbol{k}'} a_{\boldsymbol{p}-\boldsymbol{k}+\boldsymbol{k}'} c_{\boldsymbol{k}} + a^\dagger_{\boldsymbol{p}} a_{\boldsymbol{p}-\boldsymbol{k}+\boldsymbol{k}'} \delta(\boldsymbol{k}'-\boldsymbol{k}) \\ =\; & a^\dagger_{\boldsymbol{q}} a^\dagger_{\boldsymbol{p}} c^\dagger_{\boldsymbol{k}'} a_{\boldsymbol{p}-\boldsymbol{k}} a_{\boldsymbol{q}+\boldsymbol{k}'} c_{\boldsymbol{k}} + a^\dagger_{\boldsymbol{q}-\boldsymbol{k}} a^\dagger_{\boldsymbol{p}+\boldsymbol{k}} a_{\boldsymbol{p}} a_{\boldsymbol{q}} \\ & + a^\dagger_{\boldsymbol{p}} c^\dagger_{\boldsymbol{k}'} a_{\boldsymbol{p}-\boldsymbol{k}+\boldsymbol{k}'} c_{\boldsymbol{k}} + a^\dagger_{\boldsymbol{p}} a_{\boldsymbol{p}}.\end{aligned} \tag{2.19}$$

The resulting four terms are shown in Figures 2.2 (d)–(g). Obviously, these are the same terms that would result from the application of Wick's theorem to (2.17),[6] i. e.,

$$\begin{aligned}(a^\dagger_{\boldsymbol{p}} a_{\boldsymbol{p}-\boldsymbol{k}} c_{\boldsymbol{k}})\underline{(a^\dagger_{\boldsymbol{q}} c^\dagger_{\boldsymbol{k}'} a_{\boldsymbol{q}+\boldsymbol{k}'})} =\; & :(a^\dagger_{\boldsymbol{p}} a_{\boldsymbol{p}-\boldsymbol{k}} c_{\boldsymbol{k}})\underline{(a^\dagger_{\boldsymbol{q}} c^\dagger_{\boldsymbol{k}'} a_{\boldsymbol{q}+\boldsymbol{k}'})}: + (a^\dagger_{\boldsymbol{p}} a_{\boldsymbol{p}-\boldsymbol{k}} \overbracket{c_{\boldsymbol{k}})\underline{(a^\dagger_{\boldsymbol{q}} c^\dagger_{\boldsymbol{k}'}} a_{\boldsymbol{q}+\boldsymbol{k}'})} \\ & + (a^\dagger_{\boldsymbol{p}} \underbracket{a_{\boldsymbol{p}-\boldsymbol{k}} c_{\boldsymbol{k}})\underline{(a^\dagger_{\boldsymbol{q}}} c^\dagger_{\boldsymbol{k}'} a_{\boldsymbol{q}+\boldsymbol{k}'})} + (a^\dagger_{\boldsymbol{p}} \underbracket{a_{\boldsymbol{p}-\boldsymbol{k}} \overbracket{c_{\boldsymbol{k}})\underline{(a^\dagger_{\boldsymbol{q}}} c^\dagger_{\boldsymbol{k}'}} a_{\boldsymbol{q}+\boldsymbol{k}'})}.\end{aligned}$$

2.1.4 Reading diagrams in toy model

In the course of diagrammatic manipulations described in the previous section, we do not really need to keep track of the momentum labels of particle lines. A complete algebraic expression is easy to reconstruct from an unlabeled drawing using the following steps:

(I) Assign different momentum labels to all external lines except the one whose momentum is obtained from the conservation condition.[7]

(II) Assign momentum labels to all internal lines so that the momentum conservation law is observed at each individual vertex. If the diagram has *loops*, then for each independent loop you have to introduce an additional *loop momentum*.[8]

(III) Read external lines from top to bottom and write the corresponding particle operators from left to right.[9]

---

6 To complete the calculation of $\underline{V_1 V_1}$, we need to consider three more products, 2.1 (a) × 2.2 (a) + 2.1 (a) × 2.2 (b) + 2.1 (b) × 2.2 (b). They do not contribute to the electron–electron scattering, and we leave them as an exercise for the reader.

7 The sum of momenta of all incoming (annihilated) particles should be equal to the sum of momenta of all outgoing (created) particles.

8 See diagram 2.2 (g), where $\boldsymbol{k}$ is the loop momentum.

9 If the diagram is normally ordered, then the outgoing lines (creation operators) are followed by the incoming lines (annihilation operators).

(IV) For each rectangle, write the factor $(-1)/(E_f - E_i)$, where E_i is the sum of energies of the particles (lines) entering the rectangle and E_f is the sum of energies of the outgoing particles.

(V) For each vertex, write the factor $\frac{e\hbar c}{\sqrt{2(2\pi\hbar)^3 \varepsilon_k}}$, where $\boldsymbol{k}$ is the momentum of the photon line attached to the vertex.

(VI) Integrate the resulting expression with respect to all independent external and loop momenta.

With certain practice, diagram calculations can be performed with considerable speed and accuracy.

2.1.5 Scattering in second order

Let us now analyze the four second-order diagrams calculated above: 2.2 (d)–(g). The diagram 2.2 (d) is *disconnected*, since it consists of two unrelated pieces. Such diagrams describe independent scattering processes,[10] so they should not interest us, and we will simply ignore them here.

The term 2.2 (g) has the structure $a^\dagger a$, i. e., it belongs to the *renorm* type. We will discuss it in Section 2.2.

The potential 2.2 (f) $a^\dagger c^\dagger ac$ annihilates the pair electron + photon in the initial state and recreates the same particles in the final state. Hence, it describes the elastic electron–photon (Compton) scattering.

Let us consider in more detail the electron–electron scattering term $(a^\dagger a^\dagger aa)$ described by the diagram in Figure 2.2 (e). In accordance with rules (I)–(VI), its algebraic expression is

$$\text{Figure 2.2 (e)} = -\frac{\hbar^2 e^2 c^2}{2(2\pi\hbar)^3} \int \frac{d\boldsymbol{p}d\boldsymbol{q}d\boldsymbol{k}}{\varepsilon_{\boldsymbol{k}}(\varepsilon_{\boldsymbol{k}} + \omega_{\boldsymbol{q}-\boldsymbol{k}} - \omega_{\boldsymbol{q}})} a^\dagger_{\boldsymbol{q}-\boldsymbol{k}} a^\dagger_{\boldsymbol{p}+\boldsymbol{k}} a_{\boldsymbol{p}} a_{\boldsymbol{q}}.$$

To take the t-integral we will use formula (1.66). We have

$$\underbrace{\text{Figure 2.2 (e)}} = \frac{\pi i e^2 \hbar^2 c^2}{(2\pi\hbar)^3} \int d\boldsymbol{p}d\boldsymbol{q}d\boldsymbol{k} \frac{\delta(\omega_{\boldsymbol{q}-\boldsymbol{k}} + \omega_{\boldsymbol{p}+\boldsymbol{k}} - \omega_{\boldsymbol{p}} - \omega_{\boldsymbol{q}})}{\varepsilon_{\boldsymbol{k}}(\varepsilon_{\boldsymbol{k}} + \omega_{\boldsymbol{p}-\boldsymbol{k}} - \omega_{\boldsymbol{p}})} a^\dagger_{\boldsymbol{q}-\boldsymbol{k}} a^\dagger_{\boldsymbol{p}+\boldsymbol{k}} a_{\boldsymbol{p}} a_{\boldsymbol{q}}, \quad (2.20)$$

where the delta function expresses the conservation of energy in the scattering event.

In the nonrelativistic approximation $(p, q, k \ll m_e c)$,

$$\omega_{\boldsymbol{p}} \equiv \sqrt{p^2c^2 + m_e^2 c^4} \approx m_e c^2 + \frac{p^2}{2m_e}, \quad (2.21)$$

10 Possibly occurring in different corners of the universe.

$$\begin{aligned}\varepsilon_{\boldsymbol{k}}(\varepsilon_{\boldsymbol{k}} + \omega_{\boldsymbol{p}-\boldsymbol{k}} - \omega_{\boldsymbol{p}}) &\approx \varepsilon_{\boldsymbol{k}}\left(\varepsilon_{\boldsymbol{k}} + m_e c^2 + \frac{(\boldsymbol{p}-\boldsymbol{k})^2}{2m_e} - m_e c^2 - \frac{p^2}{2m_e}\right)\\ &\approx \varepsilon_{\boldsymbol{k}}^2 = \lambda^2 c^4 + k^2 c^2.\end{aligned} \tag{2.22}$$

Substituting this result in (2.20), we obtain the desired second-order contribution to the S-operator,

$$S_2[a^\dagger a^\dagger aa] \approx \frac{ie^2}{8\pi^2\hbar}\int d\boldsymbol{p}d\boldsymbol{q}d\boldsymbol{k}\frac{\delta(\omega_{\boldsymbol{q}-\boldsymbol{k}} + \omega_{\boldsymbol{p}+\boldsymbol{k}} - \omega_{\boldsymbol{p}} - \omega_{\boldsymbol{q}})}{\lambda^2 c^2 + k^2}a^\dagger_{\boldsymbol{q}-\boldsymbol{k}}a^\dagger_{\boldsymbol{p}+\boldsymbol{k}}a_{\boldsymbol{p}}a_{\boldsymbol{q}}. \tag{2.23}$$

As expected, this operator is nonzero only for particle momenta lying on the *energy shell*, which is the solution of the equation

$$\omega_{\boldsymbol{q}-\boldsymbol{k}} + \omega_{\boldsymbol{p}+\boldsymbol{k}} = \omega_{\boldsymbol{q}} + \omega_{\boldsymbol{p}}. \tag{2.24}$$

Knowing coefficient functions of the S-operator on its energy shell, it is not difficult to obtain scattering amplitudes and cross sections by usual formulas of collision theory [7]. It appears that our "naïve" theory is in a fair agreement with experiments on electron–electron and photon–electron scattering. However, this agreement is limited only to the lowest (second) order of perturbation theory. Higher-order diagrams inevitably contain loops, and integrals with respect to loop momenta quite often diverge. In Section 2.2, we will see how it is possible to "repair" the theory by introducing renormalization.

2.2 Renormalization in toy model

2.2.1 Renormalization of electron self-scattering in second order

According to our diagram rules, the *renorm* term in Figure 2.2 (g) is

$$\Sigma_2^{\text{ren}} = (V_1\underline{V_1})^{\text{ren}} = -\frac{e^2\hbar^2c^2}{2(2\pi\hbar)^3}\int d\boldsymbol{q}\left(\int\frac{d\boldsymbol{k}}{\varepsilon_{\boldsymbol{k}}(\omega_{\boldsymbol{q}-\boldsymbol{k}} - \omega_{\boldsymbol{q}} + \varepsilon_{\boldsymbol{k}})}\right)a^\dagger_{\boldsymbol{q}}a_{\boldsymbol{q}}. \tag{2.25}$$

The presence of this term has catastrophic consequences for the theory. First, the integral with respect to $\boldsymbol{k}$ diverges. However, this divergence is not even the most disturbing feature. Even if this integral converged, we would have to deal with a t-independent renorm term Σ_2^{ren} whose contribution to the S-operator is given by the divergent t-integral

$$\begin{aligned}S_2[a^\dagger a]^{\text{ren}} &= \underbrace{(V_1\underline{V_1})^{\text{ren}}}\\ &= -\frac{i}{\hbar}\int\limits_{-\infty}^{\infty} dt\left(-\frac{e^2\hbar^2c^2}{2(2\pi\hbar)^3}\int\frac{d\boldsymbol{q}d\boldsymbol{k}}{\varepsilon_{\boldsymbol{k}}(\omega_{\boldsymbol{q}-\boldsymbol{k}} - \omega_{\boldsymbol{q}} + \varepsilon_{\boldsymbol{k}})}a^\dagger_{\boldsymbol{q}}a_{\boldsymbol{q}}\right) = \infty.\end{aligned} \tag{2.26}$$

The deep reason for this divergence is that our interaction (2.7), in fact, is an "persistent" self-action. It acts constantly even on a free electron. We can say that in our naïve theory, the electron is permanently scattered on itself by emitting and absorbing "virtual photons." This contradicts the fundamental assumption of scattering theory that interactions between particles are activated only during short collision intervals. So, strictly speaking, scattering theory cannot be applied to our model. It would be natural to simply declare our theory untenable and discard it. However, in the mid-20th century, having met similar problems in QED, physicists decided in a different way. Instead of discarding this divergent theory, they decided to fix it, trying to remove the divergences. This repair of quantum field theories is called the *renormalization*. Here we will consider the simplest version of such a procedure.

According to the general renormalization recipe, we remove the divergence (2.26) by changing our Hamiltonian. This change is achieved by adding to H^n a second-order renormalization *counterterm*,

$$Q_2 \equiv -(V_1 \underline{V_1})^{\text{ren}} = \frac{e^2\hbar^2 c^2}{2(2\pi\hbar)^3} \int \frac{d\boldsymbol{q}d\boldsymbol{k}}{\varepsilon_{\boldsymbol{k}}(\omega_{\boldsymbol{q}-\boldsymbol{k}} - \omega_{\boldsymbol{q}} + \varepsilon_{\boldsymbol{k}})} a^\dagger_{\boldsymbol{q}} a_{\boldsymbol{q}}. \tag{2.27}$$

Then the Hamiltonian, renormalized up to the second order, takes the form[11]

$$H^c = H_0 + V_1 + Q_2.$$

To make sure that the counterterm (2.27) does cancel the undesirable divergent contribution (2.25), we substitute the new interaction

$$V^c = V_1 + Q_2 \tag{2.28}$$

into the standard formula (2.8) for the Σ-operator. Then we have

$$\begin{aligned}(\Sigma^c)^{\text{ren}} &= [V_1 + Q_2]^{\text{ren}} + [(V_1 + Q_2)\underline{(V_1 + Q_2)}]^{\text{ren}} + \Big[(V_1 + Q_2)\underline{(V_1 + Q_2)\underline{(V_1 + Q_2)}}\Big]^{\text{ren}} \\ &\quad + \Big[(V_1 + Q_2)\underline{(V_1 + Q_2)\underline{(V_1 + Q_2)\underline{(V_1 + Q_2)}}}\Big]^{\text{ren}} + \cdots. \end{aligned} \tag{2.29}$$

Restricting ourselves to terms of the second order, we get

$$(\Sigma^c_2)^{\text{ren}} = Q_2 + (V_1 \underline{V_1})^{\text{ren}} = 0.$$

Thus, the S-operator contribution from the counterterm Q_2 cancels exactly with the undesirable *renorm* term (2.26).[12]

---

11 The subscript "c" will label the Hamiltonian with counterterms.

12 In fact, the operators Q_2 and $(V_1 \underline{V_1})^{\text{ren}}$ are infinite, so our reasoning about the cancellation is formal and nonrigorous. A more consistent approach should include *regularization* of divergent integrals, for example, as explained in Subsection 4.1.5.

Now we can understand why theorists often say that the renormalization “sweeps the divergences under the rug.” The role of the rug is played by the Hamiltonian. We decided for ourselves that we will seek physical observable effects (scattering cross sections, decay probabilities, etc.) in the S-matrix. Therefore, we wanted to free the S-matrix from divergences. We did this by merely adding compensating divergences to the Hamiltonian. One would say: “How is this possible?! After all, the Hamiltonian is an important physical operator representing an observable quantity (energy) and generating the time evolution of states and observables. What right do we have to add divergent counterterms to it?” This is a good question, and we will try to answer it in Volume 3 of our book.

2.2.2 Renormalization of electron self-scattering in fourth order

In the previous subsection, we considered renormalization in the second order of perturbation theory. In the full theory, this procedure should be repeated also in higher orders. Third-order contributions to the S-operator have odd numbers of particle creation and annihilation operators, so there are no suspicious terms $a^\dagger a$. However, such terms reappear in the fourth order. Indeed, consider the fourth-order term in (2.29),

$$\begin{aligned}\Sigma_4[a^\dagger a]^{\mathrm{ren}} &= [Q_2\underline{Q_2}]^{\mathrm{ren}} + [V_1\underline{V_1 Q_2}]^{\mathrm{ren}} + \left[V_1\underline{Q_2\underline{V_1}}\right]^{\mathrm{ren}} \\ &\quad + \left[Q_2\underline{V_1\underline{V_1}}\right]^{\mathrm{ren}} + \left[V_1\underline{V_1\underline{V_1\underline{V_1}}}\right]^{\mathrm{ren}}. \end{aligned} \tag{2.30}$$

Let us explain in some detail how to evaluate such expressions. For example, omitting numerical factors and integral signs, the last term in (2.30) takes the form

$$\begin{aligned}&(a^\dagger_{p1}a_{q1}c^\dagger_{k1} + a^\dagger_{p1}a_{q1}c_{k1}) \\ &\quad \times \underline{(a^\dagger_{p2}a_{q2}c^\dagger_{k2} + a^\dagger_{p2}a_{q2}c_{k2})\underline{(a^\dagger_{p3}a_{q3}c^\dagger_{k3} + a^\dagger_{p3}a_{q3}c_{k3})\underline{(a^\dagger_{p4}a_{q4}c^\dagger_{k4} + a^\dagger_{p4}a_{q4}c_{k4})}}}. \end{aligned} \tag{2.31}$$

To extract renorm contributions, it is necessary to do three things:
(1) expand the four parentheses in this product;
(2) convert the resulting terms to the normal order by the Wick theorem;
(3) collect terms with the operator structure $a^\dagger a$.

After step (1) we obtain a sum of 16 “primary” terms, each of which is a product of $3 \times 4 = 12$ particle operators in a mixed order. The normal ordering (step (2)) would lead to dozens of terms. However, only few of them are of interest to us. It is not difficult to see that $a^\dagger a$ terms can appear only from the following contractions[13]:

$$(\cdots + a^\dagger ac)(\cdots + a^\dagger ac)(a^\dagger ac^\dagger + \cdots)(a^\dagger ac^\dagger + \cdots), \tag{2.32}$$

$$(\cdots + a^\dagger ac)(\cdots + a^\dagger ac)(a^\dagger ac^\dagger + \cdots)(a^\dagger ac^\dagger + \cdots), \tag{2.33}$$

$$(\cdots + a^\dagger ac)(a^\dagger ac^\dagger + \cdots)(\cdots + a^\dagger ac)(a^\dagger ac^\dagger + \cdots), \tag{2.34}$$

whose results are conveniently represented by three diagrams in Figure 2.3.[14]

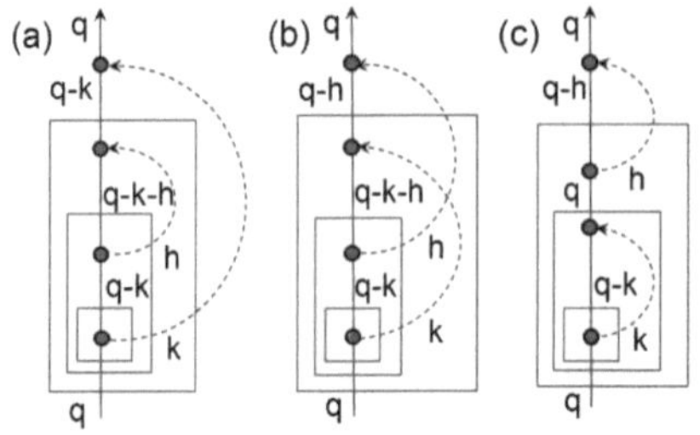

Figure 2.3: Diagrams corresponding to the three terms (2.32)–(2.34).

Next consider the first four terms on the right-hand side of (2.30). Taking into account the graphical representation of the counterterm vertex (2.27) in Figure 2.4, we get the four corresponding diagrams in Figures 2.5 (a)–(d).

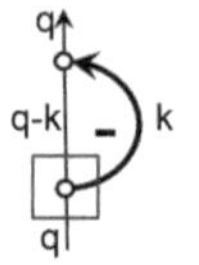

Figure 2.4: Diagram of the counterterm $Q_2 = -(\underline{V_1 V_1})^{\text{ren}}$. For easier identification of this fragment in complex diagrams, the internal photon arc is shown by a full bold line. The total "minus" sign is placed inside the loop.

Obviously the pairs of diagrams (2.3 (c) + 2.5 (a)) and (2.5 (c) + 2.5 (d)) cancel out.[15] To advance, we have to translate the three remaining diagrams 2.3 (a) + 2.3 (b) + 2.5 (b) into algebraic expressions. Applying rules from Subsection 2.1.4, we get the fourth-order *renorm* contribution

$$\Sigma_4[a^\dagger a]^{\text{ren}} = \int d\boldsymbol{q}\sigma_4(\boldsymbol{q})a^\dagger_{\boldsymbol{q}} a_{\boldsymbol{q}}$$

13 To avoid cluttering, we omitted the "underline" signs.

14 These diagrams were constructed by the usual rules. The only exception is that we no longer draw the "order axis" and do not connect fermion external lines to it.

15 Diagrams in these pairs have the same topology, but the numbers of "minus" signs differ by one.

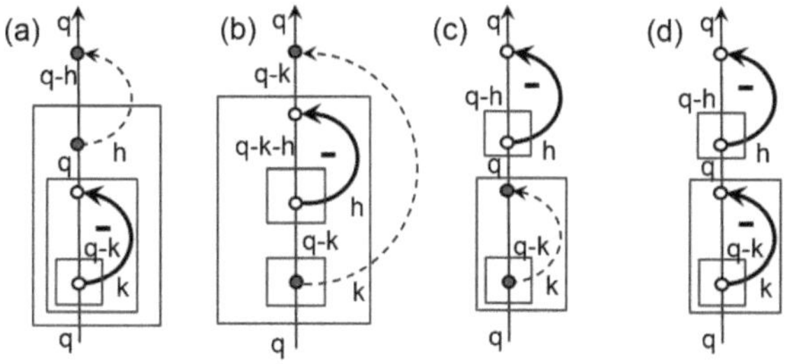

Figure 2.5: Contributions to Σ_4^{ren} from the first four terms in (2.30).

with the coefficient function

$$\begin{aligned}
\sigma_4(\boldsymbol{q}) &= \frac{e^4\hbar^4c^4}{4(2\pi\hbar)^6}\int\frac{d\boldsymbol{k}d\boldsymbol{h}}{\varepsilon_{\boldsymbol{k}}\varepsilon_{\boldsymbol{h}}} \\
&\quad\times\left[-\frac{1}{(\omega_{\boldsymbol{q}-\boldsymbol{k}}-\omega_{\boldsymbol{q}}+\varepsilon_{\boldsymbol{k}})^2(\omega_{\boldsymbol{q}-\boldsymbol{h}-\boldsymbol{k}}-\omega_{\boldsymbol{q}}+\varepsilon_{\boldsymbol{k}}+\varepsilon_{\boldsymbol{h}})}\right. \\
&\quad-\frac{1}{(\omega_{\boldsymbol{q}-\boldsymbol{k}}-\omega_{\boldsymbol{q}}+\varepsilon_{\boldsymbol{k}})(\omega_{\boldsymbol{q}-\boldsymbol{h}}-\omega_{\boldsymbol{q}}+\varepsilon_{\boldsymbol{h}})(\omega_{\boldsymbol{q}-\boldsymbol{h}-\boldsymbol{k}}-\omega_{\boldsymbol{q}}+\varepsilon_{\boldsymbol{k}}+\varepsilon_{\boldsymbol{h}})} \\
&\quad\left.+\frac{1}{(\omega_{\boldsymbol{q}-\boldsymbol{k}}-\omega_{\boldsymbol{q}}+\varepsilon_{\boldsymbol{k}})^2(\omega_{\boldsymbol{q}-\boldsymbol{h}-\boldsymbol{k}}-\omega_{\boldsymbol{q}-\boldsymbol{k}}+\varepsilon_{\boldsymbol{h}})}\right] \\
&= \frac{e^4\hbar^4c^4}{4(2\pi\hbar)^6}\int\frac{d\boldsymbol{k}d\boldsymbol{h}}{\varepsilon_{\boldsymbol{k}}\varepsilon_{\boldsymbol{h}}}\big[(\omega_{\boldsymbol{q}-\boldsymbol{k}}-\omega_{\boldsymbol{q}}+\varepsilon_{\boldsymbol{k}})^{-1} \\
&\quad\times(\omega_{\boldsymbol{q}-\boldsymbol{h}-\boldsymbol{k}}-\omega_{\boldsymbol{q}}+\varepsilon_{\boldsymbol{k}}+\varepsilon_{\boldsymbol{h}})^{-1}(\omega_{\boldsymbol{q}-\boldsymbol{h}-\boldsymbol{k}}-\omega_{\boldsymbol{q}-\boldsymbol{k}}+\varepsilon_{\boldsymbol{h}})^{-1} \\
&\quad-(\omega_{\boldsymbol{q}-\boldsymbol{k}}-\omega_{\boldsymbol{q}}+\varepsilon_{\boldsymbol{k}})^{-1}(\omega_{\boldsymbol{q}-\boldsymbol{h}}-\omega_{\boldsymbol{q}}+\varepsilon_{\boldsymbol{h}})^{-1}(\omega_{\boldsymbol{q}-\boldsymbol{h}-\boldsymbol{k}}-\omega_{\boldsymbol{q}}+\varepsilon_{\boldsymbol{k}}+\varepsilon_{\boldsymbol{h}})^{-1}\big] \\
&= \frac{e^4\hbar^4c^4}{4(2\pi\hbar)^6}\int\frac{d\boldsymbol{k}d\boldsymbol{h}}{\varepsilon_{\boldsymbol{k}}\varepsilon_{\boldsymbol{h}}}(\omega_{\boldsymbol{q}-\boldsymbol{h}}-\omega_{\boldsymbol{q}}-\omega_{\boldsymbol{q}-\boldsymbol{h}-\boldsymbol{k}}+\omega_{\boldsymbol{q}-\boldsymbol{k}})(\omega_{\boldsymbol{q}-\boldsymbol{k}}-\omega_{\boldsymbol{q}}+\varepsilon_{\boldsymbol{k}})^{-1} \\
&\quad\times(\omega_{\boldsymbol{q}-\boldsymbol{h}-\boldsymbol{k}}-\omega_{\boldsymbol{q}}+\varepsilon_{\boldsymbol{k}}+\varepsilon_{\boldsymbol{h}})^{-1}(\omega_{\boldsymbol{q}-\boldsymbol{h}-\boldsymbol{k}}-\omega_{\boldsymbol{q}-\boldsymbol{k}}+\varepsilon_{\boldsymbol{h}})^{-1}(\omega_{\boldsymbol{q}-\boldsymbol{h}}-\omega_{\boldsymbol{q}}+\varepsilon_{\boldsymbol{h}})^{-1}.
\end{aligned}$$

As we already explained, this contribution is unacceptable in the S-operator, so we have to compensate for it by adding a fourth-order counterterm Q_4 to the Hamiltonian. Hence, finally, the Hamiltonian renormalized up to the fourth order has the form

$$H^c = H_0 + V_1 + Q_2 + Q_4 + \cdots, \tag{2.35}$$

where

$$\begin{aligned}
Q_4[a^\dagger a]^{\text{ren}} &\equiv -\frac{e^4\hbar^4c^4}{4(2\pi\hbar)^6}\int d\boldsymbol{q} \\
&\quad\times\left[\int\frac{d\boldsymbol{k}d\boldsymbol{h}}{\varepsilon_{\boldsymbol{k}}\varepsilon_{\boldsymbol{h}}}(\omega_{\boldsymbol{q}-\boldsymbol{h}}-\omega_{\boldsymbol{q}}-\omega_{\boldsymbol{q}-\boldsymbol{h}-\boldsymbol{k}}+\omega_{\boldsymbol{q}-\boldsymbol{k}})(\omega_{\boldsymbol{q}-\boldsymbol{k}}-\omega_{\boldsymbol{q}}+\varepsilon_{\boldsymbol{k}})^{-1}\right.
\end{aligned}$$

$$\times (\omega_{q-h-k} - \omega_q + \varepsilon_k + \varepsilon_h)^{-1}(\omega_{q-h-k} - \omega_{q-k} + \varepsilon_h)^{-1}(\omega_{q-h} - \omega_q + \varepsilon_h)^{-1}\Big]$$
$$\times a_q^\dagger a_q. \tag{2.36}$$

2.3 Diagrams in general theory

2.3.1 Products of diagrams

The diagram technique developed above can be easily extended to general interactions (1.57)–(1.58), which are sums of potentials $V^{(i)}$. Each potential $V^{(i)}$ has $N^{(i)}$ creation and $M^{(i)}$ annihilation operators. It can be represented as a vertex with $N^{(i)}$ outgoing and $M^{(i)}$ incoming lines. When calculating scattering operators (2.8), we will encounter products of such potentials,

$$Y = V^{(1)}V^{(2)}\dots V^{(\mathcal{V})}, \tag{2.37}$$

where $\mathcal{V}$ is the number of potentials in the product. Each factor $V^{(i)}$ in (2.37) has $N^{(i)} + M^{(i)}$ external momenta, so the product Y initially has

$$\mathcal{N} = \sum_{i=1}^{\mathcal{V}}(N^{(i)} + M^{(i)}) \tag{2.38}$$

3D integrals and independent integration variables. The operator Y also contains a product of $\mathcal{V}$ delta functions expressing momentum conservation in each factor $V^{(i)}$.

Next we have to translate the product (2.37) into a normally ordered form, that is, the sum of terms $y^{(j)}$,

$$Y = \sum_j y^{(j)}, \tag{2.39}$$

represented as diagrams with $\mathcal{V}$ vertices.

According to the process described in Subsection 2.1.3, in the course of normal ordering, a certain number of pairs of external lines in the factors $V^{(i)}$ will connect and form internal lines. Each such contraction adds one momentum delta function to the product. Let us denote $\mathcal{I}$ the number of such internal lines and their delta functions in $y^{(j)}$. Then the total number of delta functions in $y^{(j)}$ is equal to

$$N_\delta = \mathcal{V} + \mathcal{I} \tag{2.40}$$

and the number of external lines is

$$\mathcal{E} = \mathcal{N} - 2\mathcal{I}. \tag{2.41}$$

This information will be useful to us when discussing cluster separability in Subsection 2.4.2.

2.3.2 Connected and disconnected diagrams

The terms $y^{(j)}$ in the normally ordered product (2.39) can be either *connected* or *disconnected*. In the former case, there is a continuous sequence (path) of internal lines connecting any two vertices. In the latter case, such path does not exist for all pairs, and the diagram splits into several disjointed pieces.

We already met a disconnected term as a part of the product $V_1 \underline{V_1}$ in the diagram 2.2 (d). It is not difficult to verify that this term does not depend on the order of the factors: $(V_1\underline{V_1})_{\text{disc}} = (\underline{V_1}V_1)_{\text{disc}}$. It turns out that the same property holds for the product of any two bosonic operators.[16]

Lemma 2.1. *In QED, a normally ordered disconnected part of the product of two connected bosonic operators does not depend on the order of the factors, i. e.,*

$$(V^{(1)}V^{(2)})_{\text{disc}} = (V^{(2)}V^{(1)})_{\text{disc}}. \tag{2.42}$$

Proof. The products $V^{(1)}V^{(2)}$ and $V^{(2)}V^{(1)}$ differ only in the order of particle operators. Hence, after converting to the normal form, the disconnected parts $(V^{(1)}V^{(2)})_{\text{disc}}$ and $(V^{(2)}V^{(1)})_{\text{disc}}$ can differ, at most, by a sign. Our goal is to show that both expressions have the same sign.

When performing the normal ordering, swapping positions of boson particle operators does not affect the sign of the product, so in our proof we shall focus on the order of fermion operators in the products $V^{(1)}V^{(2)}$ and $V^{(2)}V^{(1)}$. For simplicity, we assume that only electron and positron operators are present in $V^{(1)}$ and $V^{(2)}$.[17] For each of the two factors $V^{(i)}$ (where $i = 1, 2$) we denote $N_{\text{el}}^{(i)}$ the number of electron creation operators, $N_{\text{po}}^{(i)}$ the number of positron creation operators, $M_{\text{el}}^{(i)}$ the number of electron annihilation operators and $M_{\text{po}}^{(i)}$ the number of positron annihilation operators. Taking into account that the factors $V^{(i)}$ are already normally ordered, we can formally write

$$V^{(1)} \propto [N_{\text{el}}^{(1)}][N_{\text{po}}^{(1)}][M_{\text{el}}^{(1)}][M_{\text{po}}^{(1)}],$$

$$V^{(2)} \propto [N_{\text{el}}^{(2)}][N_{\text{po}}^{(2)}][M_{\text{el}}^{(2)}][M_{\text{po}}^{(2)}],$$

where the bracket $[N_{\text{el}}^{(1)}]$ denotes a product of $N_{\text{el}}^{(1)}$ electron creation operators from the factor $V^{(1)}$, the bracket $[N_{\text{po}}^{(1)}]$ is a product of $N_{\text{po}}^{(1)}$ positron creation operators from the factor $V^{(1)}$, etc. Then the two sides of (2.42) can be written as

$$V^{(1)}V^{(2)} \propto [N_{\text{el}}^{(1)}][N_{\text{po}}^{(1)}][M_{\text{el}}^{(1)}][M_{\text{po}}^{(1)}][N_{\text{el}}^{(2)}][N_{\text{po}}^{(2)}][M_{\text{el}}^{(2)}][M_{\text{po}}^{(2)}], \tag{2.43}$$

$$V^{(2)}V^{(1)} \propto [N_{\text{el}}^{(2)}][N_{\text{po}}^{(2)}][M_{\text{el}}^{(2)}][M_{\text{po}}^{(2)}][N_{\text{el}}^{(1)}][N_{\text{po}}^{(1)}][M_{\text{el}}^{(1)}][M_{\text{po}}^{(1)}]. \tag{2.44}$$

16 As we established in Subsection 1.2.2, all potentials considered in this book are bosonic.

17 Proton and antiproton operators can be analyzed in the same way.

Let us now bring particle operators on the right-hand side of (2.44) to the same order as on the right-hand side of (2.43). First, we move $N_{el}^{(1)}$ electron creation operators to the extreme left position in the product. This step includes $N_{el}^{(1)}M_{el}^{(2)}$ permutations with electron annihilation operators in the factor $V^{(2)}$ and $N_{el}^{(1)}N_{el}^{(2)}$ permutations with electron creation operators in the factor $V^{(2)}$.[18] Each of these permutations changes the sign of our disconnected term, so that in the end it acquires the multiplier $(-1)^{N_{el}^{(1)}(N_{el}^{(2)}+M_{el}^{(2)})}$.

Next, we have to move the factor $[N_{po}^{(1)}]$ to the second position on the left. This leads to the multiplier $(-1)^{N_{po}^{(1)}(N_{po}^{(2)}+M_{po}^{(2)})}$. Then we move the factors $[M_{el}^{(1)}]$ and $[M_{po}^{(1)}]$ to the third and fourth positions, respectively. After all these permutations, the total numerical multiplier accumulated in the expression $(V^{(2)}V^{(1)})_{disc}$ becomes

$$f = (-1)^{K_{el}^{(1)}K_{el}^{(2)}+K_{po}^{(1)}K_{po}^{(2)}}, \tag{2.45}$$

where we denote

$$K_{el}^{(i)} \equiv N_{el}^{(i)} + M_{el}^{(i)},$$
$$K_{po}^{(i)} \equiv N_{po}^{(i)} + M_{po}^{(i)}$$

full (= creation + annihilation) numbers of electron and positron operators in the factor $V^{(i)}$. Let us now prove that the power of (−1) in (2.45) is an even number, so that

$$f = 1, \tag{2.46}$$

as promised. Indeed, consider the case when $K_{el}^{(1)}$ is even, but $K_{el}^{(2)}$ is odd, and the product $K_{el}^{(1)}K_{el}^{(2)}$ is odd. From the bosonic character of $V^{(1)}$ and $V^{(2)}$ it follows that $K_{el}^{(1)} + K_{po}^{(1)}$ and $K_{el}^{(2)} + K_{po}^{(2)}$ are even numbers. Thus, $K_{po}^{(1)}$ is even, and $K_{po}^{(2)}$ is odd, which means that the product $K_{po}^{(1)}K_{po}^{(2)}$ is odd and the full power of (−1) in (2.45) is even.

The same result is obtained with any other combination of even/odd characters of $K_{el}^{(1)}$ and $K_{el}^{(2)}$. This proves equalities (2.46) and (2.42). □

Theorem 2.2. *Each multiple commutator of connected bosonic potentials is connected.*

Proof. First consider a simple commutator of two potentials $V^{(1)}$ and $V^{(2)}$,

$$V^{(1)}V^{(2)} - V^{(2)}V^{(1)}. \tag{2.47}$$

According to Lemma 2.1, after normal ordering, the disconnected terms $(V^{(1)}V^{(2)})_{disc}$ and $(V^{(2)}V^{(1)})_{disc}$ in the commutator (2.47) cancel out, and all remaining terms are connected. This proves the theorem for the simple commutator (2.47). Since this commutator is a bosonic operator, repeating previous arguments by induction, we conclude that all multiple commutators of bosonic operators are connected. □

18 We ignore contractions, because here we are interested only in the disconnected term.

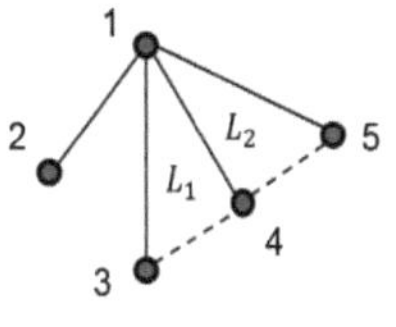

Figure 2.6: To the proof of Lemma 2.3. In a diagram with five vertices and six internal lines there are $6 - (5 - 1) = 2$ independent loops L_1 and L_2.

Lemma 2.3. *In a connected diagram with $\mathcal{V}$ vertices and $\mathcal{I}$ internal lines, the number of independent loops is equal to*

$$\mathcal{L} = \mathcal{I} - \mathcal{V} + 1. \tag{2.48}$$

Proof. $\mathcal{V}$ vertices can be joined together without forming loops by $\mathcal{V}-1$ internal lines.[19] Each additional internal line will lead to the formation of one new independent loop. Therefore, the number of independent loops is $\mathcal{I} - (\mathcal{V} - 1)$. □

Figure 2.7 shows an example of a connected diagram in an imaginary theory.[20] This diagram has $\mathcal{V} = 4$ vertices, $\mathcal{E} = 5$ external lines, $\mathcal{I} = 7$ internal lines and $\mathcal{L} = 4$ independent loops. It describes a nine-fold momentum integral. Five integration momenta correspond to the external lines of the diagram: two incoming momenta $\boldsymbol{p}_1$, $\boldsymbol{p}_2$ and three outgoing momenta $\boldsymbol{p}_3$, $\boldsymbol{p}_4$ and $\boldsymbol{p}_5$. These five integrals are a part of the standard expression (1.58) for any regular operator. Four additional integrals are performed with respect to loop momenta $\boldsymbol{p}_6$, $\boldsymbol{p}_7$, $\boldsymbol{p}_8$ and $\boldsymbol{p}_9$. These integrals are included

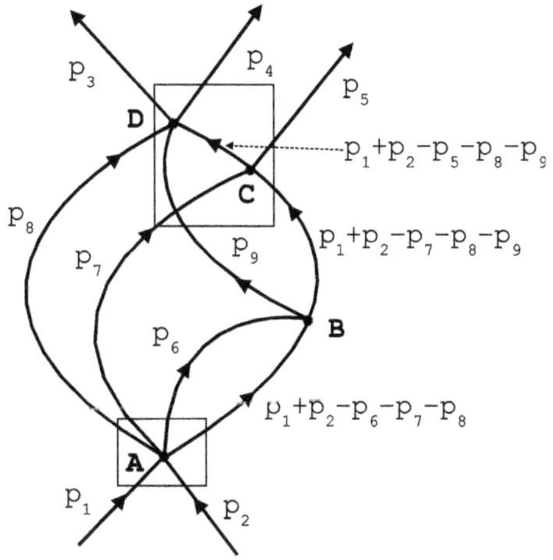

Figure 2.7: A diagram representing one fourth-order term in the S-operator of a hypothetical theory.

19 For example, you can select one vertex and draw lines from it to all remaining $\mathcal{V} - 1$ vertices. They are shown by solid lines in Figure 2.6.

20 We do not draw the order axis, as in Subsection 2.1.3. Instead, we emphasize the normal ordering by depicting all outgoing lines in the top of the diagram and all incoming lines in the bottom.

in the definition of the coefficient function

$$
\begin{aligned}
D_{3,2}&(\boldsymbol{p}_3, \boldsymbol{p}_4, \boldsymbol{p}_5; \boldsymbol{p}_1, \boldsymbol{p}_2) \\
&= \int d\boldsymbol{p}_6 d\boldsymbol{p}_7 d\boldsymbol{p}_8 d\boldsymbol{p}_9 \\
&\quad \times D_A(\boldsymbol{p}_6, \boldsymbol{p}_7, \boldsymbol{p}_8, \boldsymbol{p}_1 + \boldsymbol{p}_2 - \boldsymbol{p}_6 - \boldsymbol{p}_7 - \boldsymbol{p}_8; \boldsymbol{p}_1, \boldsymbol{p}_2) \\
&\quad \times D_B(\boldsymbol{p}_9, \boldsymbol{p}_1 + \boldsymbol{p}_2 - \boldsymbol{p}_7 - \boldsymbol{p}_8 - \boldsymbol{p}_9; \boldsymbol{p}_6, \boldsymbol{p}_1 + \boldsymbol{p}_2 - \boldsymbol{p}_6 - \boldsymbol{p}_7 - \boldsymbol{p}_8) \\
&\quad \times D_C(\boldsymbol{p}_5, \boldsymbol{p}_1 + \boldsymbol{p}_2 - \boldsymbol{p}_5 - \boldsymbol{p}_8 - \boldsymbol{p}_9; \boldsymbol{p}_7, \boldsymbol{p}_1 + \boldsymbol{p}_2 - \boldsymbol{p}_7 - \boldsymbol{p}_8 - \boldsymbol{p}_9) \\
&\quad \times D_D(\boldsymbol{p}_3, \boldsymbol{p}_4; \boldsymbol{p}_8, \boldsymbol{p}_9, \boldsymbol{p}_1 + \boldsymbol{p}_2 - \boldsymbol{p}_5 - \boldsymbol{p}_8 - \boldsymbol{p}_9) \frac{1}{\mathcal{E}_A(\mathcal{E}_C + \mathcal{E}_D)},
\end{aligned}
\tag{2.49}
$$

where $\mathcal{E}_A$, $\mathcal{E}_C$ and $\mathcal{E}_D$ are energy functions of the corresponding vertices, and D_A, D_B, D_C, D_D are coefficient functions at the vertices. The factor $\mathcal{E}_A^{-1}(\mathcal{E}_C + \mathcal{E}_D)^{-1}$ is shown by two rectangles enclosing the vertices A and $C + D$, respectively.

2.3.3 Divergence of loop integrals

In the preceding subsection we showed that S-operator terms often include loop integrals like (2.49), and there is no guarantee that these integrals converge. This problem is present even in our toy model, where in the diagram 2.2 (g) the loop integral with respect to $\boldsymbol{k}$,

$$
(V_1 \underline{V_1})^{\text{ren}} = \frac{e^2\hbar^2c^2}{2(2\pi\hbar)^3} \int d\boldsymbol{p} \left[\int \frac{d\boldsymbol{k}}{(\omega_{\boldsymbol{p}} - \omega_{\boldsymbol{p}-\boldsymbol{k}} - \varepsilon_{\boldsymbol{k}})\varepsilon_{\boldsymbol{k}}} \right] a^\dagger_{\boldsymbol{p}} a_{\boldsymbol{p}}, \tag{2.50}
$$

is divergent.[21] As we made clear in Section 2.2, the presence of this term makes the S-operator meaningless and the entire theory unacceptable.

The appearance of divergences in perturbative calculations of the S-operator is a common phenomenon in QFT. In this subsection we formulate a sufficient condition for the convergence of loop integrals. As a rule, this condition is not satisfied in the traditional QFT. However, we will appreciate its usefulness when discussing our new approach in the third volume of this book.

Let us review the diagram in Figure 2.7. There are three reasons why loop integrals can diverge:

(I) The coefficient functions $D_A, D_B, \ldots$ of the interaction vertices (2.49) can contain nonintegrable singularities.[22] Such singularities often arise in the presence

21 For large values of k, we have $\omega_{\boldsymbol{p}-\boldsymbol{k}} \to ck$, $\varepsilon_{\boldsymbol{k}} \to ck$ and the integrand in (2.50) tends to $-1/(2c^2k^2)$. Hence, in the limit of large momentum integration cutoff ($\Lambda \to \infty$) the integral with respect to k grows in proportion to Λ.

22 One such example is interaction (2.7), which is singular for $\lambda \to 0$, $\boldsymbol{k} \to \boldsymbol{0}$.

of massless particles (such as photons). They are often responsible for the so-called *infrared divergences* of loop integrals. We will discuss them in more detail in Chapter 4 and in Volume 3.

(II) In addition, singularities can appear due to zeros of energy denominators, such as $\mathcal{E}_A^{-1}$ and $(\mathcal{E}_C+\mathcal{E}_D)^{-1}$. These denominators can be rendered harmless by applying the adiabatic switching recipe from Subsection **1**-7.1.4. This leads to the appearance of small imaginary additions to denominators, which remove the singularities.

(III) The coefficient functions $D_A, D_B, \ldots$ may decrease too slowly as the loop integration momenta tend to infinity. Then the loop integrals may diverge because of the infinite integration volume. These *ultraviolet divergences* represent the greatest danger, which we are going to discuss here in more detail.

In particular, we want to prove the following.

Theorem 2.4. *If coefficient functions of the potentials decrease rapidly enough (for example, exponentially), when their arguments move away from the energy shell, then all loop integrals converge.*

Idea of the proof. Formula (2.49) is an integral in the 12-dimensional space of the four loop momenta $\boldsymbol{p}_6, \boldsymbol{p}_7, \boldsymbol{p}_8$ and $\boldsymbol{p}_9$. We denote this space by Ξ. Consider, for example, the dependence of the integrand in (2.49) on the loop momentum $\boldsymbol{p}_9$ when $\boldsymbol{p}_9 \to \infty$ and all other variables are fixed.[23] Taking into account that for large momenta $\omega_{\boldsymbol{p}} \approx cp$, in the limit $\boldsymbol{p}_9 \to \infty$ we obtain

$$
\begin{aligned}
&\mathcal{E}_A \to \text{const},\\
&\mathcal{E}_B = \omega_{\boldsymbol{p}_1+\boldsymbol{p}_2-\boldsymbol{p}_7-\boldsymbol{p}_8-\boldsymbol{p}_9} + \omega_{\boldsymbol{p}_9} - \omega_{\boldsymbol{p}_6} - \omega_{\boldsymbol{p}_1+\boldsymbol{p}_2-\boldsymbol{p}_6-\boldsymbol{p}_7-\boldsymbol{p}_8} \approx 2cp_9 \to \infty,\\
&\mathcal{E}_C = \omega_{\boldsymbol{p}_1+\boldsymbol{p}_2-\boldsymbol{p}_5-\boldsymbol{p}_8-\boldsymbol{p}_9} + \omega_{\boldsymbol{p}_5} - \omega_{\boldsymbol{p}_7} - \omega_{\boldsymbol{p}_1+\boldsymbol{p}_2-\boldsymbol{p}_7-\boldsymbol{p}_8-\boldsymbol{p}_9} \to \text{const},\\
&\mathcal{E}_D = \omega_{\boldsymbol{p}_3} + \omega_{\boldsymbol{p}_4} - \omega_{\boldsymbol{p}_8} - \omega_{\boldsymbol{p}_9} - \omega_{\boldsymbol{p}_1+\boldsymbol{p}_2-\boldsymbol{p}_5-\boldsymbol{p}_8-\boldsymbol{p}_9} \approx -2cp_9 \to \infty.
\end{aligned}
$$

The loop B–D–C–B has a *bottom vertex* B, a *top vertex* D and an *intermediate vertex* C.[24] When the loop momentum goes to infinity, the energy functions of the top and bottom vertices also tend to infinity, i. e., the integration momentum moves away from the energy surface. So, in accordance with the condition of the theorem, in this limit the coefficient function of the vertex C tends to a finite constant while coefficient functions of the vertices B and D rapidly tend to zero. This ensures fast decay of the product $D_B D_C D_D$. Hence, the loop integral with respect to $\boldsymbol{p}_9$ converges.

23 Notice that in Figure 2.7 we have chosen integration variables in such a way that each loop momentum is present only in the internal lines forming the corresponding loop. For example, the momentum $\boldsymbol{p}_9$ is present in the loop $BDCB$, and the energy function $\mathcal{E}_A$ of the vertex A (outside this loop) is independent of $\boldsymbol{p}_9$. Such a choice of integration variables can be made in any diagram.

24 For a general loop there can be several intermediate vertices or none at all.

The above analysis is also suitable for other loop variables $\boldsymbol{p}_6$, $\boldsymbol{p}_7$ and $\boldsymbol{p}_8$. Consequently, choosing an arbitrary direction in the space Ξ, we shall see that along this direction at least one loop momentum tends to infinity, and at least one energy function ($\mathcal{E}_A$, $\mathcal{E}_B$, $\mathcal{E}_C$ or $\mathcal{E}_D$) grows linearly, while others do not change (in the worst case) or grow too. Therefore, in accordance with the condition of the theorem, the integrand rapidly decreases along this direction. Thus, the product of coefficient functions in the integrand rapidly tends to zero in all directions in Ξ, hence the integral (2.49) converges in the limit of large loop momenta. □

In Chapter 4, we will see that in QED the asymptotic decrease of coefficient functions is not fast enough, so our Theorem 2.4 is not applicable, and loop integrals often diverge. These divergences have to be fixed by the traditional renormalization method.

2.4 Cluster separability

2.4.1 Cluster separability of interaction

In Subsection 1.2.7, we showed how to express familiar interparticle potentials through creation and annihilation operators. The question is, under what conditions do these potentials have the property of cluster separability?

Let us now consider an example in which the electron–proton potential

$$V = \int d\boldsymbol{p}d\boldsymbol{p}'d\boldsymbol{k}D(\boldsymbol{p},\boldsymbol{p}',\boldsymbol{k})d^{\dagger}_{\boldsymbol{p}+\boldsymbol{k}}a^{\dagger}_{\boldsymbol{p}'-\boldsymbol{k}}d_{\boldsymbol{p}}a_{\boldsymbol{p}'} \tag{2.51}$$

acts in the three-particle (1 proton + 2 electrons) sector $\mathscr{H}(2,0,1,0,0)$ of the Fock space, where state vectors have the form

$$|\Psi\rangle = \int d\boldsymbol{q}d\boldsymbol{q}_1d\boldsymbol{q}_2\psi(\boldsymbol{q};\boldsymbol{q}_1,\boldsymbol{q}_2)d^{\dagger}_{\boldsymbol{q}}a^{\dagger}_{\boldsymbol{q}_1}a^{\dagger}_{\boldsymbol{q}_2}|\text{vac}\rangle. \tag{2.52}$$

Applying operator (2.51) to this state, we get

$$V|\Psi\rangle = \int d\boldsymbol{p}d\boldsymbol{p}'d\boldsymbol{k}\int d\boldsymbol{q}d\boldsymbol{q}_1d\boldsymbol{q}_2D(\boldsymbol{p},\boldsymbol{p}',\boldsymbol{k})\psi(\boldsymbol{q};\boldsymbol{q}_1,\boldsymbol{q}_2)d^{\dagger}_{\boldsymbol{p}+\boldsymbol{k}}a^{\dagger}_{\boldsymbol{p}'-\boldsymbol{k}}d_{\boldsymbol{p}}a_{\boldsymbol{p}'}d^{\dagger}_{\boldsymbol{q}}a^{\dagger}_{\boldsymbol{q}_1}a^{\dagger}_{\boldsymbol{q}_2}|\text{vac}\rangle. \tag{2.53}$$

Then we perform the normal ordering, leaving only terms that do not have annihilation operators on the right, i. e.,

$$\begin{aligned}
&d^{\dagger}_{\boldsymbol{p}+\boldsymbol{k}}a^{\dagger}_{\boldsymbol{p}'-\boldsymbol{k}}d_{\boldsymbol{p}}a_{\boldsymbol{p}'}d^{\dagger}_{\boldsymbol{q}}a^{\dagger}_{\boldsymbol{q}_1}a^{\dagger}_{\boldsymbol{q}_2} \\
&\quad = d^{\dagger}_{\boldsymbol{p}+\boldsymbol{k}}a^{\dagger}_{\boldsymbol{p}'-\boldsymbol{k}}d_{\boldsymbol{p}}a_{\boldsymbol{p}'}d^{\dagger}_{\boldsymbol{q}}a^{\dagger}_{\boldsymbol{q}_1}a^{\dagger}_{\boldsymbol{q}_2} + d^{\dagger}_{\boldsymbol{p}+\boldsymbol{k}}a^{\dagger}_{\boldsymbol{p}'-\boldsymbol{k}}d_{\boldsymbol{p}}a_{\boldsymbol{p}'}d^{\dagger}_{\boldsymbol{q}}a^{\dagger}_{\boldsymbol{q}_1}a^{\dagger}_{\boldsymbol{q}_2} + \cdots \\
&\quad = d^{\dagger}_{\boldsymbol{p}+\boldsymbol{k}}a^{\dagger}_{\boldsymbol{p}'-\boldsymbol{k}}a^{\dagger}_{\boldsymbol{q}_2}\delta(\boldsymbol{p}-\boldsymbol{q})\delta(\boldsymbol{p}'-\boldsymbol{q}_1) - d^{\dagger}_{\boldsymbol{p}+\boldsymbol{k}}a^{\dagger}_{\boldsymbol{p}'-\boldsymbol{k}}a^{\dagger}_{\boldsymbol{q}_1}\delta(\boldsymbol{p}-\boldsymbol{q})\delta(\boldsymbol{p}'-\boldsymbol{q}_2) + \cdots.
\end{aligned}$$

Substituting this result in (2.53), we get

$$
\begin{aligned}
V|\Psi\rangle = &- \int d\boldsymbol{p}d\boldsymbol{p}'d\boldsymbol{k} \int d\boldsymbol{q}d\boldsymbol{q}_1 d\boldsymbol{q}_2 D(\boldsymbol{p},\boldsymbol{p}',\boldsymbol{k})\psi(\boldsymbol{q};\boldsymbol{q}_1,\boldsymbol{q}_2) \\
&\times \delta(\boldsymbol{q}-\boldsymbol{p})\delta(\boldsymbol{q}_2-\boldsymbol{p}')d^\dagger_{\boldsymbol{p}+\boldsymbol{k}}a^\dagger_{\boldsymbol{p}'-\boldsymbol{k}}a^\dagger_{\boldsymbol{q}_1}|\mathrm{vac}\rangle \\
&+ \int d\boldsymbol{p}d\boldsymbol{p}'d\boldsymbol{k} \int d\boldsymbol{q}d\boldsymbol{q}_1 d\boldsymbol{q}_2 D(\boldsymbol{p},\boldsymbol{p}',\boldsymbol{k})\psi(\boldsymbol{q};\boldsymbol{q}_1,\boldsymbol{q}_2) \\
&\times \delta(\boldsymbol{q}-\boldsymbol{p})\delta(\boldsymbol{q}_1-\boldsymbol{p}')d^\dagger_{\boldsymbol{p}+\boldsymbol{k}}a^\dagger_{\boldsymbol{p}'-\boldsymbol{k}}a^\dagger_{\boldsymbol{q}_2}|\mathrm{vac}\rangle \\
= &- \int d\boldsymbol{p}d\boldsymbol{p}'d\boldsymbol{k}d\boldsymbol{q}_1 D(\boldsymbol{p},\boldsymbol{p}',\boldsymbol{k})\psi(\boldsymbol{p};\boldsymbol{q}_1,\boldsymbol{p}')d^\dagger_{\boldsymbol{p}+\boldsymbol{k}}a^\dagger_{\boldsymbol{p}'-\boldsymbol{k}}a^\dagger_{\boldsymbol{q}_1}|\mathrm{vac}\rangle \\
&+ \int d\boldsymbol{p}d\boldsymbol{p}'d\boldsymbol{k}d\boldsymbol{q}_2 D(\boldsymbol{p},\boldsymbol{p}',\boldsymbol{k})\psi(\boldsymbol{p};\boldsymbol{p}',\boldsymbol{q}_2)d^\dagger_{\boldsymbol{p}+\boldsymbol{k}}a^\dagger_{\boldsymbol{p}'-\boldsymbol{k}}a^\dagger_{\boldsymbol{q}_2}|\mathrm{vac}\rangle \\
= &\int d\boldsymbol{p}d\boldsymbol{q}_2 d\boldsymbol{q}_1 \Big(\int d\boldsymbol{k} D(\boldsymbol{p}-\boldsymbol{k},\boldsymbol{q}_2+\boldsymbol{k},\boldsymbol{k})\psi(\boldsymbol{p}-\boldsymbol{k};\boldsymbol{q}_1,\boldsymbol{q}_2+\boldsymbol{k}) \\
&+ D(\boldsymbol{p}-\boldsymbol{k},\boldsymbol{q}_1+\boldsymbol{k},\boldsymbol{k})\psi(\boldsymbol{p}-\boldsymbol{k};\boldsymbol{q}_1+\boldsymbol{k},\boldsymbol{q}_2)\Big) d^\dagger_{\boldsymbol{p}}a^\dagger_{\boldsymbol{q}_1}a^\dagger_{\boldsymbol{q}_2}|\mathrm{vac}\rangle.
\end{aligned}
$$

Comparison with (2.52) shows that V acts on the three-particle wave function as follows:

$$
\begin{aligned}
\hat{V}\psi(\boldsymbol{p};\boldsymbol{q}_1,\boldsymbol{q}_2) = &\int d\boldsymbol{k} D(\boldsymbol{p}-\boldsymbol{k},\boldsymbol{q}_2+\boldsymbol{k},\boldsymbol{k})\psi(\boldsymbol{p}-\boldsymbol{k};\boldsymbol{q}_1,\boldsymbol{q}_2+\boldsymbol{k}) \\
&+ \int d\boldsymbol{k} D(\boldsymbol{p}-\boldsymbol{k},\boldsymbol{q}_1+\boldsymbol{k},\boldsymbol{k})\psi(\boldsymbol{p}-\boldsymbol{k};\boldsymbol{q}_1+\boldsymbol{k},\boldsymbol{q}_2).
\end{aligned}
$$

As expected, the two-particle interaction in the three-particle sector is split into two terms. One term acts on the pair of variables $(\boldsymbol{p},\boldsymbol{q}_2)$, and the second term acts on variables $(\boldsymbol{p},\boldsymbol{q}_1)$.

We can remove electron 2 to infinity by multiplying the original momentum wave function $\psi(\boldsymbol{p};\boldsymbol{q}_1,\boldsymbol{q}_2)$ in (2.52) by the factor $\exp(\frac{i}{\hbar}\boldsymbol{q}_2\cdot\boldsymbol{a})$, where $\boldsymbol{a}\to\infty$. The action of V on this state is

$$
\begin{aligned}
&\lim_{a\to\infty} \hat{V}\big[\psi(\boldsymbol{p};\boldsymbol{q}_1,\boldsymbol{q}_2)e^{\frac{i}{\hbar}\boldsymbol{q}_2\cdot\boldsymbol{a}}\big] \\
&= \lim_{a\to\infty}\Big[\int d\boldsymbol{k} D(\boldsymbol{p}-\boldsymbol{k},\boldsymbol{q}_2+\boldsymbol{k},\boldsymbol{k})\psi(\boldsymbol{p}-\boldsymbol{k};\boldsymbol{q}_1,\boldsymbol{q}_2+\boldsymbol{k})e^{\frac{i}{\hbar}(\boldsymbol{q}_2+\boldsymbol{k})\cdot\boldsymbol{a}} \\
&\quad+ \int d\boldsymbol{k} D(\boldsymbol{p}-\boldsymbol{k},\boldsymbol{q}_1+\boldsymbol{k},\boldsymbol{k})\psi(\boldsymbol{p}-\boldsymbol{k};\boldsymbol{q}_1+\boldsymbol{k},\boldsymbol{q}_2)e^{\frac{i}{\hbar}\boldsymbol{q}_2\cdot\boldsymbol{a}}\Big].
\end{aligned}
$$

In the limit $\boldsymbol{a}\to\infty$, the exponent in the first integral is a rapidly oscillating function of $\boldsymbol{k}$. If $D(\boldsymbol{p},\boldsymbol{q},\boldsymbol{k})$ is a smooth function of $\boldsymbol{k}$, then the integral with respect to $\boldsymbol{k}$ tends to zero by the *Riemann–Lebesgue lemma* A.1. Therefore, only the second term gives a nonvanishing contribution

$$
\int d\boldsymbol{k} D(\boldsymbol{p}-\boldsymbol{k},\boldsymbol{q}_1+\boldsymbol{k},\boldsymbol{k})\psi(\boldsymbol{p}-\boldsymbol{k};\boldsymbol{q}_1+\boldsymbol{k},\boldsymbol{q}_2)e^{\frac{i}{\hbar}\boldsymbol{q}_2\cdot\boldsymbol{a}}.
$$

Thus, the spatial translation of electron 2 leads to a state in which the free remote electron 2 coexists with the interacting subsystem "proton + electron 1". This proves that the potential $\hat{V}$ with a smooth coefficient function is cluster-separable.

For more complex potentials of the general form (1.57)–(1.58), the above arguments can be repeated. If all potentials have smooth coefficient functions and a group of particles is removed to infinity, then such potentials automatically separate into a sum of independent terms, as required by the condition of separability. These arguments give us the right to formulate the following theorem.

Theorem 2.5 (cluster separability). *The cluster separability of the general interaction* (1.57)–(1.58) *is guaranteed if coefficient functions D_{NM} of the potentials V_{NM} are smooth functions of particle momenta.*

The power and utility of this statement come from the fact that by expressing interaction potentials as polynomials in creation and annihilation operators, we get a very simple criterion for the cluster separability: *coefficient functions of the potentials must be smooth.*[25] Recall that in Subsection **1**-6.4.6 it was very difficult to ensure the cluster separability of even the simplest three-particle interaction potentials written in terms of particle observables.

2.4.2 Cluster separability of *S*-operator

In this book, we will consider only cluster-separable interactions. Is it true that the S-operator, calculated with such interactions, is also separable? Of course, the answer is "yes." However, the proof of this statement is not so straightforward. From formulas of perturbation theory, we know that, generally speaking, S is a sum of products of interaction potentials, as in (2.37). Cluster separability of interactions means that coefficient functions of the potentials $V^{(i)}$ in the product (2.37) are smooth. According to Theorem 2.5, the cluster separability of this product would be guaranteed if the normally ordered terms on the right-hand side of (2.39) are also smooth operators. However, the question of their smoothness is not so simple, because the normal ordering requires permutations of particle operators, which lead to the appearance of singular delta functions.

The following theorem establishes an important connection between the smoothness of terms on the right-hand side of (2.39) and the connectedness of the corresponding diagrams.

Theorem 2.6. *Each term $y^{(j)}$ in the normal-order decomposition* (2.39) *of a product of smooth potentials is smooth if and only if it is represented by a connected diagram.*

25 This is why we called cluster-separable potentials *smooth* in Subsection **1**-6.4.3.

Proof. First, assume that $y^{(j)}$ is a connected diagram. We will establish the smoothness of $y^{(j)}$ by proving that it can be represented in the general form (1.58), where the integrand contains only one delta function required by the momentum conservation law and the coefficient function D_{NM} is smooth. Indeed, the initial number of momentum integrals in each term $y^{(j)}$ is $\mathcal{N}$ from equation (2.38). The integrals corresponding to $\mathcal{E}$ external lines are parts of the standard form (1.58), and the integrals corresponding to $\mathcal{L}$ loops are parts of the definition of the coefficient function $y^{(j)}$. The number of remaining integrals can be obtained from (2.41) and (2.48), i. e.,

$$\mathcal{N}' = \mathcal{N} - \mathcal{E} - \mathcal{L} = \mathcal{I} + \mathcal{V} - 1. \tag{2.54}$$

This is just the sufficient number of integrals to absorb all momentum delta functions (2.40), except one that is needed to ensure the conservation of the total momentum. This proves the smoothness of the term $y^{(i)}$.

Conversely, suppose that $y^{(j)}$ is represented by a disconnected diagram with $\mathcal{V}$ vertices and $\mathcal{I}$ internal lines. Then the number of independent loops $\mathcal{L}$ is less than the number $\mathcal{I}-\mathcal{V}+1$ characteristic for connected diagrams. This also means that the number of integrations $\mathcal{N}'$ in the equality (2.54) is less than $\mathcal{I} + \mathcal{V} - 1$, and the number of delta functions remaining in the integrand $(\mathcal{N}' - \mathcal{N}_\delta)$ is greater than 1. Therefore, the coefficient function of the term $y^{(j)}$ has a singular delta-function multiplier, and the corresponding operator is not smooth. □

Theorem 2.6 establishes that smooth operators are represented by connected diagrams. In the future we will use the terms *smooth* and *connected* as synonyms, when applied to operators in the Fock space.

Putting together Theorems 2.2 and 2.6, we immediately arrive at the next important observation.

Theorem 2.7. *All terms in a normally ordered multiple commutator of smooth bosonic potentials are smooth.*

Now everything is ready for discussing the cluster separability of the S-operator. Let us write it in the Magnus form (**1**-7.18)

$$S = e^{\underbrace{\Phi}}, \tag{2.55}$$

where Φ is a sum of multiple commutators (**1**-7.19) of smooth bosonic potentials from V. By Theorem 2.7, the operators Φ and $\underbrace{\Phi}$ are also smooth and cluster-separable. Hence, if all colliding particles are broken up into spatially separated groups (1) and (2), then the phase operator in (2.55) takes the form of the sum,

$$\underbrace{\Phi} \to \underbrace{\Phi^{(1)}} + \underbrace{\Phi^{(2)}},$$

where $\underbrace{\Phi}^{(1)}$ acts only on particle variables in group (1) and $\underbrace{\Phi}^{(2)}$ acts only on particle variables in group (2). These two terms commute with each other, and the S-operator separates into the product of two independent factors, i. e.,

$$S \to \exp(\underbrace{\Phi}^{(1)} + \underbrace{\Phi}^{(2)}) = \exp(\underbrace{\Phi}^{(1)}) \exp(\underbrace{\Phi}^{(2)}) = S^{(1)} S^{(2)}.$$

In other words, the total scattering amplitude for spatially separated collisions is equal to the product of individual amplitudes in accordance with the physical meaning of cluster separability: the result of scattering in each of the subsystems does not depend on what happens in another subsystem.

3 Quantum electrodynamics

I have worked like a galley slave throughout these eight years, morning till night, and I have given all I could to this work. I am happy with the results.
Vladimir V. Putin

Up to now we have been developing a general formalism of quantum theory in the Fock space. We considered several model examples, but they were purely academic and had a very remote relationship to the real systems observed in nature. One of the reasons for this inadequacy is that our models were unable to satisfy at once all three requirements[1] listed in the preface.

For example, in Section **1**-6.4 we formulated a model that clearly satisfied the requirement of relativistic invariance. We also managed to ensure its cluster separability in the three-particle sector. In principle, this approach can be extended to interactions in all n-particle sectors [9]. There is even the possibility of describing systems with a variable number of particles [12]. However, the resulting mathematical formalism is very complex and practically applicable only to model systems.

In Section 2.1 we considered an example of a toy theory based on the formalism of creation and annihilation operators. The huge advantage of this model is that the cluster separability condition is conveniently expressed in terms of the smoothness of interaction potentials (see Theorem 2.5). The hard part is to ensure the Poincaré invariance. We did not even try to make our toy model relativistic.

As we mentioned in the preface, to date, there is only one universally recognized approach which fulfills all three requirements. This approach is called *quantum field theory* (QFT).

A special version of QFT for describing interactions between electrically charged particles and photons is called *quantum electrodynamics* (QED). We will discuss QED in the rest of this volume and in the most part of Volume 3. In Section 3.1, we will explicitly write down QED operators of potential energy V and potential boost $\boldsymbol{Z}$. The relativistic invariance of this approach will be proved in Appendix E.2. Section 3.2 will show how to calculate the simplest elements of the S-matrix in the lowest nonvanishing order of perturbation theory.

3.1 Interaction in QED

Our goal here is to construct a realistic interacting representation $U(\Lambda; \tilde{a})$ of the Poincaré group in the Fock space (1.1) populated by electrons, positrons, protons, antiprotons and photons.

1 Relativistic invariance, cluster separability and variable number of particles.

https://doi.org/10.1515/9783110493207-003

3.1.1 Why do we need quantum fields?

From our discussion in Chapters 1 and 2 it should be clear that interaction operators V and $\boldsymbol{Z}$ should be sought as polynomials (1.57)–(1.58) in creation and annihilation operators of particles. However, it is very difficult to define these polynomials in such a way as to satisfy commutators (**1**-6.26)–(**1**-6.30) of the Poincaré Lie algebra. This problem is simplified by the introduction of quantum fields $\phi_\alpha(\tilde{x})$, which are nothing but special linear combinations of particle operators.[2] These combinations are chosen in such a way that polynomials composed of quantum fields produce relativistically invariant interactions almost automatically.

We do not offer any physical interpretation for quantum fields. For us, these are simply abstract multicomponent functions from the four-dimensional Minkowski space–time $\mathcal{M}$ to Fock space operators. In our approach, the only role assigned to quantum fields is to provide convenient "building blocks" for constructing Poincaré-invariant interactions V and $\boldsymbol{Z}$. Similarly, we refuse to identify the Minkowski "space–time" coordinates $(t, \boldsymbol{x})$[3] with physical positions and times of events measured in real experiments. We understand the "space–time" $\mathcal{M}$ as an abstract four-dimensional manifold with a pseudo-Euclidean metric. In Section 8.6 of the third volume, we will discuss in more detail our attitude to quantum fields and their arguments $(t, \boldsymbol{x})$, and we will analyze our differences with the prevailing tradition.

3.1.2 Simple quantum field theories

Before turning to the full-fledged QED, it is useful to warm up on simpler quantum field theories. In such simple theories, the relativistic interaction is constructed in three steps [21, 19].

Step 1. For each particle species[4] participating in the theory, we build a *quantum field*, which is a multicomponent operator-valued function[5] $\phi_\alpha(t, \boldsymbol{x})$ defined on the Minkowski space–time $\mathcal{M}$ and having the following properties:

(I) The operator $\phi_\alpha(t, \boldsymbol{x})$ contains only terms linear in creation and annihilation operators of the particle and antiparticle.

2 Definitions of quantum fields for fermions and bosons are given in Appendices B.4 and C.1, respectively.

3 Which are arguments of quantum fields.

4 In our terminology, a particle and its antiparticle belong to the same species; therefore in our version of QED we will need three fields: electron–positron, proton–antiproton and photon.

5 This means that for each value of its arguments $(t, \boldsymbol{x})$ and index α the symbol $\phi_\alpha(t, \boldsymbol{x})$ defines an operator acting in the Fock space.

(II) The quantum field $\phi_\alpha(t, \boldsymbol{x})$ should have a simple transformation law[6]

$$U_0(\Lambda; \tilde{a})\phi_\alpha(\tilde{x})U_0^{-1}(\Lambda; \tilde{a}) = \sum_\beta D_{\alpha\beta}(\Lambda^{-1})\phi_\beta(\Lambda(\tilde{x} + \tilde{a})) \tag{3.1}$$

with respect to the noninteracting representation (1.40) $U_0(\Lambda; \tilde{a})$ of the Poincaré group in the Fock space, where Λ is a boost/rotation, $\tilde{a}$ is a space–time translation and $D_{\alpha\beta}$ is some finite-dimensional representation[7] of the Lorentz group.

(III) We require the following *anticommutators* for fermion fields (that is, fields of particles with half-integer spin) for the same values of t:

$$\{\psi_\alpha(t, \boldsymbol{x}), \psi_\beta^\dagger(t, \boldsymbol{y})\} = \delta(\boldsymbol{x} - \boldsymbol{y})\delta_{\alpha\beta}, \tag{3.2}$$

$$\{\psi_\alpha(t, \boldsymbol{x}), \psi_\beta(t, \boldsymbol{y})\} = \{\psi_\alpha^\dagger(t, \boldsymbol{x}), \psi_\beta^\dagger(t, \boldsymbol{y})\} = 0. \tag{3.3}$$

(IV) For boson fields (i. e., fields of particles with integer spin or helicity) the following equal-time *commutators* are postulated:

$$[\phi_\alpha(t, \boldsymbol{x}), \phi_\beta^\dagger(t, \boldsymbol{y})] = \delta(\boldsymbol{x} - \boldsymbol{y})\delta_{\alpha\beta}, \tag{3.4}$$

$$[\phi_\alpha(t, \boldsymbol{x}), \phi_\beta(t, \boldsymbol{y})] = [\phi_\alpha^\dagger(t, \boldsymbol{x}), \phi_\beta^\dagger(t, \boldsymbol{y})] = 0. \tag{3.5}$$

Step 2. Having at our disposal quantum fields $\phi_\alpha(\tilde{x}), \psi_\beta(\tilde{x}), \chi_\gamma(\tilde{x}), \ldots$ for all particle species, we build an operator of the *potential energy density*, i. e.,

$$V(\tilde{x}) \equiv V(t, \boldsymbol{x}) = \sum_j V^{(j)}(t, \boldsymbol{x}), \tag{3.6}$$

in the form of a polynomial, where each term is a (local) product of several fields at the same point $(t, \boldsymbol{x})$, i. e.,

$$V^{(j)}(t, \boldsymbol{x}) = \sum_{\alpha,\beta,\gamma\ldots} G^{(j)}_{\alpha\beta\gamma\ldots}\phi_\alpha(t, \boldsymbol{x})\psi_\beta(t, \boldsymbol{x})\chi_\gamma(t, \boldsymbol{x})\cdots, \tag{3.7}$$

and the numerical coefficients $G^{(j)}_{\alpha\beta\gamma\ldots}$ are selected in such a way that $V(\tilde{x})$

(I) is a bosonic[8] Hermitian operator function on the Minkowski space–time $\mathcal{M}$;

6 Our definition (3.1) differs slightly from (5.16)–(5.17) in [21]. This difference can be traced to our convention (**1**-2.6) about the order of factors in the definition of a general inertial transformation $(\Lambda; \tilde{a})$ = (boost) × (rotation) × (translation).

7 The representation $D_{\alpha\beta}$ is definitely nonunitary, since the Lorentz group is noncompact and it is well known that noncompact groups do not have finite-dimensional unitary representations. This does not imply any contradiction, because we are not going to draw any parallels between quantum fields and position–space wave functions. Therefore, no unitarity is required here.

8 That is, each term in (3.7) contains an even number of fermion fields.

(II) transforms as a scalar with respect to the noninteracting representation of the Poincaré group, i. e.,

$$U_0(\Lambda;\tilde{a})V(\tilde{x})U_0^{-1}(\Lambda;\tilde{a}) = V(\Lambda(\tilde{x}+\tilde{a})). \tag{3.8}$$

From the above properties it is not difficult to prove that $V(\tilde{x})$ commutes with itself at space-like intervals, i. e.,

$$[V(t,\boldsymbol{x}), V(t,\boldsymbol{y})] = 0, \quad \text{if } \boldsymbol{x} \neq \boldsymbol{y}. \tag{3.9}$$

Step 3. The required interacting generators of the Poincaré group representation are constructed in the instant form of dynamics. This means that the generators of spatial translations $\boldsymbol{P}_0$ and rotations $\boldsymbol{J}_0$ remain in their noninteracting forms (1.34) and (1.37), respectively, while the generators of time translations H and boosts $\boldsymbol{K}$ are obtained from their noninteracting counterparts (1.32) and (1.39) by adding integrals of the potential energy density (3.6) at $t = 0$, namely

$$H = H_0 + V = H_0 + \int d\boldsymbol{x} V(0,\boldsymbol{x}), \tag{3.10}$$

$$\boldsymbol{K} = \boldsymbol{K}_0 + \boldsymbol{Z} = \boldsymbol{K}_0 - \frac{1}{c^2}\int d\boldsymbol{x}\boldsymbol{x} V(0,\boldsymbol{x}). \tag{3.11}$$

Based on these definitions, one can show (see Appendix E.1) that the above operators $\{\boldsymbol{P}_0, \boldsymbol{J}_0, H, \boldsymbol{K}\}$ form a nontrivial interacting representation of the Poincaré Lie algebra in the Fock space. By Theorem 2.5, the smoothness of coefficient functions of the potentials (3.7) guarantees their cluster separability. Interactions that change the number of particles are obtained in a natural way. Hence, all three conditions of a successful theory, listed in the preface, are fairly easily satisfied in QFT. This explains the astonishing success of QFT in describing realistic physical systems.

3.1.3 Interaction operators in QED

Unfortunately, simple formulas (3.10) and (3.11) work only in the simplest QFT models. More interesting theories, such as QED, require the introduction of some modifications to this scheme. Such modifications are necessary, for example, in theories containing photons. Indeed, the Poincaré transformation (C.32) of the photon field contains the additional term $\Omega_\mu(\tilde{x},\boldsymbol{\theta})$, so it does not obey the simple formula (3.1).[9] Hence the above algorithm breaks down. In this book, we do not attempt to derive QED interactions

9 Despite it not being a 4-vector, we mark the photon quantum field with the tilde $\tilde{\mathcal{A}}(\tilde{x})$ to emphasize its four-component nature.

from first principles.[10] We simply borrow from the standard theory expressions for four interacting Poincaré generators $H = H_0 + V$ and $\boldsymbol{K} = \boldsymbol{K}_0 + \boldsymbol{Z}$ in terms of quantum fields for electrons/positrons $\psi_a(\tilde{x})$, protons/antiprotons $\Psi_a(\tilde{x})$ and photons $A_\mu(\tilde{x})$.

The full QED Hamiltonian has the familiar form

$$H^n = H_0 + V^n, \tag{3.12}$$

where the noninteracting part H_0 is taken from (1.32), and the interaction V^n is composed of two terms,[11]

$$V^n = V_1 + V_2. \tag{3.13}$$

The first-order interaction V_1 is a pseudo-scalar product of two four-component operators. One of them is the 4-vector of the fermion current density $\tilde{\mathrm{j}}(\tilde{x})$ defined in (D.1). The other is the photon quantum field $\tilde{\mathcal{A}}(\tilde{x})$, i. e.,[12]

$$V_1 = \int d\boldsymbol{x}\tilde{\mathrm{j}}(0,\boldsymbol{x}) \cdot \tilde{\mathcal{A}}(0,\boldsymbol{x}) \equiv \int d\boldsymbol{x}\mathrm{j}^\mu(0,\boldsymbol{x})\mathcal{A}_\mu(0,\boldsymbol{x}) = -\int d\boldsymbol{x}\boldsymbol{\mathrm{j}}(0,\boldsymbol{x}) \cdot \boldsymbol{\mathcal{A}}(0,\boldsymbol{x}). \tag{3.14}$$

The second-order interaction is

$$V_2 = \int d\boldsymbol{x}d\boldsymbol{y}\frac{\mathrm{j}^0(0,\boldsymbol{x})\mathrm{j}^0(0,\boldsymbol{y})}{8\pi|\boldsymbol{x}-\boldsymbol{y}|}. \tag{3.15}$$

The interaction term in the boost operator

$$\boldsymbol{K}^n = \boldsymbol{K}_0 + \boldsymbol{Z}^n \tag{3.16}$$

is defined as

$$\begin{aligned}\boldsymbol{Z}^n = &-\frac{1}{c^2}\int d\boldsymbol{x}\boldsymbol{x}(\boldsymbol{\mathrm{j}}(0,\boldsymbol{x}) \cdot \boldsymbol{\mathcal{A}}(0,\boldsymbol{x})) - \frac{1}{c^2}\int d\boldsymbol{x}d\boldsymbol{y}\frac{\boldsymbol{x}\mathrm{j}^0(0,\boldsymbol{x})\mathrm{j}^0(0,\boldsymbol{y})}{8\pi|\boldsymbol{x}-\boldsymbol{y}|}\\ &-\frac{1}{c^2}\int d\boldsymbol{x}\mathrm{j}^0(0,\boldsymbol{x})\boldsymbol{C}(0,\boldsymbol{x}),\end{aligned} \tag{3.17}$$

where the operator function $\boldsymbol{C}(t,\boldsymbol{x})$ is defined as[13]

$$\boldsymbol{C}(\tilde{x}) \equiv \frac{i\hbar^2\sqrt{c}}{\sqrt{2(2\pi\hbar)^3}}\int\frac{d\boldsymbol{p}}{p^{3/2}}\sum_\tau(e^{-\frac{i}{\hbar}\tilde{p}\cdot\tilde{x}}\boldsymbol{e}(\boldsymbol{p},\tau)c_{\boldsymbol{p}\tau} - e^{\frac{i}{\hbar}\tilde{p}\cdot\tilde{x}}\boldsymbol{e}^*(\boldsymbol{p},\tau)c^\dagger_{\boldsymbol{p}\tau}). \tag{3.18}$$

10 The traditional derivation uses the formalism of canonical quantization. It can be found in Chapter 8 of Weinberg's book [21]; see also our discussion in Subsection **3**-1.3.1.

11 The subscript "n" comes from the word "naïve," because, as we will see in Chapter 4, this interaction is unable to describe the physics of charged particles and should be modified by adding renormalization counterterms, just as we did in our toy theory in Section 2.2.

12 The last equality in (3.14) follows from equation (C.12).

13 See the last formula on page 461 in [19] and (B.7) in [20].

The above operators H^n and $\boldsymbol{K}^n$ correspond to QED in the so-called *Coulomb gauge* [21, 19].[14] We will also need an expression of the interaction V^n through creation and annihilation operators of particles. For this, one can substitute quantum field expansions (D.1), (B.34) and (C.2) into equations (3.14) and (3.15). The resulting formulas are rather long and cumbersome, so we moved them to Appendix D.

3.2 *S*-operator in QED

So, we have at our disposal all 10 generators $\{H, \boldsymbol{P}_0, \boldsymbol{J}_0, \boldsymbol{K}\}$ of the Poincaré group representation in the Fock space. Therefore, in principle, nothing should prevent us from calculating any physical quantity related to systems of charged particles and photons. However, this statement is too optimistic, and the point here is not only and not so much that performing such calculations is a rather complex mathematical task. In Chapter 4 we will see that the theory outlined above has a number of serious fundamental difficulties and internal contradictions. In fact, this theory is only suitable for calculations in the lowest order of perturbation theory. An example of such a calculation will be presented in this section, where we obtain a portion of the S-operator for the electron–proton scattering in the second perturbation order. To go to higher orders, we will have to modify our theory by introducing renormalization in Chapter 4. In the third volume we will discuss further modifications of QED that will help us to go beyond the elementary theory of scattering and to achieve an adequate description of bound states and the time evolution.

3.2.1 *S*-operator in second order

We are interested in S-operator terms having the form $d^\dagger a^\dagger da$, i. e., describing the electron–proton scattering. It will be convenient to begin this calculation with the expansion of the scattering phase operator (**1**-7.21) in powers of the coupling constant. Substituting the interaction operator (3.13) in (**1**-7.21) instead of V, we get

$$
\begin{aligned}
\Phi &= \Phi_1 + \Phi_2 + \cdots, \\
\Phi_1 &= V_1, \\
\Phi_2 &= V_2 - \frac{1}{2}[\underline{V_1}, V_1], \qquad (3.19) \\
&\cdots .
\end{aligned}
$$

14 As we shall see in Subsection 3.2.3, the Coulomb gauge formulation is not convenient for S-matrix calculations, which become much easier in the Feynman gauge. Nevertheless, we begin our story precisely with the Coulomb gauge because in this case there is a fairly simple proof of relativistic invariance; see Appendix E.2.

Taking into account that operator V_1 is *unphys* (see Appendix D.2), and due to (1.76) $\underbrace{\Phi_1} = \underbrace{V_1} = 0$, we obtain the following expansion for the S-operator:

$$
\begin{aligned}
S &= e^{\underbrace{\Phi}} \\
&= 1 + \underbrace{\Phi} + \frac{1}{2!}\underbrace{\Phi}\,\underbrace{\Phi} + \cdots \\
&= 1 + \underbrace{\Phi_1} + \underbrace{\Phi_2} + \frac{1}{2!}\underbrace{\Phi_1}\,\underbrace{\Phi_1} + \frac{1}{2!}\underbrace{\Phi_2}\,\underbrace{\Phi_1} + \frac{1}{2!}\underbrace{\Phi_1}\,\underbrace{\Phi_2} + \underbrace{\Phi_3} + \cdots \\
&= 1 + \underbrace{\Phi_2} + \underbrace{\Phi_3} + \cdots \\
&= 1 + \underbrace{V_2} - \frac{1}{2}\underbrace{[\underline{V_1}, V_1]} + \underbrace{\Phi_3} + \cdots .
\end{aligned}
\tag{3.20}
$$

Here our goal is to calculate the second and third terms in this expression.

Let us first calculate the contribution $-\frac{1}{2}[\underline{V_1}, V_1]$ in (3.20). Since we are only interested in terms of the type $d^\dagger a^\dagger da$, it will suffice to consider only the first four terms in (D.9), namely[15]

$$
\begin{aligned}
V_1 = &-\frac{e}{(2\pi\hbar)^{3/2}}\int d\boldsymbol{k}d\boldsymbol{p}\overline{A}^\dagger_a(\boldsymbol{p}+\boldsymbol{k})A_b(\boldsymbol{p})\mathcal{C}_{ab}(\boldsymbol{k}) \\
&-\frac{e}{(2\pi\hbar)^{3/2}}\int d\boldsymbol{k}d\boldsymbol{p}\overline{A}^\dagger_a(\boldsymbol{p}-\boldsymbol{k})A_b(\boldsymbol{p})\mathcal{C}^\dagger_{ab}(\boldsymbol{k}) \\
&+\frac{e}{(2\pi\hbar)^{3/2}}\int d\boldsymbol{k}d\boldsymbol{p}\overline{D}^\dagger_a(\boldsymbol{p}+\boldsymbol{k})D_b(\boldsymbol{p})\mathcal{C}_{ab}(\boldsymbol{k}) \\
&+\frac{e}{(2\pi\hbar)^{3/2}}\int d\boldsymbol{k}d\boldsymbol{p}\overline{D}^\dagger_a(\boldsymbol{p}-\boldsymbol{k})D_b(\boldsymbol{p})\mathcal{C}^\dagger_{ab}(\boldsymbol{k}) + \cdots .
\end{aligned}
$$

According to (1.65), the corresponding terms in $\underline{V_1}$ are

$$
\begin{aligned}
\underline{V_1} = &\frac{e}{(2\pi\hbar)^{3/2}}\int d\boldsymbol{k}d\boldsymbol{p}\overline{A}^\dagger_a(\boldsymbol{p}+\boldsymbol{k})A_b(\boldsymbol{p})\mathcal{C}_{ab}(\boldsymbol{k})\frac{1}{\omega_{\boldsymbol{p}+\boldsymbol{k}} - \omega_{\boldsymbol{p}} - ck} \\
&+\frac{e}{(2\pi\hbar)^{3/2}}\int d\boldsymbol{k}d\boldsymbol{p}\overline{A}^\dagger_a(\boldsymbol{p}-\boldsymbol{k})A_b(\boldsymbol{p})\mathcal{C}^\dagger_{ab}(\boldsymbol{k})\frac{1}{\omega_{\boldsymbol{p}-\boldsymbol{k}} - \omega_{\boldsymbol{p}} + ck} \\
&-\frac{e}{(2\pi\hbar)^{3/2}}\int d\boldsymbol{k}d\boldsymbol{p}\overline{D}^\dagger_a(\boldsymbol{p}+\boldsymbol{k})D_b(\boldsymbol{p})\mathcal{C}_{ab}(\boldsymbol{k})\frac{1}{\Omega_{\boldsymbol{p}+\boldsymbol{k}} - \Omega_{\boldsymbol{p}} - ck} \\
&-\frac{e}{(2\pi\hbar)^{3/2}}\int d\boldsymbol{k}d\boldsymbol{p}\overline{D}^\dagger_a(\boldsymbol{p}-\boldsymbol{k})D_b(\boldsymbol{p})\mathcal{C}^\dagger_{ab}(\boldsymbol{k})\frac{1}{\Omega_{\boldsymbol{p}-\boldsymbol{k}} - \Omega_{\boldsymbol{p}} + ck} \\
&+\cdots .
\end{aligned}
\tag{3.21}
$$

In order to get the desired terms of the type $D^\dagger A^\dagger DA$ in $[\underline{V_1}, V_1]$, we should consider four commutators:

(1) [first term in $\underline{V_1}$, fourth term in V_1];
(2) [second term in $\underline{V_1}$, third term in V_1];

15 The operators A, D and $\mathcal{C}$ are defined in (B.52)–(B.53), (B.56)–(B.57) and (C.16), respectively.

(3) [third term in $\underline{V}_1$, second term in V_1];
(4) [fourth term in $\underline{V}_1$, first term in V_1].

Then using equation (C.18), we obtain

$$
\begin{aligned}
-\frac{1}{2}[\underline{V}_1, V_1] = &-\frac{e^2}{2(2\pi\hbar)^3}\int d\boldsymbol{k}d\boldsymbol{p}d\boldsymbol{k}'d\boldsymbol{p}'\overline{A}_a^\dagger(\boldsymbol{p}+\boldsymbol{k})A_b(\boldsymbol{p})\overline{D}_c^\dagger(\boldsymbol{p}'-\boldsymbol{k}')D_d(\boldsymbol{p}')\frac{[C_{ab}(\boldsymbol{k}), C_{cd}^\dagger(\boldsymbol{k}')]}{\omega_{\boldsymbol{p}+\boldsymbol{k}}-\omega_{\boldsymbol{p}}-ck}\\
&-\frac{e^2}{2(2\pi\hbar)^3}\int d\boldsymbol{k}d\boldsymbol{p}d\boldsymbol{k}'d\boldsymbol{p}'\overline{A}_a^\dagger(\boldsymbol{p}-\boldsymbol{k})A_b(\boldsymbol{p})\overline{D}_c^\dagger(\boldsymbol{p}'+\boldsymbol{k}')D_d(\boldsymbol{p}')\frac{[C_{ab}^\dagger(\boldsymbol{k}), C_{cd}(\boldsymbol{k}')]}{\omega_{\boldsymbol{p}-\boldsymbol{k}}-\omega_{\boldsymbol{p}}+ck}\\
&-\frac{e^2}{2(2\pi\hbar)^3}\int d\boldsymbol{k}d\boldsymbol{p}d\boldsymbol{k}'d\boldsymbol{p}'\overline{D}_a^\dagger(\boldsymbol{p}+\boldsymbol{k})D_b(\boldsymbol{p})\overline{A}_c^\dagger(\boldsymbol{p}'-\boldsymbol{k}')A_d(\boldsymbol{p}')\frac{[C_{ab}(\boldsymbol{k}), C_{cd}^\dagger(\boldsymbol{k}')]}{\Omega_{\boldsymbol{p}+\boldsymbol{k}}-\Omega_{\boldsymbol{p}}-ck}\\
&-\frac{e^2}{2(2\pi\hbar)^3}\int d\boldsymbol{k}d\boldsymbol{p}d\boldsymbol{k}'d\boldsymbol{p}'\overline{D}_a^\dagger(\boldsymbol{p}-\boldsymbol{k})D_b(\boldsymbol{p})\overline{A}_c^\dagger(\boldsymbol{p}'+\boldsymbol{k}')A_d(\boldsymbol{p}')\frac{[C_{ab}^\dagger(\boldsymbol{k}), C_{cd}(\boldsymbol{k}')]}{\Omega_{\boldsymbol{p}-\boldsymbol{k}}-\Omega_{\boldsymbol{p}}+ck}\\
&+\cdots\\
= &\frac{e^2\hbar^2 c}{4(2\pi\hbar)^3}\int \frac{d\boldsymbol{k}d\boldsymbol{p}d\boldsymbol{q}}{k}\gamma_{ab}^\mu\gamma_{cd}^\nu h_{\mu\nu}(\boldsymbol{k})\\
&\times\Bigg(-\overline{D}_c^\dagger(\boldsymbol{p}-\boldsymbol{k})D_d(\boldsymbol{p})\overline{A}_a^\dagger(\boldsymbol{q}+\boldsymbol{k})A_b(\boldsymbol{q})\frac{1}{\omega_{\boldsymbol{q}+\boldsymbol{k}}-\omega_{\boldsymbol{q}}-ck}\\
&+\overline{D}_c^\dagger(\boldsymbol{p}+\boldsymbol{k})D_d(\boldsymbol{p})\overline{A}_a^\dagger(\boldsymbol{q}-\boldsymbol{k})A_b(\boldsymbol{q})\frac{1}{\omega_{\boldsymbol{q}-\boldsymbol{k}}-\omega_{\boldsymbol{q}}+ck}\\
&-\overline{D}_a^\dagger(\boldsymbol{p}+\boldsymbol{k})D_b(\boldsymbol{p})\overline{A}_c^\dagger(\boldsymbol{q}-\boldsymbol{k})A_d(\boldsymbol{q})\frac{1}{\Omega_{\boldsymbol{p}+\boldsymbol{k}}-\Omega_{\boldsymbol{p}}-ck}\\
&+\overline{D}_a^\dagger(\boldsymbol{p}-\boldsymbol{k})D_b(\boldsymbol{p})\overline{A}_c^\dagger(\boldsymbol{q}+\boldsymbol{k})A_d(\boldsymbol{q})\frac{1}{\Omega_{\boldsymbol{p}-\boldsymbol{k}}-\Omega_{\boldsymbol{p}}+ck}+\cdots\Bigg)\\
= &\frac{e^2\hbar^2 c}{4(2\pi\hbar)^3}\int \frac{d\boldsymbol{k}d\boldsymbol{p}d\boldsymbol{q}}{k}\gamma_{ab}^\mu\gamma_{cd}^\nu h_{\mu\nu}(\boldsymbol{k})\\
&\times\Bigg(-\overline{D}_a^\dagger(\boldsymbol{p}-\boldsymbol{k})D_b(\boldsymbol{p})\overline{A}_c^\dagger(\boldsymbol{q}+\boldsymbol{k})A_d(\boldsymbol{q})\frac{1}{\omega_{\boldsymbol{q}+\boldsymbol{k}}-\omega_{\boldsymbol{q}}-ck}\\
&+\overline{D}_a^\dagger(\boldsymbol{p}-\boldsymbol{k})D_b(\boldsymbol{p})\overline{A}_c^\dagger(\boldsymbol{q}+\boldsymbol{k})A_d(\boldsymbol{q})\frac{1}{\omega_{\boldsymbol{q}+\boldsymbol{k}}-\omega_{\boldsymbol{q}}+ck}\\
&-\overline{D}_a^\dagger(\boldsymbol{p}-\boldsymbol{k})D_b(\boldsymbol{p})\overline{A}_c^\dagger(\boldsymbol{q}+\boldsymbol{k})A_d(\boldsymbol{q})\frac{1}{\Omega_{\boldsymbol{p}-\boldsymbol{k}}-\Omega_{\boldsymbol{p}}-ck}\\
&+\overline{D}_a^\dagger(\boldsymbol{p}-\boldsymbol{k})D_b(\boldsymbol{p})\overline{A}_c^\dagger(\boldsymbol{q}+\boldsymbol{k})A_d(\boldsymbol{q})\frac{1}{\Omega_{\boldsymbol{p}-\boldsymbol{k}}-\Omega_{\boldsymbol{p}}+ck}+\cdots\Bigg)\\
= &-\frac{e^2\hbar^2c^2}{2(2\pi\hbar)^3}\int d\boldsymbol{k}d\boldsymbol{p}d\boldsymbol{q}\gamma_{ab}^\mu\gamma_{cd}^\nu\frac{h_{\mu\nu}(\boldsymbol{k})}{(\omega_{\boldsymbol{q}+\boldsymbol{k}}-\omega_{\boldsymbol{q}})^2-c^2k^2}\\
&\times\overline{D}_a^\dagger(\boldsymbol{p}-\boldsymbol{k})\overline{A}_c^\dagger(\boldsymbol{q}+\boldsymbol{k})D_b(\boldsymbol{p})A_d(\boldsymbol{q})
\end{aligned}
$$

$$-\frac{e^2\hbar^2c^2}{2(2\pi\hbar)^3}\int d\boldsymbol{k}d\boldsymbol{p}d\boldsymbol{q}\gamma^{\mu}_{ab}\gamma^{\nu}_{cd}\frac{h_{\mu\nu}(\boldsymbol{k})}{(\Omega_{\boldsymbol{p}-\boldsymbol{k}}-\Omega_{\boldsymbol{p}})^2-c^2k^2}$$
$$\times\overline{D}^{\dagger}_a(\boldsymbol{p}-\boldsymbol{k})\overline{A}^{\dagger}_c(\boldsymbol{q}+\boldsymbol{k})D_b(\boldsymbol{p})A_d(\boldsymbol{q}). \tag{3.22}$$

Next we take into account that we are interested only in the S-operator behavior near the energy shell, where

$$\Omega_{\boldsymbol{p}-\boldsymbol{k}}-\Omega_{\boldsymbol{p}}=\omega_{\boldsymbol{q}}-\omega_{\boldsymbol{q}+\boldsymbol{k}}. \tag{3.23}$$

Let us use the notation (B.64)–(B.65) in which

$$\overline{A}^{\dagger}(\boldsymbol{q}+\boldsymbol{k})\gamma^{\nu}A(\boldsymbol{q})=\frac{m_ec^2}{\sqrt{\omega_{\boldsymbol{q}+\boldsymbol{k}}\omega_{\boldsymbol{q}}}}\sum_{\sigma\sigma'}\mathcal{U}^{\nu}((\boldsymbol{q}+\boldsymbol{k})\sigma',\boldsymbol{q}\sigma)a^{\dagger}_{(\boldsymbol{q}+\boldsymbol{k})\sigma'}a_{\boldsymbol{q}\sigma},$$
$$\overline{D}^{\dagger}(\boldsymbol{p}-\boldsymbol{k})\gamma^{\mu}D(\boldsymbol{p})=\frac{m_pc^2}{\sqrt{\Omega_{\boldsymbol{p}-\boldsymbol{k}}\Omega_{\boldsymbol{p}}}}\sum_{\tau\tau'}\mathcal{W}^{\mu}((\boldsymbol{p}-\boldsymbol{k})\tau',\boldsymbol{p}\tau)d^{\dagger}_{(\boldsymbol{p}-\boldsymbol{k})\tau'}d_{\boldsymbol{p}\tau}$$

and equation (C.20). Then

$$-\frac{1}{2}\underline{[V_1,V_1]}\approx-\frac{e^2\hbar^2c^2}{(2\pi\hbar)^3}\sum_{\sigma\tau\sigma'\tau'}\int d\boldsymbol{k}d\boldsymbol{p}d\boldsymbol{q}\frac{m_pm_ec^4}{\sqrt{\omega_{\boldsymbol{q}+\boldsymbol{k}}\omega_{\boldsymbol{q}}}\sqrt{\Omega_{\boldsymbol{p}-\boldsymbol{k}}\Omega_{\boldsymbol{p}}}}$$
$$\times\frac{h_{\mu\nu}(\boldsymbol{k})\mathcal{U}^{\nu}((\boldsymbol{q}+\boldsymbol{k})\sigma',\boldsymbol{q}\sigma)\mathcal{W}^{\mu}((\boldsymbol{p}-\boldsymbol{k})\tau',\boldsymbol{p}\tau)}{(\omega_{\boldsymbol{q}+\boldsymbol{k}}-\omega_{\boldsymbol{q}})^2-c^2k^2}d^{\dagger}_{(\boldsymbol{p}-\boldsymbol{k})\tau'}a^{\dagger}_{(\boldsymbol{q}+\boldsymbol{k})\sigma'}d_{\boldsymbol{p}\tau}a_{\boldsymbol{q}\sigma}$$
$$=-\frac{e^2\hbar^2c^2}{(2\pi\hbar)^3}\sum_{\sigma\tau\sigma'\tau'}\int d\boldsymbol{k}d\boldsymbol{p}d\boldsymbol{q}\frac{m_pm_ec^4}{\sqrt{\omega_{\boldsymbol{q}+\boldsymbol{k}}\omega_{\boldsymbol{q}}\Omega_{\boldsymbol{p}-\boldsymbol{k}}\Omega_{\boldsymbol{p}}}}$$
$$\times\left[\frac{\mathcal{U}((\boldsymbol{q}+\boldsymbol{k})\sigma',\boldsymbol{q}\sigma)\cdot\mathcal{W}((\boldsymbol{p}-\boldsymbol{k})\tau',\boldsymbol{p}\tau)}{(\omega_{\boldsymbol{q}+\boldsymbol{k}}-\omega_{\boldsymbol{q}})^2-c^2k^2}\right.$$
$$\left.-\frac{(\boldsymbol{k}\cdot\mathcal{U}((\boldsymbol{q}+\boldsymbol{k})\sigma',\boldsymbol{q}\sigma))(\boldsymbol{k}\cdot\mathcal{W}((\boldsymbol{p}-\boldsymbol{k})\tau',\boldsymbol{p}\tau))}{k^2((\omega_{\boldsymbol{q}+\boldsymbol{k}}-\omega_{\boldsymbol{q}})^2-c^2k^2)}\right]d^{\dagger}_{(\boldsymbol{p}-\boldsymbol{k})\tau'}a^{\dagger}_{(\boldsymbol{q}+\boldsymbol{k})\sigma'}d_{\boldsymbol{p}\tau}a_{\boldsymbol{q}\sigma}.$$

Combining this expression with the term $D^{\dagger}A^{\dagger}DA$ in V_2,[16] we see that the operator Φ_2 in (3.19) takes the form

$$\Phi_2=\sum_{\sigma\tau\sigma'\tau'}\int d\boldsymbol{p}d\boldsymbol{q}d\boldsymbol{p}'d\boldsymbol{q}'\phi_2(\boldsymbol{p}'\tau',\boldsymbol{q}'\sigma';\boldsymbol{p}\tau,\boldsymbol{q}\sigma)\delta(\boldsymbol{p}+\boldsymbol{q}-\boldsymbol{p}'-\boldsymbol{q}')d^{\dagger}_{\boldsymbol{p}'\tau'}a^{\dagger}_{\boldsymbol{q}'\sigma'}d_{\boldsymbol{p}\tau}a_{\boldsymbol{q}\sigma}, \tag{3.24}$$

where the coefficient function is

$$\phi_2(\boldsymbol{p}'\tau',\boldsymbol{q}'\sigma';\boldsymbol{p}\tau,\boldsymbol{q}\sigma)$$
$$=\frac{e^2\hbar^2c^2}{(2\pi\hbar)^3}\frac{m_pm_ec^4}{\sqrt{\omega_{\boldsymbol{q}+\boldsymbol{k}}\omega_{\boldsymbol{q}}\Omega_{\boldsymbol{p}-\boldsymbol{k}}\Omega_{\boldsymbol{p}}}}\left[-\frac{\mathcal{U}((\boldsymbol{q}+\boldsymbol{k})\sigma',\boldsymbol{q}\sigma)\cdot\mathcal{W}((\boldsymbol{p}-\boldsymbol{k})\tau',\boldsymbol{p}\tau)}{(\omega_{\boldsymbol{q}+\boldsymbol{k}}-\omega_{\boldsymbol{q}})^2-c^2k^2}\right.$$

16 The third term on the right-hand side of (D.12).

$$+\frac{(\boldsymbol{k}\cdot\mathcal{U}((\boldsymbol{q}+\boldsymbol{k})\sigma',\boldsymbol{q}\sigma))(\boldsymbol{k}\cdot\mathcal{W}((\boldsymbol{p}-\boldsymbol{k})\tau',\boldsymbol{p}\tau))}{k^2((\omega_{\boldsymbol{q}+\boldsymbol{k}}-\omega_{\boldsymbol{q}})^2-c^2k^2)}$$
$$-\frac{\mathcal{U}^0((\boldsymbol{q}+\boldsymbol{k})\sigma',\boldsymbol{q}\sigma)\mathcal{W}^0((\boldsymbol{p}-\boldsymbol{k})\tau',\boldsymbol{p}\tau)}{c^2k^2}\Bigg]. \tag{3.25}$$

In the $(v/c)^2$ approximation, we apply formula (G.3) to obtain[17]

$$\phi_2(\boldsymbol{p}'\tau',\boldsymbol{q}'\sigma';\boldsymbol{p}\tau,\boldsymbol{q}\sigma)\approx\frac{ie^2\delta_{\tau\tau'}\delta_{\sigma\sigma'}}{(-2\pi i)(2\pi)^2\hbar}\left(\frac{1}{k^2}-\frac{1}{8m_e^2c^2}\right)-\frac{\alpha\delta_{\tau\tau'}}{(-2\pi i)4\pi m_e^2c}\chi^\dagger_{\sigma'}\frac{\boldsymbol{\sigma}_{\mathrm{el}}\cdot[\boldsymbol{k}\times\boldsymbol{q}]}{k^2}\chi_\sigma. \tag{3.26}$$

As the next step we insert this result into formula (3.20) for the S-operator. According to (1.66), in order to perform the integration with respect to t from $-\infty$ to ∞, we should simply multiply the coefficient function by $-2\pi i\delta(\mathcal{E}(\boldsymbol{p},\boldsymbol{q},\boldsymbol{k}))$, where

$$\mathcal{E}(\boldsymbol{p},\boldsymbol{q},\boldsymbol{k})=\Omega_{\boldsymbol{p}-\boldsymbol{k}}+\omega_{\boldsymbol{q}+\boldsymbol{k}}-\Omega_{\boldsymbol{p}}-\omega_{\boldsymbol{q}}$$

is the energy function. Finally, the second-order scattering operator is represented in the standard form (1.58), i. e.,

$$S_2[d^\dagger a^\dagger da]=(-2\pi i)\sum_{\sigma\tau\sigma'\tau'}\int d\boldsymbol{p}d\boldsymbol{q}d\boldsymbol{p}'d\boldsymbol{q}'\phi_2(\boldsymbol{p}'\tau',\boldsymbol{q}'\sigma';\boldsymbol{p}\tau,\boldsymbol{q}\sigma)\delta^4(\tilde{p}+\tilde{q}-\tilde{p}'-\tilde{q}')$$
$$\times d^\dagger_{\boldsymbol{p}'\tau'}a^\dagger_{\boldsymbol{q}'\sigma'}d_{\boldsymbol{p}\tau}a_{\boldsymbol{q}\sigma}.$$

In the extreme nonrelativistic approximation, we can omit terms in (3.26) having c in the denominators and obtain

$$S_2[d^\dagger a^\dagger da]\approx\frac{ie^2}{(2\pi)^2\hbar}\sum_{\sigma\tau}\int d\boldsymbol{p}d\boldsymbol{q}d\boldsymbol{k}\frac{\delta(\mathcal{E}(\boldsymbol{p},\boldsymbol{q},\boldsymbol{k}))}{k^2}d^\dagger_{(\boldsymbol{p}-\boldsymbol{k})\tau}a^\dagger_{(\boldsymbol{q}+\boldsymbol{k})\sigma}d_{\boldsymbol{p}\tau}a_{\boldsymbol{q}\sigma}, \tag{3.27}$$

which is consistent with our toy model result (2.23) if we set $\lambda=0$ there.[18]

3.2.2 Covariant form of S-operator

Our expression (3.25) for the scattering phase can be simplified by taking into account that the coefficient function outside the energy shell may be chosen arbitrarily.

Note that the 4-vector of the transferred momentum can be written in two equivalent forms, i. e.,

---

17 Here $\alpha\equiv e^2/(4\pi\hbar c)\approx 1/137$ is the *fine structure constant* and $\boldsymbol{k}\equiv\boldsymbol{p}-\boldsymbol{p}'=\boldsymbol{q}'-\boldsymbol{q}$ is the *transferred momentum.*

18 The additional factor of 1/2 in (2.23) is due to the fact that this expression refers to the scattering of indistinguishable particles.

$$\tilde{k} = \tilde{p} - \tilde{p}' = (\Omega_{\boldsymbol{p}} - \Omega_{\boldsymbol{p}-\boldsymbol{k}}, c\boldsymbol{p} - c(\boldsymbol{p}-\boldsymbol{k})) = (\Omega_{\boldsymbol{p}} - \Omega_{\boldsymbol{p}-\boldsymbol{k}}, c\boldsymbol{k}),$$
$$\tilde{k} = \tilde{q}' - \tilde{q} = (\omega_{\boldsymbol{q}+\boldsymbol{k}} - \omega_{\boldsymbol{q}}, c(\boldsymbol{q}+\boldsymbol{k}) - c\boldsymbol{q}) = (\omega_{\boldsymbol{q}+\boldsymbol{k}} - \omega_{\boldsymbol{q}}, c\boldsymbol{k}).$$

Then from (B.95)–(B.96) and (3.23) we obtain on the energy shell

$$\begin{aligned}(\boldsymbol{k}\cdot\boldsymbol{\mathcal{W}}((\boldsymbol{p}-\boldsymbol{k})\tau',\boldsymbol{p}\tau)) &= \frac{\Omega_{\boldsymbol{p}} - \Omega_{\boldsymbol{p}-\boldsymbol{k}}}{c}\mathcal{W}^0((\boldsymbol{p}-\boldsymbol{k})\tau',\boldsymbol{p}\tau)\\ &= \frac{\omega_{\boldsymbol{q}+\boldsymbol{k}} - \omega_{\boldsymbol{q}}}{c}\mathcal{W}^0((\boldsymbol{p}-\boldsymbol{k})\tau',\boldsymbol{p}\tau),\\ (\boldsymbol{k}\cdot\boldsymbol{\mathcal{U}}((\boldsymbol{q}+\boldsymbol{k})\sigma',\boldsymbol{q}\sigma)) &= \frac{\omega_{\boldsymbol{q}+\boldsymbol{k}} - \omega_{\boldsymbol{q}}}{c}\mathcal{U}^0((\boldsymbol{q}+\boldsymbol{k})\sigma',\boldsymbol{q}\sigma)\end{aligned}$$

and

$$\begin{aligned}&\frac{(\boldsymbol{k}\cdot\boldsymbol{\mathcal{U}})(\boldsymbol{k}\cdot\boldsymbol{\mathcal{W}})}{k^2((\omega_{\boldsymbol{q}+\boldsymbol{k}} - \omega_{\boldsymbol{q}})^2 - c^2k^2)} - \frac{\mathcal{U}^0\mathcal{W}^0}{c^2k^2}\\ &= \frac{(\omega_{\boldsymbol{q}+\boldsymbol{k}} - \omega_{\boldsymbol{q}})^2\mathcal{U}^0\mathcal{W}^0}{c^2k^2((\omega_{\boldsymbol{q}+\boldsymbol{k}} - \omega_{\boldsymbol{q}})^2 - c^2k^2)} - \frac{[(\omega_{\boldsymbol{q}+\boldsymbol{k}} - \omega_{\boldsymbol{q}})^2 - c^2k^2]\mathcal{U}^0\mathcal{W}^0}{c^2k^2((\omega_{\boldsymbol{q}+\boldsymbol{k}} - \omega_{\boldsymbol{q}})^2 - c^2k^2)}\\ &= \frac{\mathcal{U}^0\mathcal{W}^0}{(\omega_{\boldsymbol{q}+\boldsymbol{k}} - \omega_{\boldsymbol{q}})^2 - c^2k^2}.\end{aligned}$$

Substituting these results in (3.25), we obtain a manifestly covariant coefficient function of the S-operator, i. e.,

$$\begin{aligned}&s_2(\boldsymbol{p}'\tau',\boldsymbol{q}'\sigma';\boldsymbol{p}\tau,\boldsymbol{q}\sigma)\\ &= (-2\pi i)\phi_2(\boldsymbol{p}'\tau',\boldsymbol{q}'\sigma';\boldsymbol{p}\tau,\boldsymbol{q}\sigma)\\ &= \frac{e^2\hbar^2c^2(-2\pi i)}{(2\pi\hbar)^3((\omega_{\boldsymbol{q}+\boldsymbol{k}} - \omega_{\boldsymbol{q}})^2 - c^2k^2)}\frac{m_pm_ec^4}{\sqrt{\omega_{\boldsymbol{q}+\boldsymbol{k}}\omega_{\boldsymbol{q}}\Omega_{\boldsymbol{p}-\boldsymbol{k}}\Omega_{\boldsymbol{p}}}}\\ &\quad\times[\mathcal{U}^0((\boldsymbol{q}+\boldsymbol{k})\sigma',\boldsymbol{q}\sigma)\mathcal{W}^0((\boldsymbol{p}-\boldsymbol{k})\tau',\boldsymbol{p}\tau) - \boldsymbol{\mathcal{U}}((\boldsymbol{q}+\boldsymbol{k})\sigma',\boldsymbol{q}\sigma)\cdot\boldsymbol{\mathcal{W}}((\boldsymbol{p}-\boldsymbol{k})\tau',\boldsymbol{p}\tau)]\\ &= \frac{e^2\hbar^2c^2(-2\pi i)}{(2\pi\hbar)^3}\frac{m_pm_ec^4}{\sqrt{\omega_{\boldsymbol{q}+\boldsymbol{k}}\omega_{\boldsymbol{q}}\Omega_{\boldsymbol{p}-\boldsymbol{k}}\Omega_{\boldsymbol{p}}}}\frac{\mathcal{U}_\mu((\boldsymbol{q}+\boldsymbol{k})\sigma',\boldsymbol{q}\sigma)\mathcal{W}^\mu((\boldsymbol{p}-\boldsymbol{k})\tau',\boldsymbol{p}\tau)}{(\tilde{q}-\tilde{q}')^2}.\end{aligned} \tag{3.28}$$

Note also that this coefficient function can be regarded as a matrix element of the S-operator between two-particle momentum–spin basis states, i. e.,

$$\begin{aligned}&\langle\text{vac}|a_{\boldsymbol{q}'\sigma'}d_{\boldsymbol{p}'\tau'}S_2[d^\dagger a^\dagger da]d^\dagger_{\boldsymbol{p}\tau}a^\dagger_{\boldsymbol{q}\sigma}|\text{vac}\rangle\\ &= \langle\text{vac}|a_{\boldsymbol{q}'\sigma'}d_{\boldsymbol{p}'\tau'}\bigg[\sum_{\pi\rho\pi'\rho'}\int d\boldsymbol{s}d\boldsymbol{t}d\boldsymbol{s}'d\boldsymbol{t}'s_2(\boldsymbol{s}'\pi',\boldsymbol{t}'\rho';\boldsymbol{s}\pi,\boldsymbol{t}\rho)\\ &\quad\times\delta(\boldsymbol{s}+\boldsymbol{t}-\boldsymbol{s}'-\boldsymbol{t}')\delta(\Omega_{\boldsymbol{s}} + \omega_{\boldsymbol{t}} - \Omega_{\boldsymbol{s}'} - \omega_{\boldsymbol{t}'})d^\dagger_{\boldsymbol{s}'\pi'}a^\dagger_{\boldsymbol{t}'\rho'}d_{\boldsymbol{s}\pi}a_{\boldsymbol{t}\rho}\bigg]d^\dagger_{\boldsymbol{p}\tau}a^\dagger_{\boldsymbol{q}\sigma}|\text{vac}\rangle\end{aligned}$$

$$= \sum_{\pi\rho\pi'\rho'} \int d\mathbf{s}d\mathbf{t}d\mathbf{s}'d\mathbf{t}' s_2(\mathbf{s}'\pi', \mathbf{t}'\rho'; \mathbf{s}\pi, \mathbf{t}\rho)\delta(\mathbf{s} + \mathbf{t} - \mathbf{s}' - \mathbf{t}')\delta(\Omega_{\mathbf{s}'} + \omega_{\mathbf{t}'} - \Omega_{\mathbf{s}} - \omega_{\mathbf{t}})$$
$$\times \delta(\mathbf{s} - \mathbf{p})\delta_{\pi\tau}\delta(\mathbf{t} - \mathbf{q})\delta_{\rho\sigma}\delta(\mathbf{s}' - \mathbf{p}')\delta_{\pi'\tau'}\delta(\mathbf{t}' - \mathbf{q}')\delta_{\rho'\sigma'}$$
$$= s_2(\mathbf{p}'\tau', \mathbf{q}'\sigma'; \mathbf{p}\tau, \mathbf{q}\sigma)\delta(\tilde{p} + \tilde{q} - \tilde{p}' - \tilde{q}'). \tag{3.29}$$

3.2.3 Feynman gauge

The coefficient function (3.28) is manifestly covariant, and it is true that the S-operator calculated in this way does satisfy the Poincaré invariance condition (**1**-7.7). However, this result looks almost accidental. Indeed, in our interacting Hamiltonian $H_0 + V_1 + V_2$ it is not easy to recognize the zero component of a 4-momentum energy vector. We also used the noncovariant form (C.21) of the function $h_{\mu\nu}(\mathbf{k})$ and observed a successful cancellation of noncovariant terms when calculating the S-operator. Was it a lucky coincidence or a predictable inevitability?

Of course, under more detailed examination, this result does not look surprising, because in Appendix E.2 we proved that the Coulomb gauge QED meets all the requirements of Poincaré invariance. However, we would like to have a method in which the relativistic invariance remains explicit at all stages of the calculation.

An additional incentive to find an alternative approach is that calculations of scattering amplitudes with interaction (3.13) $V_1 + V_2$ are very labor-intensive. As we see in Appendices D.2 and D.3, operators V_1 and V_2 have rather complex expressions in terms of creation and annihilation operators. In high perturbation orders, calculations of their multiple normally ordered products and/or commutators become almost impossible.

Here, the *gauge invariance* of electrodynamics comes to our aid. It turns out that one can change the gauge and redefine quantum fields and the Hamiltonian in such a way that observable quantities, such as scattering cross sections, remain unchanged. The theory looks most simple and invariant in the so-called *Feynman gauge*. The recipe for the transition to this formulation is as follows:

(1) Instead of the noncovariant interaction $V_1 + V_2$ in (3.13), leave only the "Lorentz scalar" operator V_1 from (3.14), i. e.,

$$V_1 = \int d\mathbf{x} j^\mu(0, \mathbf{x})\mathcal{A}_\mu(0, \mathbf{x})$$
$$= -e \int d\mathbf{x}\overline{\psi}(\tilde{x})\gamma_\mu\psi(\tilde{x})\mathcal{A}^\mu(\tilde{x}) + e \int d\mathbf{x}\overline{\Psi}(\tilde{x})\gamma_\mu\Psi(\tilde{x})\mathcal{A}^\mu(\tilde{x}). \tag{3.30}$$

(2) In formula (C.23) for the photon propagator, use a covariant expression $-\eta_{\mu\nu}$ instead of the matrix $h_{\mu\nu}(\mathbf{k})$.

If, in addition to these rules, one computes the S-operator with the help of the time-ordered perturbation series (**1**-7.17), then calculations become much easier.

We omit the (nontrivial) proof that the Feynman gauge approach works in all orders of perturbation theory and yields results identical to the original Coulomb gauge theory. This proof is most elegant within the functional integral approach [21, 10], the discussion of which is beyond the scope of our book. Here we just illustrate our claims by repeating our calculation of the S-matrix element (3.28), this time using the Feynman gauge rules. We use formulas (**1**-7.17), (3.29) and (D.1) to obtain

$$
\begin{aligned}
&s_2(\boldsymbol{p}'\tau',\boldsymbol{q}'\sigma';\boldsymbol{p}\tau,\boldsymbol{q}\sigma)\delta(\tilde{p}+\tilde{q}-\tilde{p}'-\tilde{q}')\\
&= \langle \mathrm{vac}|a_{\boldsymbol{q}'\sigma'}d_{\boldsymbol{p}'\tau'}S_2 d^\dagger_{\boldsymbol{p}\tau}a^\dagger_{\boldsymbol{q}\sigma}|\mathrm{vac}\rangle\\
&= -\langle \mathrm{vac}|a_{\boldsymbol{q}'\sigma'}d_{\boldsymbol{p}'\tau'}\left(\frac{1}{2!\hbar^2}\int_{-\infty}^{+\infty} dt_1dt_2 T[V_1(t_1)V_1(t_2)]\right)d^\dagger_{\boldsymbol{p}\tau}a^\dagger_{\boldsymbol{q}\sigma}|\mathrm{vac}\rangle\\
&= -\frac{1}{2!\hbar^2}\int d^4x_1d^4x_2\langle \mathrm{vac}|a_{\boldsymbol{q}'\sigma'}d_{\boldsymbol{p}'\tau'}\\
&\quad\times T[(j^\mu_{ep}(\tilde{x}_1)\mathcal{A}_\mu(\tilde{x}_1)+j^\mu_{pa}(\tilde{x}_1)\mathcal{A}_\mu(\tilde{x}_1))(j^\mu_{ep}(\tilde{x}_2)\mathcal{A}_\mu(\tilde{x}_2)+j^\mu_{pa}(\tilde{x}_2)\mathcal{A}_\mu(\tilde{x}_2))]d^\dagger_{\boldsymbol{p}\tau}a^\dagger_{\boldsymbol{q}\sigma}|\mathrm{vac}\rangle\\
&= -\frac{1}{2!\hbar^2}\int d^4x_1d^4x_2\langle \mathrm{vac}|a_{\boldsymbol{q}'\sigma'}d_{\boldsymbol{p}'\tau'}\\
&\quad\times (T[j^\mu_{ep}(\tilde{x}_1)\mathcal{A}_\mu(\tilde{x}_1)j^\mu_{pa}(\tilde{x}_2)\mathcal{A}_\mu(\tilde{x}_2)]+T[j^\mu_{pa}(\tilde{x}_1)\mathcal{A}_\mu(\tilde{x}_1)j^\mu_{ep}(\tilde{x}_2)\mathcal{A}_\mu(\tilde{x}_2)])d^\dagger_{\boldsymbol{p}\tau}a^\dagger_{\boldsymbol{q}\sigma}|\mathrm{vac}\rangle\\
&= \frac{e^2}{\hbar^2}\int d^4x_1d^4x_2\langle \mathrm{vac}|a_{\boldsymbol{q}'\sigma'}d_{\boldsymbol{p}'\tau'}\\
&\quad\times T[\overline{\psi}(\tilde{x}_1)\gamma^\mu\psi(\tilde{x}_1)\mathcal{A}_\mu(\tilde{x}_1)\overline{\Psi}(\tilde{x}_2)\gamma^\nu\Psi(\tilde{x}_2)\mathcal{A}_\nu(\tilde{x}_2)]d^\dagger_{\boldsymbol{p}\tau}a^\dagger_{\boldsymbol{q}\sigma}|\mathrm{vac}\rangle. \qquad (3.31)
\end{aligned}
$$

In the integrand we have the operator $a_{\boldsymbol{q}'\sigma'}d_{\boldsymbol{p}'\tau'}T[\cdots]d^\dagger_{\boldsymbol{p}\tau}a^\dagger_{\boldsymbol{q}\sigma}$, sandwiched by vacuum vectors $\langle\mathrm{vac}|\cdots|\mathrm{vac}\rangle$. This operator should be normally ordered. Then, only the c-number term will contribute to the matrix element. To provide such a number, the operator under the T-order symbol must have the structure $d^\dagger a^\dagger da$. From expressions (B.34) and (B.37) for quantum fields ψ and Ψ, we conclude that the operator $d^\dagger$ can originate only from the field $\overline{\Psi}$, the operator $a^\dagger$ comes from the field $\overline{\psi}$ and operators d and a come from factors Ψ and ψ, respectively. The photon operators in the fields $\mathcal{A}_\mu$ and $\mathcal{A}_\nu$ have to be consumed by Wick's contractions, explained in Subsection 1.1.9. In the process of bringing the entire product to the normal order, the fermion creation (annihilation) operators inside the T-symbol should be contracted with the corresponding annihilation (creation) operators outside this symbol. After contractions we obtain expressions of the form (momentum delta function) × (Kronecker delta symbol of spin components) × (numerical factor). The delta function and the Kronecker delta disappear after integration (summation), and only the numerical factor remains. For example, the contraction of the electron creation operator from $\overline{\psi}_a$ with the $a_{\boldsymbol{q}'\sigma'}$ annihilation operator outside the T-symbol produces the following numerical factor:

$$
\sqrt{\frac{m_ec^2}{(2\pi\hbar)^3\omega_{\boldsymbol{q}'}}}\exp\left(\frac{i}{\hbar}\tilde{q}'\cdot\tilde{x}_1\right)\overline{u}_a(\boldsymbol{q}',\sigma'). \qquad (3.32)
$$

After all these routine manipulations, the coefficient function takes the form[19]

$$\begin{aligned}
&s_2(\boldsymbol{p}'\tau',\boldsymbol{q}'\sigma';\boldsymbol{p}\tau,\boldsymbol{q}\sigma)\delta(\tilde{p}+\tilde{q}-\tilde{p}'-\tilde{q}')\\
&\approx \int d^4x_1d^4x_2\frac{e^2m_pm_ec^4}{\hbar^2(2\pi\hbar)^6\sqrt{\omega_{\boldsymbol{q}}\omega_{\boldsymbol{q}'}\Omega_{\boldsymbol{p}}\Omega_{\boldsymbol{p}'}}}\exp\left(\frac{i}{\hbar}\tilde{q}'\cdot\tilde{x}_1\right)\\
&\times\exp\left(-\frac{i}{\hbar}\tilde{q}\cdot\tilde{x}_1\right)\exp\left(\frac{i}{\hbar}\tilde{p}'\cdot\tilde{x}_2\right)\exp\left(-\frac{i}{\hbar}\tilde{p}\cdot\tilde{x}_2\right)\\
&\times\overline{u}(\boldsymbol{q}',\sigma')\gamma^\mu u(\boldsymbol{q},\sigma)\overline{w}(\boldsymbol{p}',\tau')\gamma^\nu w(\boldsymbol{p},\tau)\langle\mathrm{vac}|T[\mathcal{A}_\mu(\tilde{x}_1)\mathcal{A}_\nu(\tilde{x}_2)]|\mathrm{vac}\rangle\\
&=\frac{1}{2\pi i}\int d^4x_1d^4x_2d^4s\frac{e^2c^2m_pm_ec^4}{(2\pi\hbar)^9\sqrt{\omega_{\boldsymbol{q}}\omega_{\boldsymbol{q}'}\Omega_{\boldsymbol{p}}\Omega_{\boldsymbol{p}'}}}\\
&\times\exp\left(\frac{i}{\hbar}(\tilde{q}'-\tilde{q}+\tilde{s})\cdot\tilde{x}_1\right)\exp\left(\frac{i}{\hbar}(\tilde{p}'-\tilde{p}-\tilde{s})\cdot\tilde{x}_2\right)\\
&\times\overline{u}_b(\boldsymbol{q}',\sigma')\gamma^\mu_{ba}u_a(\boldsymbol{q},\sigma)\frac{\eta_{\mu\nu}}{\tilde{s}^2}\overline{w}_d(\boldsymbol{p}',\tau')\gamma^\nu_{dc}w_c(\boldsymbol{p},\tau)\\
&=-\frac{ie^2c^2m_pm_ec^4}{4\pi^2\hbar\sqrt{\omega_{\boldsymbol{q}}\omega_{\boldsymbol{q}'}\Omega_{\boldsymbol{p}}\Omega_{\boldsymbol{p}'}}}\delta^4(\tilde{p}+\tilde{q}-\tilde{p}'-\tilde{q}')\\
&\times\overline{u}_b(\boldsymbol{q}',\sigma')\gamma^\mu_{ba}u_a(\boldsymbol{q},\sigma)\frac{\eta_{\mu\nu}}{(\tilde{q}-\tilde{q}')^2}\overline{w}_d(\boldsymbol{p}',\tau')\gamma^\nu_{dc}w_c(\boldsymbol{p},\tau) \qquad (3.33)\\
&=-\frac{ie^2m_pm_ec^6\delta^4(\tilde{p}+\tilde{q}-\tilde{p}'-\tilde{q}')\mathcal{U}^\mu(\boldsymbol{q}'\sigma',\boldsymbol{q}\sigma)\mathcal{W}_\mu(\boldsymbol{p}'\tau',\boldsymbol{p}\tau)}{4\pi^2\hbar\sqrt{\omega_{\boldsymbol{q}}\omega_{\boldsymbol{q}'}\Omega_{\boldsymbol{p}}\Omega_{\boldsymbol{p}'}}(\tilde{q}-\tilde{q}')^2}. \qquad (3.34)
\end{aligned}$$

This formula, as expected, coincides with the result (3.28) of the noncovariant approach.

3.2.4 Feynman diagrams

The above calculation of the simplest scattering amplitude is still rather involved. However, it can be greatly simplified by noticing that amplitudes are always constructed by simple rules from a small number of standard factors. Feynman came up with a convenient way to represent these rules and factors in the form of pictures – diagrams. Derivations of the *Feynman rules* can be found in any QFT textbook. Here we will simply list the rules for drawing and interpreting diagrams in QED.

So, to write down the matrix element of the S-operator in the $\mathcal{V}$th perturbation order, it is necessary to perform the following steps:

19 The matrix element $\langle\mathrm{vac}|T[\mathcal{A}^\mu(\tilde{x}_1)\mathcal{A}^\nu(\tilde{x}_2)]|\mathrm{vac}\rangle$ is called the *photon propagator*. We took its value from (C.24).

Table 3.1: Correspondence between elements of Feynman diagrams and factors in S-matrix elements ($e^{\mp}$ = electron/positron, $p^{\pm}$ = proton/antiproton, γ = photon).

Element of diagram	Numerical factor	Physical interpretation
outgoing line (thin) e^-	$\sqrt{\frac{m_e c^2}{(2\pi\hbar)^3 \omega_p}} \overline{u}_a(\mathbf{p},\sigma)$	e^- in state $\lvert\mathbf{p}\sigma\rangle$ at $t = +\infty$
incoming line (thin) e^-	$\sqrt{\frac{m_e c^2}{(2\pi\hbar)^3 \omega_p}} u_a(\mathbf{p},\sigma)$	e^- in state $\lvert\mathbf{p}\sigma\rangle$ at $t = -\infty$
outgoing line (thick) p^+	$\sqrt{\frac{m_p c^2}{(2\pi\hbar)^3 \Omega_p}} \overline{w}_a(\mathbf{p},\sigma)$	p^+ in state $\lvert\mathbf{p}\sigma\rangle$ at $t = +\infty$
incoming line (thick) p^+	$\sqrt{\frac{m_p c^2}{(2\pi\hbar)^3 \Omega_p}} w_a(\mathbf{p},\sigma)$	p^+ in state $\lvert\mathbf{p}\sigma\rangle$ at $t = -\infty$
outgoing line (wavy) γ	$\frac{\sqrt{c}}{\sqrt{(2\pi\hbar)^3 2p}} e^*_\mu(\mathbf{p},\tau)$	γ in state $\lvert\mathbf{p}\rangle^\tau$ at $t = +\infty$
incoming line (wavy) γ	$\frac{\sqrt{c}}{\sqrt{(2\pi\hbar)^3 2p}} e_\mu(\mathbf{p},\tau)$	γ in state $\lvert\mathbf{p}\rangle^\tau$ at $t = -\infty$
internal line (thin) $e^{\mp}$	$\frac{(\not{p}+m_e c^2)_{ab}}{(2\pi i)(2\pi\hbar)^3(\tilde{p}^2 - m_e^2 c^4)}$	no interpretation
internal line (wavy) γ	$\frac{\hbar c^2 \eta_{\mu\nu}}{(2\pi i)(2\pi\hbar)^3 \tilde{p}^2}$	no interpretation
interaction vertex $e^{\mp}\gamma$	$-i(2\pi\hbar)^4 e\gamma^\mu_{ab}$	no interpretation
interaction vertex $p^{\pm}\gamma$	$i(2\pi\hbar)^4 e\gamma^\mu_{ab}$	no interpretation

(1) Draw a connected[20] Feynman diagram with $\mathcal{V}$ vertices, $\mathcal{I}$ internal lines and $\mathcal{L} = \mathcal{I} - \mathcal{V} + 1$ independent loops. Each vertex should be connected to two fermion lines (electron or proton, external or internal) and one photon line (external or internal). External incoming (outgoing) lines correspond to the initial (final) configuration of the colliding particles. Momenta and spins of the particles in these asymptotic states are assumed to be given.

(2) Assign any 4-momentum labels to $\mathcal{L}$ internal lines in independent loops.

(3) Respecting the 4-momentum conservation rule at each vertex, assign 4-momentum labels to all remaining internal lines.

(4) The integrand is now formed from the factors corresponding to all lines and vertices in the diagram according to the rules from Table 3.1. Internal lines correspond to the photon (C.24) and electron (B.100) *propagators*. Each vertex factor γ^μ_{ab} has three summation indices corresponding to the three lines converging at the vertex. Two Dirac indices a and b are coupled with the indices of the fermion lines, and the 4-vector index μ is coupled with the photon line index.

(5) Integrate the resulting expression with respect to all loop 4-momenta.

---

20 As we explained in Subsection 2.4.2, we should ignore disconnected diagrams, because they correspond to spatially separated and independent collisions.

(6) Multiply by $(-1)^{\mathcal{F}}$, where $\mathcal{F}$ is the number of closed fermion loops.[21]
(7) Multiply this expression by the 4D delta function, which expresses the conservation of the total energy–momentum in the scattering process.
(8) Repeat steps (1)–(7) for all possible topologies of diagrams describing the process of interest in the given perturbation order $\mathcal{V}$. Add the resulting expressions.

Using these rules, it is not difficult to verify that equation (3.33) for the electron–proton S-matrix element is represented by the Feynman diagram in Figure 3.1.

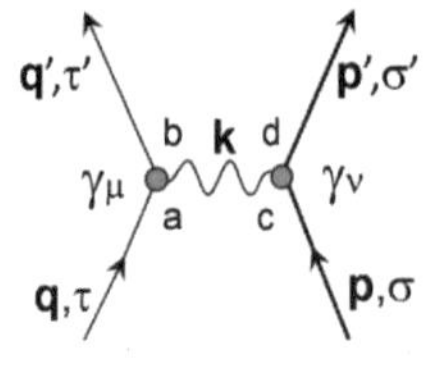

Figure 3.1: Feynman diagram for the electron–proton scattering in the second perturbation order (thin full line = electron; thick full line = proton).

3.2.5 Compton scattering

As another application of the Feynman rules, let us consider the *Compton scattering* electron + photon → electron + photon. Two diagrams describing this process in the second perturbation order are shown in Figures 3.2 (a) and (b). In accordance with the Feynman rules, the scattering amplitude is

$$
\begin{aligned}
&\langle \mathrm{vac}|a_{\boldsymbol{p}'\sigma'}c_{\boldsymbol{q}'\tau'}S_2[a^\dagger c^\dagger ac]a^\dagger_{\boldsymbol{p}\sigma}c^\dagger_{\boldsymbol{q}\tau}|\mathrm{vac}\rangle \\
&= \left[\sqrt{\frac{m_ec^2}{(2\pi\hbar)^3\omega_{\boldsymbol{p}'}}}\bar{u}_d(\boldsymbol{p}',\sigma')\right][-i(2\pi\hbar)^4 e\gamma^\mu_{dc}]\left[\frac{\sqrt{c}e^*_\mu(\boldsymbol{q}',\tau')}{\sqrt{(2\pi\hbar)^3 2q'}}\right] \\
&\quad\times\left[\frac{(\not{p}+\not{q}+m_ec^2)_{cb}}{(2\pi i)(2\pi\hbar)^3((\tilde{p}+\tilde{q})^2-m_e^2c^4)}\right][-i(2\pi\hbar)^4 e\gamma^\nu_{ba}] \\
&\quad\times\left[\frac{\sqrt{c}e_\nu(\boldsymbol{q},\tau)}{\sqrt{(2\pi\hbar)^3 2q}}\right]\left[\sqrt{\frac{m_ec^2}{(2\pi\hbar)^3\omega_{\boldsymbol{p}}}}u_a(\boldsymbol{p},\sigma)\right]\delta^4(\tilde{p}+\tilde{q}-\tilde{p}'-\tilde{q}') \\
&\quad+\left[\sqrt{\frac{m_ec^2}{(2\pi\hbar)^3\omega_{\boldsymbol{p}'}}}\bar{u}_d(\boldsymbol{p}',\sigma')\right][-i(2\pi\hbar)^4 e\gamma^\nu_{dc}]\left[\frac{\sqrt{c}e_\nu(\boldsymbol{q},\tau)}{\sqrt{(2\pi\hbar)^3 2q}}\right] \\
&\quad\times\left[\frac{(\not{p}-\not{q}'+m_ec^2)_{cb}}{(2\pi i)(2\pi\hbar)^3((\tilde{p}-\tilde{q}')^2-m_e^2c^4)}\right][-i(2\pi\hbar)^4 e\gamma^\mu_{ba}] \\
&\quad\times\left[\frac{\sqrt{c}e^*_\mu(\boldsymbol{q}',\tau')}{\sqrt{(2\pi\hbar)^3 2q'}}\right]\left[\sqrt{\frac{m_ec^2}{(2\pi\hbar)^3\omega_{\boldsymbol{p}}}}u_a(\boldsymbol{p},\sigma)\right]\delta^4(\tilde{p}+\tilde{q}-\tilde{p}'-\tilde{q}').
\end{aligned}
$$

21 See page 120 in [10].

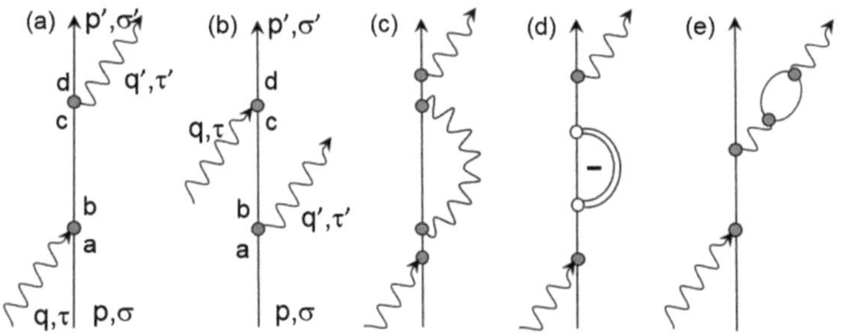

Figure 3.2: Feynman diagrams for the electron–photon (Compton) scattering. (a), (b) Second-order terms. (c), (d), (e) Selected fourth-order terms discussed in Section 4.2.

From this expression, it is not difficult to obtain the cross section for the elastic Compton scattering. We will not reproduce here this standard result,[22] but only note that in the limits of low photon and electron energies we will get exactly the *Thomson formula*, known from classical electrodynamics.[23] Since the low-energy Thomson formula was verified in detail in experiments, all higher-order corrections to this result should vanish. This observation will prove very useful in our discussion of charge renormalization in Chapter 4.

3.2.6 Virtual particles?

In this volume, we want to avoid speculations about the physical meaning of quantum fields. This issue will be discussed in Volume 3. However, we should warn the reader against a too literal interpretation of Feynman diagrams. One often hears the opinion that these diagrams represent real physical processes in which virtual photons and electrons are emitted and absorbed at interaction vertices and that these virtual particle exchanges are the true reasons for the attraction or repulsion of the observed charges.

Such interpretations are absolutely groundless and misleading. In fact, Feynman diagrams are not graphical representations of any real physical events. The lines and vertices in the diagrams have nothing to do with particle trajectories. They are simply convenient symbols for certain factors in perturbation theory integrals. There is no deeper meaning in these pictures.

As shown in Table 3.1, external lines represent physical particles in asymptotic states. In particular, 4-momenta of these lines are on the *mass shell*, which is another

22 See, for example, Section 5.5 in [10] and Section 8.7 in [21].
23 For example, formula (5.92) in [10].

way of saying that they satisfy standard energy–momentum relationships, i. e.,

$$p_0 - c^2p^2 = 0, \tag{3.35}$$

$$p_0 - c^2p^2 = m_e^2c^4, \tag{3.36}$$

$$p_0 - c^2p^2 = m_p^2c^4, \tag{3.37}$$

for photon, electron and proton external lines, respectively. Four-momenta of internal lines do not satisfy equations (3.35)–(3.37). They are out of the mass shell and lack any physical interpretation.

4 Renormalization

There is no great thing that would not be surmounted by a still greater thing. There is no thing so small that no smaller thing could fit into it.
Kozma Prutkov

In the preceding chapter, we calculated the second-order contribution to the S-operator. The obtained result (3.28) agrees rather well with experiments on electron–proton scattering. Similarly, in the second order, one can also obtain cross sections for other processes, such as Compton scattering or electron–positron annihilation, with good accuracy. Can we expect even better agreement with the experiment by including terms of higher orders in the expansion (3.20)? Unfortunately, the answer to this question is "no." As we shall see in this chapter, many high-order terms in the expansion (3.20) are not just inaccurate – they are divergent!

This is the same situation as that described in Section 2.2 for the toy model. We will apply the same renormalization methods for fixing the problem of divergences in QED.

4.1 Two renormalization conditions

In this section we will be interested in general physical principles underlying the renormalization and the removal of the so-called ultraviolet divergences. We express these principles in the form of two *renormalization conditions*, namely:

(1) the absence of self-scattering in the vacuum and one-particle states;
(2) the charge renormalization.

Note that cancellation of divergences is not required explicitly in our conditions. Nevertheless, in the renormalized theory, all scattering amplitudes turn out to be finite and perfectly agreeing with the experiment.

4.1.1 No self-scattering condition

It should be noted that divergence of loop integrals is not the most painful problem in the naïve QED. Even if all loop integrals converged, the S-operator would contain very unpleasant divergences. Let us look in more detail where these infinities come from and how to deal with them.

Recall that the QED interaction operator (3.30) has only *unphys* terms. The corresponding scattering phase Φ in (3.20) is obtained by calculating multiple commutators of V_1. Therefore, according to Table 1.2, operator Φ can contain terms of all three allowed types: *unphys*, *phys* and *renorm*. Then the most general expression for the

https://doi.org/10.1515/9783110493207-004

S-operator can be written as[1]

$$S = \exp(\underbrace{\Phi}) = \exp(\underbrace{\Phi^{\mathrm{unp}}} + \underbrace{\Phi^{\mathrm{phys}}} + \underbrace{\Phi^{\mathrm{ren}}}) = \exp(\underbrace{\Phi^{\mathrm{phys}}} + \underbrace{\Phi^{\mathrm{ren}}}). \tag{4.1}$$

Let us now apply (4.1) to the one-electron state $a^{\dagger}_{\boldsymbol{p},\sigma}|\mathrm{vac}\rangle$. From Lemma 1.4 it follows that *phys* operators yield zero when acting on the single particle. *Renorm* operators do not change the number of particles, so we can write[2]

$$\begin{aligned}
Sa^{\dagger}_{\boldsymbol{p}\sigma}|\mathrm{vac}\rangle &= \exp(\underbrace{\Phi^{\mathrm{phys}}} + \underbrace{\Phi^{\mathrm{ren}}})a^{\dagger}_{\boldsymbol{p}\sigma}|\mathrm{vac}\rangle \\
&= \left(1 + \underbrace{\Phi^{\mathrm{phys}}} + \underbrace{\Phi^{\mathrm{ren}}} + \frac{1}{2!}(\underbrace{\Phi^{\mathrm{phys}}} + \underbrace{\Phi^{\mathrm{ren}}})^2 + \cdots\right)a^{\dagger}_{\boldsymbol{p}\sigma}|\mathrm{vac}\rangle \\
&= \left(1 + \underbrace{\Phi^{\mathrm{phys}}} + \underbrace{\Phi^{\mathrm{ren}}} + \frac{1}{2!}(\underbrace{\Phi^{\mathrm{phys}}})^2 + \frac{1}{2!}\underbrace{\Phi^{\mathrm{phys}}}\,\underbrace{\Phi^{\mathrm{ren}}} + \frac{1}{2!}\underbrace{\Phi^{\mathrm{ren}}}\,\underbrace{\Phi^{\mathrm{phys}}}\right. \\
&\quad \left. + \frac{1}{2!}(\underbrace{\Phi^{\mathrm{ren}}})^2 + \cdots\right)a^{\dagger}_{\boldsymbol{p}\sigma}|\mathrm{vac}\rangle \\
&= \left(1 + \underbrace{\Phi^{\mathrm{ren}}} + \frac{1}{2!}(\underbrace{\Phi^{\mathrm{ren}}})^2 + \cdots\right)a^{\dagger}_{\boldsymbol{p}\sigma}|\mathrm{vac}\rangle \\
&= \exp(\underbrace{\Phi^{\mathrm{ren}}})a^{\dagger}_{\boldsymbol{p}\sigma}|\mathrm{vac}\rangle.
\end{aligned} \tag{4.2}$$

Similar calculations can be made for the single-photon state $c^{\dagger}_{\boldsymbol{p}\tau}|\mathrm{vac}\rangle$ and for the vacuum vector, i. e.,

$$Sc^{\dagger}_{\boldsymbol{p}\tau}|\mathrm{vac}\rangle = \exp(\underbrace{\Phi^{\mathrm{ren}}})c^{\dagger}_{\boldsymbol{p}\tau}|\mathrm{vac}\rangle, \tag{4.3}$$

$$S|\mathrm{vac}\rangle = \exp(\underbrace{\Phi^{\mathrm{ren}}})|\mathrm{vac}\rangle. \tag{4.4}$$

So, the "self-scattering" in these states depends only on the *renorm* part of Φ. We know from (1.75) that the t-integral $\underbrace{\Phi^{\mathrm{ren}}}$ is infinite even if the operator Φ^{ren} is finite. Therefore, if $\Phi^{\mathrm{ren}} \neq 0$, then the S-operator multiplies zero-particle and one-particle states by infinite phase factors, which are deemed unacceptable.

Intuitively, we expect that single-particle states and vacuum should evolve freely over the entire time interval from $t = -\infty$ to $t = +\infty$. This means that there cannot be any self-scattering in these states. This also means that the S-operator must be equivalent to the unit operator $S = 1$, when acting on such states. In other words, from physical considerations, we would like to require $\Phi^{\mathrm{ren}} = 0$. Obviously, this condition is not

1 Here we notice that the *unphys* terms in Φ do not contribute to the S-operator because of equation (1.76).

2 In fact, here we consider the result of a scattering experiment with one electron in the initial state (in the remote past). At time $t \approx 0$ this electron collides "with itself," and the products of this collision are registered in the distant future. Of course, we know in advance that the outcome of such a "scattering" should be trivial. One particle cannot collide with anything, and therefore it cannot experience scattering. The question is whether our scattering operator S can reproduce this simple result?

fulfilled in the theory presented here. Our particles are permanently self-interacting and scattering on themselves. In QFT, renormalization is introduced as a means of resolving this paradox.

The idea of renormalization is that our divergence problems come from the incorrect (naïve) choice of the interaction operator (3.30). This operator V_1 has to be modified or *renormalized* so that the self-scattering of the vacuum and single-particle states disappears. The modified interaction operator V^c will be obtained from V_1 by adding to the latter so-called renormalization counterterms whose presence will be indicated by the subscript "c." In particular, we will require that the phase operator calculated with the renormalized interaction V^c has a vanishing *renorm* part, i. e.,

$$(\Phi^c)^{\text{ren}} = 0. \tag{4.5}$$

If we achieve that, then the operator $\underbrace{\Phi^c}$ becomes purely *phys*, i. e.,

$$\underbrace{\Phi^c} = \underbrace{(\Phi^c)^{\text{phys}}},$$

and expressions (4.2)–(4.4) take physically acceptable forms

$$\begin{aligned} S^c a^\dagger_{\boldsymbol{p}\sigma}|\text{vac}\rangle &= a^\dagger_{\boldsymbol{p}\sigma}|\text{vac}\rangle, \\ S^c c^\dagger_{\boldsymbol{p}\tau}|\text{vac}\rangle &= c^\dagger_{\boldsymbol{p}\tau}|\text{vac}\rangle, \\ S^c|\text{vac}\rangle &= |\text{vac}\rangle, \end{aligned}$$

in which the self-scattering is absent. Taking into account the perturbation theory expansion $S^c = 1 + S^c_2 + S^c_3 + \cdots$, we will be able to write in each order $i = 2, 3, \ldots$

$$S^c_i a^\dagger_{\boldsymbol{p}\sigma}|\text{vac}\rangle = 0, \tag{4.6}$$

$$S^c_i c^\dagger_{\boldsymbol{p}\tau}|\text{vac}\rangle = 0, \tag{4.7}$$

$$S^c_i|\text{vac}\rangle = 0 \tag{4.8}$$

and for the S-matrix elements

$$\langle\text{vac}|a_{\boldsymbol{p}\sigma} S^c_i a^\dagger_{\boldsymbol{p}'\sigma'}|\text{vac}\rangle = 0, \tag{4.9}$$

$$\langle\text{vac}|c_{\boldsymbol{p}\tau} S^c_i c^\dagger_{\boldsymbol{p}'\tau'}|\text{vac}\rangle = 0, \tag{4.10}$$

$$\langle\text{vac}|S^c_i|\text{vac}\rangle = 0. \tag{4.11}$$

So, we summarize the above conditions as follows.

Statement 4.1 (absence of self-scattering). *The Hamiltonian of the theory must be chosen in such a way that there is no self-scattering in the vacuum and single-particle states. Nontrivial scattering is expected only in systems having at least two particles interacting with each other.*

We will deduce this statement from Postulate 2.1 in Volume 3. Later in this chapter we will see that our no-self-scattering condition is, in fact, equivalent to the traditional *mass renormalization* condition from QED textbooks.

4.1.2 Charge renormalization

It appears that the above no-self-scattering condition cannot guarantee the cancellation of all ultraviolet divergences. Additional *charge renormalization* efforts should be applied as well.

Recall that the second-order amplitude of the electron–proton scattering (3.34) has the singularity $\propto e^2/\tilde{k}^2$ at zero momentum transfer $\tilde{k} = \tilde{q}' - \tilde{q} = 0$. It is known[3] that in the position space such a singularity corresponds to the long-range Coulomb potential $-e^2/(4\pi r)$. From classical physics and experiments, we know that this potential provides a very accurate description for the interaction between charges at large distances and low energies. We may guess that to keep this agreement, all high-order corrections to the low-energy long-distance scattering should vanish. In other words, in the momentum space, components S_i $(i > 2)$ of the scattering operator should be nonsingular at $\tilde{k} = 0$.

Besides, in Subsection 3.2.5 we have seen that the second perturbation order is completely sufficient to describe the low-energy (Thomson) photon–electron scattering. Therefore, we also do not expect any corrections to this result in higher orders. Let us now raise these qualitative observations to the level of a fundamental physical principle.[4]

Postulate 4.2 (charge renormalization condition). Charge–charge and charge–photon elastic scattering cross sections at large distances and low energies are described *exactly* by the second-order term S_2 in the S-operator. All high-order contributions to these results should vanish.

4.1.3 Renormalization by counterterms

So, we have seen that the no-self-scattering and charge renormalization conditions are not fulfilled in QED with interaction (3.30). This problem must be solved by renormalization, which essentially means a transition from the naïve potential

$$V_1 = \int d\mathbf{x}[-e\overline{\psi}(\tilde{x})\gamma^\mu\psi(\tilde{x}) + e\overline{\Psi}(\tilde{x})\gamma^\mu\Psi(\tilde{x})]\mathcal{A}_\mu(\tilde{x}) \tag{4.12}$$

to the new interaction

$$V^c = V_1 + Q. \tag{4.13}$$

by adding counterterms [8], which we denote by Q. Operators in Q must be chosen in such a way as to satisfy our two renormalization conditions. In particular, the new

3 See, for example, Section 3-1.4.

4 Which is, actually, violated in the naïve QED.

scattering phase, calculated with the aid of V^c,

$$\Phi^c = V^c - \frac{1}{2}[\underline{V^c}, V^c] + \cdots,$$

should not contain *renorm* terms ($(\Phi^c)^{\text{ren}} = 0$). Moreover, higher-order contributions Φ_i^c ($i > 2$) should be nonsingular at $\tilde{k} = 0$.

Then the full renormalized S-operator will be expressed by the perturbation series (**1**-7.17):

$$S^c = 1 - \frac{i}{\hbar}\int_{-\infty}^{+\infty} dt_1 V^c(t_1) - \frac{1}{2!\hbar^2}\int_{-\infty}^{+\infty} dt_1 dt_2 T[V^c(t_1)V^c(t_2)]\cdots, \quad (4.14)$$

where each term turns out to be finite. In the rest of this chapter, we will see how the renormalization program solves the problem of ultraviolet divergences in one-loop diagrams. The problem of infrared divergences is no less difficult. We will discuss it briefly in Chapter 5 of the third volume.

4.1.4 Diagrams of electron–proton scattering

As before, we are mainly interested in the collisions of two charges – an electron and a proton. In Figure 4.1 we have collected all related Feynman diagrams of the second and fourth order. Generally speaking, these diagrams can be divided into three classes: *tree diagrams*, *loop diagrams* and *counterterm diagrams*. We have already discussed the single tree diagram 4.1 (a) in Chapter 3 and found it to be in good agreement with the experiment.

Serious problems are associated with fourth-order loop diagrams, such as in Figures 4.1 (b)–(g).[5] As we saw in Section 2.3, loops are associated with potentially divergent integrals. There are two types of divergences in QED loop diagrams. First, loop integrals in Figures 4.1 (e)–(g) diverge because the integrand is singular at small values of the argument (loop momentum). These are the so-called *infrared divergences* [21], which will be analyzed in Volume 3.

The second type is the divergence of loop integrals (in Figures 4.1 (b)–(e)) at large values of the loop momenta (see Subsection 2.3.3). This is the so-called problem of *ultraviolet divergences*. Historically, this problem was solved by *renormalization* theory developed by Tomonaga, Schwinger and Feynman in the late 1940s. As mentioned, the idea of this theory is that our "naïve" interaction operator V_1 in (3.30) is incom-

5 Here we did not show diagrams with loops formed by proton lines. Their calculation is no different from the electronic loops discussed in this chapter. However, their contribution to the scattering amplitudes is much smaller, owing to the inequality $m_e \ll m_p$.

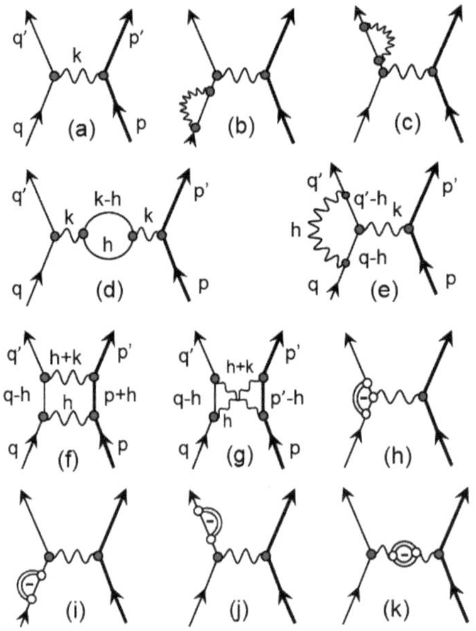

Figure 4.1: Feynman diagrams for electron–proton scattering up to the fourth perturbation order (thick solid lines = protons, thin solid lines = electrons, wavy lines = "virtual photons"). Double arcs in (h)–(k) denote counterterms, which will be discussed in Section 4.2.

plete. It has to be corrected by introducing additional interaction operators (*counterterms*) Q, which are shown in the diagrams in Figures 4.1 (h)–(k) as new interaction vertices. Formally speaking, counterterms are infinite operators. However, if they are carefully selected, then their contributions to the S-matrix cancel out the infinities from loop diagrams, so that only some finite differences remain in each perturbation order. These residual contributions are called *radiative corrections*. Taking them into account is necessary for reaching the remarkable agreement between QED and precision experimental data.

4.1.5 Regularization

As we noted above, loop integrals in QED tend to diverge. It is quite inconvenient to carry out calculations with infinite quantities, so to simplify the task it is customary to perform *regularization* of integrals. The idea is to change the theory "by hands," to force all loop integrals to converge. In this approach, all intermediate computations operate with finite quantities, and all steps are mathematically rigorous. The simplest approach to regularization, adopted in Appendix F, is to cut off the momentum integrals. Usually, such a modification depends on two parameters having the dimensions of mass: the ultraviolet cutoff Λ limits the integration volume at large loop momentum and the infrared cutoff λ controls integrals at small momenta. So, with finite Λ and λ all loop integrals are guaranteed to be finite.

Naturally, a theory with such truncated integrals cannot be exact. To obtain the final result, at the end of calculations the ultraviolet cutoff must be set to infinity $\Lambda \to \infty$.[6] If the counterterms are chosen correctly, then in this limit S-matrix elements tend to finite values agreeing with the experiment.

4.2 Counterterms

Now we proceed to the practical implementation of the renormalization program outlined above. In this section, we apply renormalization conditions 4.1–4.2 and derive explicit formulas for counterterms Q in the second and third orders of perturbation theory.

4.2.1 Electron's self-scattering

Let us see in more detail how condition (4.9) is violated in QED. There are only two connected diagrams that give second-order contributions to the electron self-scattering. They are shown in Figures 4.2 (a) and (b). Applying Feynman's rules to the diagram in Figure 4.2 (a), we get

$$\begin{aligned}
&\langle \mathrm{vac}|a_{\mathbf{p}'\sigma'} S_2^{(a)} a^{\dagger}_{\mathbf{p}\sigma}|\mathrm{vac}\rangle \\
&= -\frac{m_e e^2 c^4 \delta^4(\tilde{p}-\tilde{p}')}{(2\pi i)^2(2\pi\hbar)\sqrt{\omega_{\mathbf{p}}\omega_{\mathbf{p}'}}}\overline{u}_d(\mathbf{p}',\sigma')\left[\int d^4k \gamma_\nu^{dc}\frac{(\not{p}-\not{k}+m_e c^2)_{cb}}{(\tilde{p}-\tilde{k})^2-m_e^2c^4}\cdot\frac{\eta^{\nu\mu}}{\tilde{k}^2}\gamma_\mu^{ba}\right]u_a(\mathbf{p},\sigma) \qquad (4.15) \\
&= \frac{m_e e^2 c^4 \delta^4(\tilde{p}-\tilde{p}')}{(2\pi)^2(2\pi\hbar)\omega_{\mathbf{p}}}\overline{u}_d(\mathbf{p},\sigma)[C^{(0)}\delta_{da} + C^{(1)}(\not{p}-m_e c^2)_{da} + R_{da}(\not{p})]u_a(\mathbf{p},\sigma), \qquad (4.16)
\end{aligned}$$

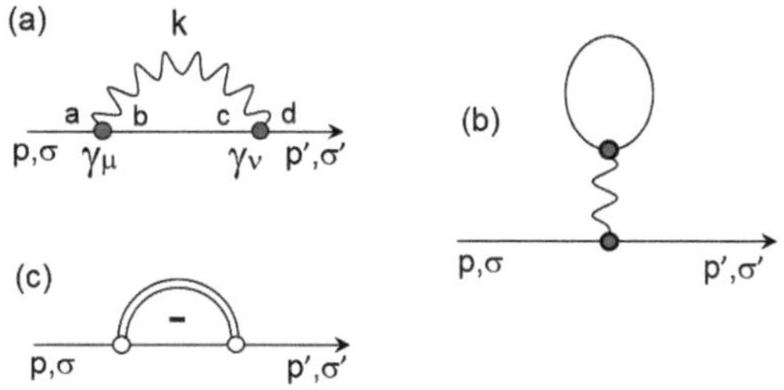

Figure 4.2: Feynman diagrams for the self-scattering electron → electron in the second perturbation order. (a) Electron–photon loop. (b) "Tadpole" loop that does not contribute to the amplitude. (c) Counterterm contribution (4.20).

6 In the complete theory, the infrared cutoff should approach zero, i. e., $\lambda \to 0$. However, in this chapter we will retain a nonzero value of λ. The limit $\lambda \to 0$ and the associated infrared divergences will be discussed in Volume 3.

where the (divergent) constants $C^{(0)}$ and $C^{(1)}$ are computed in (F.26) and (F.27), respectively:

$$C^{(0)} = -\frac{3\pi^2 m_e c^2}{2ic^3}\left(4\ln\frac{\Lambda}{m_e} + 1\right),$$
$$C^{(1)} = \frac{2\pi^2}{ic^3}\left(\ln\frac{\Lambda}{m_e} + 2\ln\frac{\lambda}{m_e} + \frac{9}{4}\right).$$

The finite[7] quantity $R(\not{p})$ includes terms that are quadratic, cubic and higher-order in $(\not{p} - m_e c^2)$; we have

$$R(\not{p}) = C^{(2)}(\not{p} - m_e c^2)^2 + C^{(3)}(\not{p} - m_e c^2)^3 + \cdots. \tag{4.17}$$

By dropping the multipliers corresponding to external electron lines and the delta function in (4.16), we get contributions from the loop itself and its two vertices:

$$\mathcal{G}^{(el)}_{loop}(\not{p}) = \hbar^2 e^2 c^2 (C^{(0)} + C^{(1)}(\not{p} - m_e c^2) + R(\not{p})). \tag{4.18}$$

If a loop like the one shown in Figure 4.2 (a) is inserted into an external electron line, then the 4-momentum $\tilde{p}$ is on the mass shell, and only the constant term survives in (4.18),[8] i. e.,

$$\mathcal{G}^{(el)}_{loop}(\not{p} = m_e c^2) = \hbar^2 e^2 c^2 C^{(0)} = 3\pi^2 i e^2 \hbar^2 m_e c\left(2\ln\frac{\Lambda}{m_e} + \frac{1}{2}\right). \tag{4.19}$$

This nonzero result contradicts the no-self-scattering condition (4.9). Moreover, this expression tends to infinity in the limit $\Lambda \to \infty$. So here we are dealing with an ultraviolet divergence, which must be suppressed by the introduction of counterterms.

For loops in internal electron lines, the 4-momentum $\tilde{p}$ is not necessarily on the mass shell, so the expression (4.18) must be taken into account in its entirety.

Now consider the second electron self-scattering diagram, in Figure 4.2 (b). We have

$$\begin{aligned}
&\langle \mathrm{vac}| a_{\mathbf{p}'\sigma'} S_2^{(b)} a^\dagger_{\mathbf{p}\sigma} |\mathrm{vac}\rangle \\
&= \frac{m_e c^2}{(2\pi\hbar)^3\sqrt{\omega_{\mathbf{p}}\omega_{\mathbf{p}'}}} \overline{u}_a(\mathbf{p}',\sigma')\gamma^\mu_{ab} u_b(\mathbf{p},\sigma)\left(-\frac{(2\pi\hbar)^8 e^2}{\hbar^2}\right)\frac{\hbar c^2 \eta_{\mu\nu}\delta^4(\tilde{p}-\tilde{p}')}{(2\pi i)^2(2\pi\hbar)^6(\tilde{p}-\tilde{p}')^2} \\
&\quad\times \int d^4k \frac{(\not{k} + m_e c^2)_{cd}\gamma^\nu_{dc}}{\tilde{k}^2 - m_e^2 c^4}.
\end{aligned}$$

7 That is, tending to a finite value in the limit $\Lambda \to \infty$.

8 This applies also to loops in the diagrams in Figures 4.1 (b) and (c). Here we can formally write the mass shell equation ($\tilde{p}^2 = m_e^2 c^4$) as $\not{p} = m_e c^2$ due to (B.19).

The integral with respect to $\tilde{k}$ vanishes due to (B.11) and (B.12), i. e.,

$$\int d^4k \frac{\mathrm{Tr}(\gamma^\nu \gamma^\rho k_\rho + \gamma^\nu m_e c^2)}{\tilde{k}^2 - m_e^2 c^4} = \int d^4k \frac{4k^\nu}{\tilde{k}^2 - m_e^2 c^4} = 0,$$

so diagrams like Figure 4.2 (b) can be ignored.

4.2.2 Electron self-scattering counterterm

In the preceding subsection we saw that the loop diagram in Figure 4.2 (a) makes a divergent contribution (4.18) to scattering amplitudes. We are going to compensate for this divergence by a second-order counterterm. This counterterm will become a part of the interaction operator, so it must satisfy all conditions formulated for such operators in Subsection 3.1.2, in particular, the relativistic invariance expressed by equation (3.8). Taking these considerations into account, we choose the following *electron self-scattering counterterm*:

$$Q_2^{(el)}(t) = (\delta m)_2 \int dx \overline{\psi}(\tilde{x})\psi(\tilde{x}) + (Z_2 - 1)_2 \int dx \overline{\psi}(\tilde{x})(-i\hbar c \gamma^\mu \partial_\mu + m_e c^2)\psi(\tilde{x}), \tag{4.20}$$

where the *4-gradient* ∂^μ is defined as

$$\partial_\mu \equiv \left(-\frac{1}{c}\frac{\partial}{\partial t}, \frac{\partial}{\partial x}, \frac{\partial}{\partial y}, \frac{\partial}{\partial z}\right) \tag{4.21}$$

and the parameters $(\delta m)_2$, $(Z_2 - 1)_2$ have to be adjusted so as to satisfy renormalization conditions.[9] The counterterm (4.20) gives the following second-order contribution to the electron → electron self-scattering amplitude[10]:

$$\begin{aligned} &\langle vac|a_{p'\sigma'} S_2^{count} a^\dagger_{p\sigma}|vac\rangle \\ &= -\frac{i(\delta m)_2}{\hbar}\langle vac|a_{p'\sigma'} \int d^4x \overline{\psi}(\tilde{x})\psi(\tilde{x}) a^\dagger_{p\sigma}|vac\rangle \end{aligned}$$

---

9 In this chapter, we follow the traditional notation and introduce renormalization constants δm, Z_1, Z_2 and Z_3. Unlike our usual convention, the subscripts of these constants are *not* related to the order of perturbation theory. Therefore, for example, we write $(Z_2 - 1)_2$ to denote the second-order contribution to the renormalization constant $Z_2 - 1$. The constant δm has the dimension of energy, and $(Z_2 - 1)$ is dimensionless. We shall see later that these quantities coincide with the traditional renormalization parameters – the mass shift and the electron wave function renormalization factor – in the conventional approaches. Note that in our interpretation we do not "shift" the electron mass and do not change the normalization of the electron–positron quantum field, as suggested in many textbooks. We simply add new terms (4.20) to the interaction Hamiltonian.

10 This formula is obtained by inserting $Q_2^{(el)}(t)$ instead of $V^c(t)$ in the second term on the right-hand side of (4.14).

$$
\begin{aligned}
&- \frac{i(Z_2-1)_2}{\hbar}\langle \mathrm{vac}|a_{\mathbf{p}'\sigma'} \int d^4x \overline{\psi}(\tilde{x})(-i\hbar c\gamma^\mu \partial_\mu + m_e c^2)\psi(\tilde{x}) a^\dagger_{\mathbf{p}\sigma}|\mathrm{vac}\rangle \\
&= -\frac{i(\delta m)_2}{\hbar} \int d^4x \frac{m_e c^2 \delta_{da}}{(2\pi\hbar)^3\sqrt{\omega_{\mathbf{p}}\omega_{\mathbf{p}'}}} e^{\frac{i}{\hbar}\tilde{p}\cdot\tilde{x}} e^{-\frac{i}{\hbar}\tilde{p}'\cdot\tilde{x}} \overline{u}_d(\mathbf{p}',\sigma') u_a(\mathbf{p},\sigma) \\
&\quad - \frac{i(Z_2-1)_2}{\hbar} \int d^4x \frac{m_e c^2}{(2\pi\hbar)^3\sqrt{\omega_{\mathbf{p}}\omega_{\mathbf{p}'}}} e^{\frac{i}{\hbar}\tilde{p}\cdot\tilde{x}} e^{-\frac{i}{\hbar}\tilde{p}'\cdot\tilde{x}} \\
&\quad \times \overline{u}_d(\mathbf{p}',\sigma')(-\not{p} + m_e c^2)_{da} u_a(\mathbf{p},\sigma) \\
&= -\frac{2\pi i(\delta m)_2 m_e c^2 \delta^4(\tilde{p}-\tilde{p}')}{\omega_{\mathbf{p}}} \overline{u}_a(\mathbf{p},\sigma) u_a(\mathbf{p},\sigma) \\
&\quad - \frac{2\pi i(Z_2-1)_2 m_e c^2 \delta^4(\tilde{p}-\tilde{p}')}{\omega_{\mathbf{p}}} \overline{u}_d(\mathbf{p},\sigma)(-\not{p} + m_e c^2)_{da} u_a(\mathbf{p},\sigma).
\end{aligned} \tag{4.22}
$$

Dropping factors corresponding to external electron lines and the momentum delta function, we obtain the pure counterterm contribution

$$
\mathcal{G}^{(\mathrm{el})}_{\mathrm{count}}(\not{p}) = -i(2\pi)^4\hbar^3 \delta m_2 + i(2\pi)^4\hbar^3 (Z_2-1)_2(\not{p} - m_e c^2). \tag{4.23}
$$

In the Feynman diagram technique, the counterterm (4.20) is described by a new interaction vertex, which is shown in Figure 4.2 (c)[11] by two hollow circles connected with a double arc and with a "–" sign in the center. This group of symbols should be understood as a single second-order vertex and cannot be divided into component parts.

4.2.3 Fitting coefficient $(\delta m)_2$

Now, consider the electron → electron self-scattering diagrams in Figures 4.2 (a) and (c). We already know that in this case the 4-momentum $\tilde{p}$ is on the mass shell ($\not{p} = m_e c^2$). Hence, the electron–photon loop in Figure 4.2 (a) gives a divergent contribution expressed by the $C^{(0)}$ term in (4.19),[12] and the counterterm's contribution is given by the first $\not{p}$-independent summand in (4.23),

$$
\mathcal{G}^{(\mathrm{el})}_{\mathrm{count}}(\not{p} = m_e c^2) = -i(2\pi)^4\hbar^3(\delta m)_2. \tag{4.24}
$$

Our task is to ensure complete cancellation of these two terms, i. e.,

$$
\mathcal{G}^{(\mathrm{el})}_{\mathrm{loop}}(\not{p} = m_e c^2) + \mathcal{G}^{(\mathrm{el})}_{\mathrm{count}}(\not{p} = m_e c^2) = 0. \tag{4.25}
$$

11 The symbol "–" inside the counterterm loop reminds about the cancellation of divergences in the sum (a) + (c); see also Figures 4.1 (i) and (j) and 3.2 (d).

12 All other terms there disappear for $\not{p} = m_e c^2$.

This can be achieved by choosing the following value of the renormalization constant:

$$(\delta m)_2 = -\frac{ie^2c^2C^{(0)}}{(2\pi)^4\hbar} = \frac{3m_ece^2}{16\pi^2\hbar}\left(\frac{1}{2} + 2\ln\frac{\Lambda}{m_e}\right). \tag{4.26}$$

This choice is sufficient to satisfy the electron's no-self-scattering condition in the second order.[13]

Now consider an arbitrary diagram with the electron–photon loop inserted in an *external* electron line (for example, Figure 4.1 (b)). This line is also on the mass shell. In a complete set of Feynman graphs, for each such loop there is a diagram (Figure 4.1 (i) in our example) in which this loop is replaced by a counterterm. According to (4.25), these two divergent diagrams cancel out *exactly*. Hence, in the renormalized QED, both types of diagrams (containing loops and counterterms in external electron/positron lines) can be simply ignored. Another example of such a cancellation is 4.1 (c) + 4.1 (j) = 0.

4.2.4 Fitting coefficient $(Z_2 - 1)_2$

Now let us consider the situation where the electron–photon loop is present in an *internal* electron line; see, for example, the diagram in Figure 3.2 (c). Again, there exists also a diagram (Figure 3.2 (d)) in which this loop is replaced by the counterterm vertex. As we have already established, the contribution $\propto C^{(0)}$ in Figure 3.2 (c) is canceled out exactly with the contribution $\propto (\delta m)_2$ in Figure 3.2 (d). However, the 4-momentum $\tilde{p}$ of the internal line is not necessarily on the mass shell, so the loop contribution $\propto C^{(1)}$ in (4.18) does not disappear; it even diverges. Obviously, to compensate for this divergence, it is necessary to select the renormalization factor $(Z_2 - 1)_2$ in (4.23) as[14]

$$(Z_2 - 1)_2 = \frac{ie^2c^2C^{(1)}}{(2\pi)^4\hbar} = \frac{e^2}{8\pi^2\hbar c}\left(\ln\frac{\Lambda}{m_e} + 2\ln\frac{\lambda}{m_e} + \frac{9}{4}\right). \tag{4.27}$$

Note that the finiteness requirement does not determine this factor uniquely. In principle, we could replace $(Z_2-1)_2$ with the value $(Z_2-1)_2+\delta$, where δ is any finite constant, and still have a finite result for the sum 3.2 (c) + 3.2 (d). The correct choice of $\delta = 0$ is explained by our desire to conform with the charge renormalization postulate, Postulate 4.2. If $\delta \neq 0$, then the sum 3.2 (c) + 3.2 (d) contains a fourth-order contribution that is singular at $\boldsymbol{k} = 0$ and, therefore, provides a correction to the classical Thomson formula for the photon–electron scattering. According to Postulate 4.2, such a correction

13 In the traditional mass renormalization method, the choice of $(\delta m)_2$ is justified by the requirement that the electron's renormalized propagator has a pole at $\not{p} = m_ec^2$, where m_e is the physical mass of the electron. Our expression (4.26) coincides with the traditional results; see, for example, equation (21) in [4], the expression immediately after (8.42) in [2] and the second equality on page 523 in [14].

14 Compare this result to (8.43) in [2] and to equation (94b) of Chapter 15 in [14].

is unacceptable. Therefore, we must set $\delta = 0$ and define the renormalization factor $(Z_2 - 1)_2$ exactly by equation (4.27).

Our choice of the renormalization constant (4.27) means that in all sums "loop + counterterm vertex" in internal electron lines (such as the sum 3.2 (c) + 3.2 (d)) only a finite and harmless R-correction is left, i. e.,

$$\mathscr{G}^{(el)}_{loop}(\not{p}) + \mathscr{G}^{(el)}_{count}(\not{p}) = \hbar^2 e^2 c^2 R(\not{p}).$$

This residual term is responsible for so-called *radiative corrections* associated with the electron's *self-energy*. These small corrections do not play any role in the physical phenomena that interest us in this book, so we will not discuss them any longer. Similarly, we omit the analysis of proton–photon loops and corresponding counterterms.

4.2.5 Photon's self-scattering

Cancellation of the photon's self-scattering in many respects repeats the steps done above for the electron. The following second-order self-scattering amplitude is obtained from the diagram in Figure 4.3 (a):

$$\begin{aligned}&\langle \mathrm{vac}| c_{\boldsymbol{p}'\tau'} S_2 c^{\dagger}_{\boldsymbol{p}\tau} |\mathrm{vac}\rangle \\ &\quad = \frac{ce^2\delta_{\tau\tau'}}{(2\pi\hbar)2p(2\pi)^2}\delta^4(\tilde{p}' - \tilde{p})e^{*}_{\mu}(\boldsymbol{p},\tau)(\tilde{p}^2\eta^{\mu\nu} - p^{\mu}p^{\nu})\Pi(\tilde{p}^2)e_{\nu}(\boldsymbol{p},\tau).\end{aligned} \tag{4.28}$$

The part of equation (4.28) associated only with the loop and its two vertices (e. g., omitting contributions from the external lines and the delta function) is given by the divergent integral[15]

$$\begin{aligned}\mathscr{G}^{(ph)}_{loop}(\tilde{p}) &= e^2\hbar^2 \int d^4k \frac{(\not{k} + m_e c^2)_{da}}{\tilde{k}^2 - m_e^2 c^4}\gamma^{\mu}_{ab}\frac{(\not{p} - \not{k} + m_e c^2)_{bc}}{(\tilde{p} - \tilde{k})^2 - m_e^2 c^4}\gamma^{\nu}_{cd} \\ &= e^2\hbar^2\Pi(\tilde{p}^2)(\tilde{p}^2\eta^{\mu\nu} - p^{\mu}p^{\nu}).\end{aligned} \tag{4.29}$$

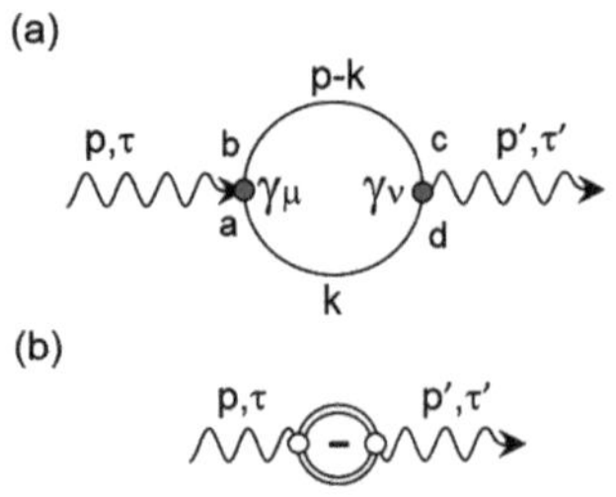

Figure 4.3: Feynman diagrams for the photon → photon self-scattering in the second perturbation order. (a) Divergent loop. (b) Counterterm contribution (4.32). The symbol "–" inside the loop reminds about the cancellation of divergences in the sum (a) + (b).

15 We omit the calculation of this integral, which can be found in many places, for example, in Section 11.2 of [21], in Section 7.5 of [10] and in Section 8.2 of [2].

It is convenient to write the divergent expression $\Pi(\tilde{p}^2)$ as a sum of its (infinite) value $\Pi(0)$ on the free photon's mass shell ($\tilde{p}^2 = 0$) and a finite remainder $\xi(\tilde{p}^2)$. Then we have

$$\Pi(\tilde{p}^2) = \Pi(0) + \xi(\tilde{p}^2).$$

By definition, the remainder vanishes on the photon's mass shell, so

$$\xi(0) = 0. \tag{4.30}$$

It is not important for us how the divergent quantity $\Pi(0)$ depends on the ultraviolet cutoff, because in Subsection 4.2.6 this contribution will be exactly compensated for by an appropriate counterterm. The function $\xi(\tilde{p}^2)$ can be represented as the integral (11.2.22) in [21], which takes the following form for small values of the 4-momentum $\tilde{p}$:

$$\xi(\tilde{p}^2) = -\frac{(2\pi)^4}{2\pi^2 i c^3} \int_0^1 x(1-x) \ln\left(1 + \frac{\tilde{p}^2 x(1-x)}{m_e^2 c^4}\right) dx \approx \frac{i(2\pi)^4 \tilde{p}^2}{60\pi^2 m_e^2 c^7}. \tag{4.31}$$

In equation (4.28), the 4-momentum $\tilde{p}$ is on the mass shell, so the loop contribution (4.29) vanishes,[16] despite the fact that the factor $\Pi(0)$ is infinite. Then the no-self-scattering condition 4.1 is fulfilled for photons without any additional efforts. The same can be said about loops in external photon lines. Diagrams with such loops can simply be ignored.[17] However, we cannot ignore loop contributions in internal photon lines (such as in Figure 4.1 (d)). In such cases, the 4-momentum $\tilde{p}$ is not necessarily on the mass shell, the expression $(\tilde{p}^2\eta^{\mu\nu} - p^\mu p^\nu)$ is nonzero and the divergent factor $\Pi(\tilde{p}^2)$ needs to be compensated for somehow.

4.2.6 Photon self-energy counterterm

Similarly to the electron's self-energy renormalization described in Subsection 4.2.2, we are going to cancel the divergence of $\Pi(\tilde{p}^2)$ by adding a new renormalization counterterm to the QED interaction operator. We define the photon self-energy counterterm as follows[18]:

$$Q_2^{(ph)}(t) = -\frac{(Z_3 - 1)_2}{4} \int d\boldsymbol{x} F^{\mu\nu}(\tilde{x}) F_{\mu\nu}(\tilde{x}), \tag{4.32}$$

16 This follows from the fact that for $\tilde{p}^2 = 0$ the second term in the parentheses $(\tilde{p}^2\eta^{\mu\nu} - p^\mu p^\nu)$ does not contribute to the full matrix element due to the property $\tilde{p} \cdot \tilde{e}(\boldsymbol{p}, \tau) = 0$, proved in (C.14).

17 See, for example, the diagram in Figure 3.2 (e), whose contribution is zero.

18 From the dimensionality (H.2) of the photon quantum field, it is easy to see that this counterterm will have the required dimensionality of energy if the factor $(Z_3 - 1)_2$ is chosen to be dimensionless. Moreover, the energy density in (4.32) explicitly satisfies the Poincaré invariance condition (3.8).

where we denoted

$$
\begin{aligned}
F_{\mu\nu} &\equiv \partial_\mu \mathcal{A}_\nu - \partial_\nu \mathcal{A}_\mu, \\
F^{\mu\nu}F_{\mu\nu} &= (\partial^\mu \mathcal{A}^\nu - \partial^\nu \mathcal{A}^\mu)(\partial_\mu \mathcal{A}_\nu - \partial_\nu \mathcal{A}_\mu) \\
&= \partial^\mu \mathcal{A}^\nu \partial_\mu \mathcal{A}_\nu - \partial^\mu \mathcal{A}^\nu \partial_\nu \mathcal{A}_\mu - \partial^\nu \mathcal{A}^\mu \partial_\mu \mathcal{A}_\nu + \partial^\nu \mathcal{A}^\mu \partial_\nu \mathcal{A}_\mu \\
&= 2\partial^\mu \mathcal{A}^\nu \partial_\mu \mathcal{A}_\nu - 2\partial^\mu \mathcal{A}^\nu \partial_\nu \mathcal{A}_\mu
\end{aligned}
$$

and $(Z_3 - 1)_2$ is a yet undefined second-order renormalization factor. Let us now calculate the effect of this counterterm on the self-scattering amplitude photon → photon. From definitions (C.2) and (4.21), we have[19]

$$
\begin{aligned}
\partial_\nu \mathcal{A}_\mu(t, \boldsymbol{x}) &= \frac{i\sqrt{c}}{(2\pi\hbar)^{3/2}} \int \frac{d\boldsymbol{q}}{\sqrt{2q}} \frac{q_\nu}{c} \sum_\tau [e^{-\frac{i}{\hbar}\tilde{q}\cdot\tilde{x}} e_\mu(\boldsymbol{q}, \tau) c_{\boldsymbol{q}\tau} - e^{\frac{i}{\hbar}\tilde{q}\cdot\tilde{x}} e^*_\mu(\boldsymbol{q}, \tau) c^\dagger_{\boldsymbol{q}\tau}], \\
\langle \text{vac}| c_{\boldsymbol{p}'\tau'} \partial_\nu \mathcal{A}_\mu(t, \boldsymbol{x}) &\to -\langle \text{vac}| \frac{i}{(2\pi\hbar)^{3/2}\sqrt{2p'c}} e^{\frac{i}{\hbar}\tilde{p}'\cdot\tilde{x}} p'_\nu e^*_\mu(\boldsymbol{p}', \tau'), \\
\partial_\nu \mathcal{A}_\mu(t, \boldsymbol{x}) c^\dagger_{\boldsymbol{p}\tau} |\text{vac}\rangle &\to \frac{i}{(2\pi\hbar)^{3/2}\sqrt{2pc}} e^{-\frac{i}{\hbar}\tilde{p}\cdot\tilde{x}} p_\nu e_\mu(\boldsymbol{p}, \tau)|\text{vac}\rangle.
\end{aligned}
$$

Then the S-matrix contribution from (4.32) is similar to (4.28) and we have

$$
\begin{aligned}
&\langle \text{vac}| c_{\boldsymbol{p}'\tau'} S_2^{\text{count}} c^\dagger_{\boldsymbol{p}\tau} |\text{vac}\rangle \\
&= \frac{i(Z_3 - 1)_2}{2\hbar} \langle \text{vac}| c_{\boldsymbol{p}'\tau'} \int d^4x (\partial^\lambda \mathcal{A}^\kappa(\tilde{x}) \partial_\lambda \mathcal{A}_\kappa(\tilde{x}) - \partial^\lambda \mathcal{A}^\kappa(\tilde{x}) \partial_\kappa \mathcal{A}_\lambda(\tilde{x})) c^\dagger_{\boldsymbol{p}\tau} |\text{vac}\rangle \\
&= \frac{i(Z_3 - 1)_2}{2\hbar c} \int d^4x \frac{e^{\frac{i}{\hbar}(\tilde{p}' - \tilde{p})\cdot\tilde{x}}}{(2\pi\hbar)^3\sqrt{4pp'}} (p')^\lambda p_\lambda e^{*\kappa}(\boldsymbol{p}', \tau') e_\kappa(\boldsymbol{p}, \tau) \\
&\quad - \frac{i(Z_3 - 1)_2}{2\hbar c} \int d^4x \frac{e^{\frac{i}{\hbar}(\tilde{p}' - \tilde{p})\cdot x}}{(2\pi\hbar)^3\sqrt{4pp'}} (p')^\lambda p_\kappa e^{*\kappa}(\boldsymbol{p}', \tau') e_\lambda(\boldsymbol{p}, \tau) \\
&= \frac{i(Z_3 - 1)_2 \pi\delta^4(\tilde{p} - \tilde{p}')}{2pc} \\
&\quad \times [p^\lambda p_\lambda e^{*\kappa}(\boldsymbol{p}, \tau) e_\kappa(\boldsymbol{p}, \tau) - p^\lambda p_\kappa e^{*\kappa}(\boldsymbol{p}, \tau) e_\lambda(\boldsymbol{p}, \tau)] \\
&= \frac{i(Z_3 - 1)_2 \pi\delta^4(\tilde{p} - \tilde{p}')}{2pc} e^*_\mu(\boldsymbol{p}, \tau)[\tilde{p}^2 \eta^{\mu\nu} - p^\mu p^\nu] e_\nu(\boldsymbol{p}, \tau).
\end{aligned}
\tag{4.33}
$$

This means that the new (counterterm) interaction vertex[20] corresponds to the multiplier[21]

---

19 Here the symbol → means only the part of the expression that is needed to compute the matrix element (4.33).

20 Such vertex is represented in diagrams as two hollow circles joined by double arches as in Figures 4.3 (b) and 4.4 (k).

21 This multiplier is obtained from (4.33) by dropping the 4-momentum delta function and factors associated with external photon lines.

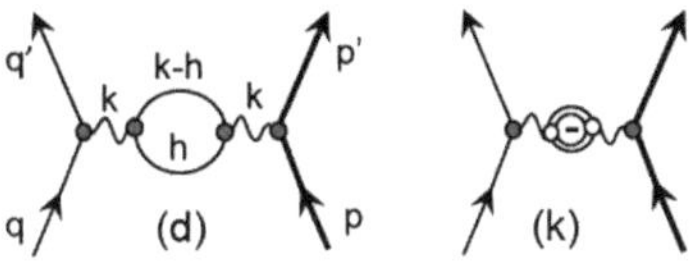

Figure 4.4: "Vacuum polarization" diagrams in electron–proton scattering. The same as in Figures 4.1 (d) and (k).

$$\mathscr{G}^{(ph)}_{count}(\tilde{p}) = \frac{8i(Z_3 - 1)_2 \pi^4 \hbar^3}{c^2} (\tilde{p}^2 \eta^{\mu\nu} - p^\mu p^\nu). \tag{4.34}$$

As before, we expect that for each diagram with an electron–positron loop, there is a similar diagram where the loop is replaced by the counterterm symbol. An example of such a pair is given by Figures 4.4 (d) and (k). We can achieve the divergence cancellation in the sum "loop + counterterm," if we choose the renormalization constant $(Z_3 - 1)_2$ such that (4.34) cancels the loop factor (4.29) near the mass shell $\tilde{p}^2 = 0$, i. e.,

$$(Z_3 - 1)_2 = \frac{ie^2 c^2 \Pi(0)}{8\pi^4 \hbar}. \tag{4.35}$$

We have already explained that, when inserted into an external photon line, the loop and the counterterm make no contributions, because the factor $(\tilde{p}^2 \eta^{\mu\nu} - p^\mu p^\nu)$ is effectively zero. However, for internal photon lines $\tilde{p}^2 \neq 0$, so the sum "loop + counterterm" is not zero, but finite, i. e.,

$$\begin{aligned} \mathscr{G}^{(ph)}_{loop}(\tilde{p}) + \mathscr{G}^{(ph)}_{count}(\tilde{p}) &= \left[e^2 \hbar^2 (\Pi(0) + \xi(\tilde{p}^2)) + \frac{8i\pi^4 \hbar^3}{c^2} \left(\frac{ie^2 c^2}{8\pi^4 \hbar} \right) \Pi(0) \right] (\tilde{p}^2 \eta^{\mu\nu} - p^\mu p^\nu) \\ &= e^2 \hbar^2 \xi(\tilde{p}^2)(\tilde{p}^2 \eta^{\mu\nu} - p^\mu p^\nu). \end{aligned} \tag{4.36}$$

This means that by adding pairs of diagrams with electron–positron loops and counterterms on photon lines, we remove ultraviolet divergences and obtain finite contributions to scattering amplitudes. These contributions are called *vacuum polarization* radiative corrections.[22]

4.2.7 Applying charge renormalization condition

If our only goal were to ensure the finiteness of the perturbation theory series for the scattering operators, then the choice of the renormalization constant (4.35) would not be unique. Indeed, we could add an arbitrary finite number δ to $\Pi(0)$ so that

$$(Z_3 - 1)_2 = -\frac{ie^2 c^2 (\Pi(0) + \delta)}{8\pi^4 \hbar}$$

22 In our understanding this traditional terminology does not make sense, because vacuum is an empty state without particles, where no physical processes, such as polarization, are possible.

and the radiative correction (4.36) would remain finite, i. e.,

$$\mathscr{G}^{(ph)}_{loop}(\tilde{p}) + \mathscr{G}^{(ph)}_{count}(\tilde{p}) = e^2\hbar^2(\xi(\tilde{p}^2) - \delta)(\tilde{p}^2\eta^{\mu\nu} - p^\mu p^\nu). \tag{4.37}$$

Why do we not do this? The answer is that such an addition would be incompatible with the charge renormalization condition in Postulate 4.2.

To see this, let us calculate the contribution to the electron–proton scattering from the diagrams in Figures 4.4 (d) and (k). Using Feynman rules and (4.37), we get[23]

$$\begin{aligned}
&\langle vac|a_{\boldsymbol{q}'\sigma'}d_{\boldsymbol{p}'\tau'}S_4^{(d)+(k)}d^\dagger_{\boldsymbol{p}\tau}a^\dagger_{\boldsymbol{q}\sigma}|vac\rangle \\
&= -\frac{e^2 m_p m_e \hbar^2 c^8}{(2\pi i)^2(2\pi\hbar)^4\sqrt{\omega_{\boldsymbol{q}}\omega_{\boldsymbol{q}'}\Omega_{\boldsymbol{p}}\Omega_{\boldsymbol{p}'}}}\delta^4(\tilde{q} - \tilde{q}' - \tilde{p}' + \tilde{p}) \\
&\quad \times e^2(\xi(\tilde{k}^2) - \delta)\mathcal{U}^\mu(\boldsymbol{q}'\sigma', \boldsymbol{q}\sigma)\frac{\eta_{\mu\nu}}{\tilde{k}^2}(\tilde{k}^2\eta^{\nu\lambda} - k^\nu k^\lambda)\frac{\eta_{\lambda\kappa}}{\tilde{k}^2}\mathcal{W}^\kappa(\boldsymbol{p}'\tau', \boldsymbol{p}\tau) \\
&= \frac{e^4 m_p m_e c^8}{\hbar^2(2\pi)^6\sqrt{\omega_{\boldsymbol{q}}\omega_{\boldsymbol{q}'}\Omega_{\boldsymbol{p}}\Omega_{\boldsymbol{p}'}}}\delta^4(\tilde{q} - \tilde{q}' - \tilde{p}' + \tilde{p})\frac{\xi(\tilde{k}^2) - \delta}{\tilde{k}^2}\mathcal{U}_\mu(\boldsymbol{q}'\sigma', \boldsymbol{q}\sigma)\mathcal{W}^\mu(\boldsymbol{p}'\tau', \boldsymbol{p}\tau) \\
&\approx \frac{e^4 c^4}{(2\pi)^6\hbar^2}\delta^4(\tilde{q} - \tilde{q}' - \tilde{p}' + \tilde{p})\frac{\xi(\tilde{k}^2) - \delta}{\tilde{k}^2}\delta_{\sigma\sigma'}\delta_{\tau\tau'}.
\end{aligned} \tag{4.38}$$

Taking into account equation (4.30), we conclude that with nonzero δ this matrix element would have a singularity $\propto -\delta/\tilde{k}^2$ for small values of $\tilde{k}$. This singularity would contribute a fourth-order correction to the long-range scattering of charged particles, and therefore violate the charge renormalization condition 4.2. The only way to avoid such a violation is to set $\delta = 0$.

4.2.8 Vertex renormalization

Let us calculate the diagram in Figure 4.5 (e). By Feynman rules we have

$$\begin{aligned}
&\langle vac|a_{\boldsymbol{q}'\sigma'}d_{\boldsymbol{p}'\tau'}S_4^{(e)}d^\dagger_{\boldsymbol{p}\tau}a^\dagger_{\boldsymbol{q}\sigma}|vac\rangle \\
&\approx -\frac{e^4 c^4 m_p m_e c^4}{(2\pi i)^4(2\pi\hbar)^2\sqrt{\omega_{\boldsymbol{q}}\omega_{\boldsymbol{q}'}\Omega_{\boldsymbol{p}}\Omega_{\boldsymbol{p}'}}}\delta^4(\tilde{q} - \tilde{q}' - \tilde{p}' + \tilde{p}) \\
&\quad \times \overline{u}(\boldsymbol{q}', \sigma')\left[\int d^4 h \gamma_\mu \frac{-\not{h} + \not{q}' + m_e c^2}{(\tilde{h} - \tilde{q}')^2 - m_e^2 c^4}\gamma_\kappa \frac{-\not{h} + \not{q} + m_e c^2}{(\tilde{h} - \tilde{q})^2 - m_e^2 c^4}\gamma^\mu \frac{1}{\tilde{h}^2}\right]u(\boldsymbol{q}, \sigma) \\
&\quad \times \frac{1}{(\tilde{q}' - \tilde{q})^2}\mathcal{W}^\kappa(\boldsymbol{p}'\tau', \boldsymbol{p}\tau).
\end{aligned} \tag{4.39}$$

23 Here we denoted $\tilde{k} \equiv \tilde{q}' - \tilde{q} = \tilde{p} - \tilde{p}'$ and used nonrelativistic approximations from Appendix B.9, formulas (B.95)–(B.96) and $\mathcal{U}^\mu\eta_{\mu\nu}k^\nu k^\lambda\eta_{\lambda\kappa}\mathcal{W}^\kappa = \mathcal{U}^\mu k_\mu k_\kappa \mathcal{W}^\kappa = 0$.

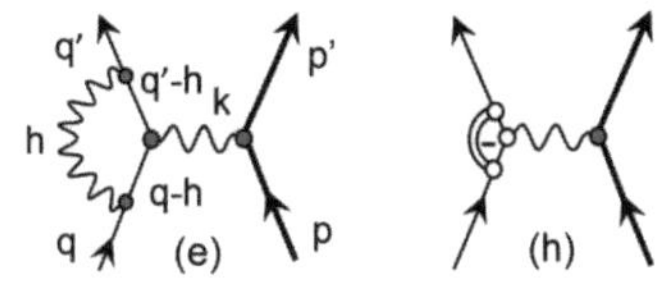

Figure 4.5: Vertex diagrams in electron–proton scattering. The same as in Figures 4.1 (e) and (h). The "–" sign inside the pseudo-loop in (h) reminds of the cancellation of divergences in the sum (e) + (h).

The integral in square brackets is calculated in (F.44), i. e.,

$$N_\kappa(\tilde{q},\tilde{q}') = -\frac{\pi^2\gamma_\kappa}{ic^3}\left(\frac{8\theta}{\tan(2\theta)}\ln\frac{\lambda}{m_e} + \frac{8}{\tan(2\theta)}\int_0^\theta x\tan x dx + \frac{1}{2} + 6\theta\cot\theta + 2\ln\frac{\Lambda}{m_e}\right)$$
$$+\frac{2\pi^2\theta(q+q')_\kappa}{im_ec^5\sin(2\theta)}. \tag{4.40}$$

Let us see if the amplitude (4.39) complies with the charge renormalization postulate, Postulate 4.2. For that we need to understand how this expression behaves at small values of the transferred momentum.[24] We use Dirac equations (B.93), (B.94) and equation (B.8) to obtain

$$\begin{aligned}
0 &= \overline{u}(\boldsymbol{q},\sigma')(\gamma_\mu(\not{q} - m_ec^2) + (\not{q} - m_ec^2)\gamma_\mu)u(\boldsymbol{q},\sigma)\\
&= \overline{u}(\boldsymbol{q},\sigma')(\gamma_\mu\gamma_\nu q^\nu + \gamma_\nu q^\nu\gamma_\mu - 2\gamma_\mu m_ec^2)u(\boldsymbol{q},\sigma)\\
&= \overline{u}(\boldsymbol{q},\sigma')(2\eta_{\mu\nu}q^\nu - 2\gamma_\mu m_ec^2)u(\boldsymbol{q},\sigma)\\
&= 2\overline{u}(\boldsymbol{q},\sigma')(q_\mu - \gamma_\mu m_ec^2)u(\boldsymbol{q},\sigma).
\end{aligned}$$

This means that the 4-vector q_μ sandwiched between $\overline{u}$ and u can be replaced by $\gamma_\mu m_ec^2$. Making the same substitution in the last term in (4.40), we set $q_\kappa \approx q'_\kappa \approx \gamma_\kappa m_ec^2$ and obtain

$$\lim_{\tilde{k}\to 0,\tilde{q}\to 0} N_\kappa(\tilde{q},\tilde{q}') = -\frac{\pi^2\gamma_\kappa}{ic^3}\left(4\ln\frac{\lambda}{m_e} + \frac{1}{2} + 6 + 2\ln\frac{\Lambda}{m_e} - 2\right) \equiv -\frac{\Gamma\pi^2\gamma_\kappa}{ic^3},$$

where we introduced an ultraviolet- and infrared-divergent constant,[25]

$$\Gamma \equiv 4\ln\frac{\lambda}{m_e} + 2\ln\frac{\Lambda}{m_e} + \frac{9}{2}.$$

Then, at low values of $\boldsymbol{q}$ and $\tilde{k} \equiv \tilde{q}' - \tilde{q}$, the scattering amplitude (4.39),

$$\langle\text{vac}|a_{\boldsymbol{q}'\sigma'}d_{\boldsymbol{p}'\tau'}S_4^{(e)}d^\dagger_{\boldsymbol{p}\tau}a^\dagger_{\boldsymbol{q}\sigma}|\text{vac}\rangle$$
$$= -\frac{ie^4c\Gamma\pi^2m_pm_ec^4\delta^4(\tilde{q}-\tilde{q}'-\tilde{p}'+\tilde{p})}{(2\pi i)^4(2\pi\hbar)^2\sqrt{\omega_{\boldsymbol{q}}\omega_{\boldsymbol{q}'}\Omega_{\boldsymbol{p}}\Omega_{\boldsymbol{p}'}}(\tilde{q}'-\tilde{q})^2}\mathcal{U}_\kappa(\boldsymbol{q}'\sigma',\boldsymbol{q}\sigma)\mathcal{W}^\kappa(\boldsymbol{p}'\tau',\boldsymbol{p}\tau), \tag{4.41}$$

24 That is, when $\tilde{q} \approx \tilde{q}'$ and, according to (F.39), $\theta \equiv \arcsin(\frac{\sqrt{\tilde{k}^2}}{2m_ec^2}) \approx 0$.

25 Compare with equation (23) in [4].

has a singularity $\propto \Gamma/\tilde{k}^2$. This means that in contradiction to Postulate 4.2, the fourth perturbation order gives a nonvanishing contribution to the electron–proton scattering at low energies (large distances). As an additional "inconvenience," this contribution is infinite in the limit $\Lambda \to \infty$.

This unacceptable situation can be corrected by adding one more (*vertex*) renormalization counterterm

$$Q_3(t) = -e(Z_1 - 1)_2 \int d\boldsymbol{x}\overline{\psi}(\tilde{x})\gamma^\mu \psi(\tilde{x})\mathcal{A}_\mu(\tilde{x}). \tag{4.42}$$

to the QED interaction. In Feynman diagrams, we denote the corresponding three-legged vertex by three hollow circles connected by lines, as shown in Figure 4.5 (h). The renormalization constant $(Z_1 - 1)_2$ has the second perturbation order, so the order of the counterterm (4.42) is three. It has the same form as the basic interaction operator (4.12) in QED, so its contribution to the diagram in Figure 4.1 (h) is easy to calculate (compare with equation (3.34)), as we have

$$\begin{aligned}&\langle \mathrm{vac}|a_{\boldsymbol{q}'\sigma'} d_{\boldsymbol{p}'\tau'} S_4^{(h)} d^\dagger_{\boldsymbol{p}\tau} a^\dagger_{\boldsymbol{q}\sigma}|\mathrm{vac}\rangle \\ &\quad = \frac{ie^2c^2(Z_1-1)_2 m_p m_e c^4 \delta^4(\tilde{q}+\tilde{p}-\tilde{q}'-\tilde{p}')}{4\pi^2\hbar\sqrt{\omega_{\boldsymbol{q}}\omega_{\boldsymbol{q}'}\Omega_{\boldsymbol{p}}\Omega_{\boldsymbol{p}'}}(\tilde{q}'-\tilde{q})^2}\mathcal{U}_\kappa(\boldsymbol{q}'\sigma',\boldsymbol{q}\sigma)\mathcal{W}^\kappa(\boldsymbol{p}'\tau',\boldsymbol{p}\tau).\end{aligned}$$

We demand that this counterterm contribution cancels the infinite and singular contribution (4.41) from the vertex loop. Therefore, our choice of the renormalization constant is[26]

$$(Z_1 - 1)_2 = \frac{e^2\Gamma}{16\pi^2 c\hbar} = \frac{e^2}{8\pi^2 c\hbar}\left(2\ln\frac{\lambda}{m_e} + \ln\frac{\Lambda}{m_e} + \frac{9}{4}\right). \tag{4.43}$$

Putting together all three counterterms (4.20), (4.32) and (4.42), the QED interaction operator, renormalized up to the third perturbation order, takes the final form[27]

$$\begin{aligned}V^c(t) &= V_1(t) + Q_2^{(\mathrm{el})}(t) + Q_2^{(\mathrm{ph})}(t) + Q_3(t) + \cdots \\ &= -e\int d\boldsymbol{x}\overline{\psi}(\tilde{x})\gamma^\mu\psi(\tilde{x})\mathcal{A}_\mu(\tilde{x}) + e\int d\boldsymbol{x}\overline{\Psi}(\tilde{x})\gamma^\mu\Psi(\tilde{x})\mathcal{A}_\mu(\tilde{x}) \\ &\quad + (\delta m)_2\int d\boldsymbol{x}\overline{\psi}(\tilde{x})\psi(\tilde{x}) + (Z_2-1)_2\int d\boldsymbol{x}\overline{\psi}(\tilde{x})(-i\hbar c\gamma^\mu\partial_\mu + m_e c^2)\psi(\tilde{x}) \\ &\quad - \frac{(Z_3-1)_2}{4}\int d\boldsymbol{x}F^{\mu\nu}(\tilde{x})F_{\mu\nu}(\tilde{x}) - e(Z_1-1)_2\int d\boldsymbol{x}\overline{\psi}(\tilde{x})\gamma^\mu\psi(\tilde{x})\mathcal{A}_\mu(\tilde{x}) + \cdots.\end{aligned} \tag{4.44}$$

---

26 Note that $(Z_1 - 1)_2$ coincides with another renormalization constant $(Z_2 - 1)_2$ in (4.27). This equality is not accidental. It is explained, for example, in Section 8.6 of [2].

27 Strictly speaking, in the full relativistic theory it is also necessary to add counterterms to the potential boost operator $\boldsymbol{Z}^n$ (3.17). However, the author is not aware of any attempts to do that.

4.3 Renormalized *S*-matrix

In this section, we are going to confirm that substituting the renormalized interaction (4.44) in the usual formula (4.14) for the S-operator, we do get ultraviolet-finite scattering amplitudes. In other words, we are going to calculate all fourth-order diagrams in Figure 4.1. As we already know, the diagrams in Figures 4.1 (b), (c), (i) and (j) cancel out exactly. Hence, we have to calculate the six remaining diagrams, which we combined into four coefficient functions s_4. We have

$$\begin{aligned}&\langle \text{vac}|a_{\mathbf{q}'\sigma'}d_{\mathbf{p}'\tau'}S_4^c d^\dagger_{\mathbf{p}\tau}a^\dagger_{\mathbf{q}\sigma}|\text{vac}\rangle \\ &\quad = (s_4^{(d)+(k)} + s_4^{(e)+(h)} + s_4^{(f)} + s_4^{(g)})\delta^4(\tilde{q} + \tilde{p} - \tilde{q}' - \tilde{p}'). \end{aligned} \tag{4.45}$$

For our purposes it will be sufficient to work in the $(v/c)^2$ approximation (see Appendix B.9), which is characteristic for low-energy collisions. In particular, the transferred momentum $\tilde{k}$ will be considered small, and the proton mass infinite ($m_p \to \infty$).

4.3.1 "Vacuum polarization" diagrams

Inserting (4.31) into (4.38) and setting $\delta = 0$, we find that in our approximations the sum of S-matrix elements in Figures 4.4 (d) and (k) does not depend on momenta and spins of the particles, i. e.,

$$s_4^{(d)+(k)} \approx \frac{ie^4}{\hbar^2(2\pi)^2 60\pi^2 m_e^2 c^3}\delta_{\sigma\sigma'}\delta_{\tau\tau'} = \frac{i\alpha^2}{15\pi^2 m_e^2 c}\delta_{\sigma\sigma'}\delta_{\tau\tau'}. \tag{4.46}$$

4.3.2 Vertex diagram

The total contribution of the electron vertex diagrams in Figures 4.5 (e) and (h) is given by equation (4.39), where the square bracket should be replaced by the ultraviolet-finite expression

$$\begin{aligned}&N_\kappa(\tilde{q},\tilde{q}') + \frac{\Gamma\pi^2}{ic^3}\gamma_\kappa \\ &\quad = -\frac{\pi^2\gamma_\kappa}{ic^3}\left(\frac{8\theta}{\tan(2\theta)}\ln\frac{\lambda}{m_e} + \frac{8}{\tan(2\theta)}\int\limits_0^\theta x\tan x dx + \frac{1}{2} + 6\theta\cot\theta + 2\ln\frac{\Lambda}{m_e}\right) \\ &\qquad + \frac{2\pi^2\theta(\tilde{q}+\tilde{q}')_\kappa}{im_e c^5\sin(2\theta)} + \frac{\pi^2\gamma_\kappa}{ic^3}\left(4\ln\frac{\lambda}{m_e} + 2\ln\frac{\Lambda}{m_e} + \frac{9}{2}\right)\end{aligned}$$

$$= -\frac{\pi^2\gamma_\kappa}{ic^3}\left(\left(\frac{8\theta}{\tan(2\theta)} - 4\right)\ln\frac{\lambda}{m_e} + \frac{8}{\tan(2\theta)}\int_0^\theta x\tan x dx - 4 + 6\theta\cot\theta\right)$$
$$+ \frac{2\pi^2\theta(\tilde{q}+\tilde{q}')_\kappa}{im_e c^5\sin(2\theta)}.$$

In the limit of low transferred momentum ($\tilde{k} \approx 0$) all components in this formula simplify, so we have[28]

$$\theta \equiv \sin^{-1}\frac{|\tilde{k}|}{2m_e c^2} \approx \frac{|\tilde{k}|}{2m_e c^2} + \frac{|\tilde{k}|^3}{48m_e^3c^6},$$
$$\tan(2\theta) \approx \frac{|\tilde{k}|}{m_e c^2} + \frac{3|\tilde{k}|^3}{8m_e^3c^6},$$
$$\frac{8\theta}{\tan(2\theta)} \approx 4\frac{1+\frac{\tilde{k}^2}{24m_e^2c^4}}{1+\frac{3\tilde{k}^2}{8m_e^2c^4}} \approx 4 - \frac{4\tilde{k}^2}{3m_e^2c^4},$$
$$\frac{8}{\tan(2\theta)}\int_0^\theta x\tan x dx \approx \frac{4}{\theta}\int_0^\theta x^2 dx = \frac{4}{\theta}\cdot\frac{\theta^3}{3} \approx \frac{\tilde{k}^2}{3m_e^2c^4},$$
$$6\theta\cot\theta \approx 6\theta\left(\frac{1}{\theta} - \frac{\theta}{3}\right) = 6 - 2\theta^2 \approx 6 - \frac{\tilde{k}^2}{2m_e^2c^4},$$
$$\frac{2\theta}{\sin(2\theta)} \approx \frac{2\theta}{2\theta - (4/3)\theta^3} \approx 1 + \frac{2\theta^2}{3} \approx 1 + \frac{\tilde{k}^2}{6m_e^2c^4},$$
$$N_\kappa(\tilde{q},\tilde{q}') + \frac{\Gamma\pi^2}{ic^3}\gamma_\kappa \approx -\frac{\pi^2}{ic^3}\left(2\gamma_\kappa\left(1 - \frac{\tilde{k}^2}{12m_e^2c^4}\right)\right.$$
$$\left. - \frac{(\tilde{q}+\tilde{q}')_\kappa}{m_e c^2}\left(1 + \frac{\tilde{k}^2}{6m_e^2c^4}\right) - \frac{4\gamma_\kappa\tilde{k}^2}{3m_e^2c^4}\ln\frac{\lambda}{m_e}\right),$$

so that the total contribution of Figures 4.5 (e) and (h) becomes

$$s_4^{(e)+(h)} \approx -\frac{ic^3\alpha^2}{4\pi^2\tilde{k}^2}\frac{m_p m_e c^4}{\sqrt{\omega_{\boldsymbol{q}}\omega_{\boldsymbol{q}'}\Omega_{\boldsymbol{p}}\Omega_{\boldsymbol{p}'}}}\mathcal{W}^\kappa(\boldsymbol{p}'\tau',\boldsymbol{p}\tau)\bar{u}(\boldsymbol{q}',\sigma')$$
$$\times\left[2\gamma_\kappa\left(1 - \frac{\tilde{k}^2}{12m_e^2c^4}\right) - \frac{(\tilde{q}+\tilde{q}')_\kappa}{m_e c^2}\left(1 + \frac{\tilde{k}^2}{6m_e^2c^4}\right) - \frac{4\gamma_\kappa\tilde{k}^2}{3m_e^2c^4}\ln\frac{\lambda}{m_e}\right]u(\boldsymbol{q},\sigma). \tag{4.47}$$

According to (G.4) and (G.6), in the $(v/c)^2$ approximation and in the limit $m_p \to \infty$ this expression is further simplified, so we have

$$s_4^{(e)+(h)} \approx -\frac{\alpha^2\delta_{\tau\tau'}}{4\pi^2m_e^2c}\chi_{\sigma'}^\dagger\frac{\boldsymbol{\sigma}_{\text{el}}\cdot[\boldsymbol{k}\times\boldsymbol{q}]}{k^2}\chi_\sigma + \frac{i\alpha^2}{3\pi^2m_e^2c}\ln\left(\frac{\lambda}{m_e}\right)\delta_{\sigma\sigma'}\delta_{\tau\tau'}. \tag{4.48}$$

28 We introduced the notation $|\tilde{k}| \equiv \sqrt{\tilde{k}^2}$.

4.3.3 Ladder diagram

Next, we study the ladder diagram in Figure 4.6. According to the Feynman rules, its contribution is[29]

$$S_4^{(f)} = \frac{e^4c^4\delta_{\sigma\sigma'}\delta_{\tau\tau'}}{(2\pi)^4(2\pi\hbar)^2}\cdot\frac{m_pm_ec^4}{(2\pi\hbar)^6\sqrt{\omega_{\boldsymbol{q}}\omega_{\boldsymbol{q}'}\Omega_{\boldsymbol{p}}\Omega_{\boldsymbol{p}'}}}\int d^4h\frac{\overline{u}(\boldsymbol{q}')\gamma_\mu(\not{q}-\not{h}+m_ec^2)\gamma^\nu u(\boldsymbol{q})}{(\tilde{q}-\tilde{h})^2-m_e^2c^4}$$
$$\times\frac{\overline{w}(\boldsymbol{p}')\gamma^\mu(\not{p}+\not{h}+m_pc^2)\gamma_\nu w(\boldsymbol{p})}{[(\tilde{p}+\tilde{h})^2-m_p^2c^4][\tilde{h}^2-\lambda^2c^4][(\tilde{h}+\tilde{k})^2-\lambda^2c^4]}.$$

We use Dirac equations (B.93)–(B.94) for the functions $u(\boldsymbol{q})$, $w(\boldsymbol{p})$ and anticommutator (B.8) of the gamma matrices to rewrite the numerator as follows:

$$\begin{aligned}
&[\overline{u}(\boldsymbol{q}')\gamma_\mu(\not{q}-\not{h}+m_ec^2)\gamma^\nu u(\boldsymbol{q})]\cdot[\overline{w}(\boldsymbol{p}')\gamma^\mu(\not{p}+\not{h}+m_pc^2)\gamma_\nu w(\boldsymbol{p})]\\
&\quad=[\overline{u}(\boldsymbol{q}')\gamma_\mu(\not{q}+m_ec^2)\gamma^\nu u(\boldsymbol{q})-\overline{u}(\boldsymbol{q}')\gamma_\mu\not{h}\gamma^\nu u(\boldsymbol{q})]\\
&\qquad\times[\overline{w}(\boldsymbol{p}')\gamma^\mu(\not{p}+m_pc^2)\gamma_\nu w(\boldsymbol{p})+\overline{w}(\boldsymbol{p}')\gamma^\mu\not{h}\gamma_\nu w(\boldsymbol{p})]\\
&\quad=[2\overline{u}(\boldsymbol{q}')\gamma_\mu q^\nu u(\boldsymbol{q})-\overline{u}(\boldsymbol{q}')\gamma_\mu\not{h}\gamma^\nu u(\boldsymbol{q})][2\overline{w}(\boldsymbol{p}')\gamma^\mu p_\nu w(\boldsymbol{p})+\overline{w}(\boldsymbol{p}')\gamma^\mu\not{h}\gamma_\nu w(\boldsymbol{p})]\\
&\quad=4(\tilde{q}\cdot\tilde{p})\overline{u}(\boldsymbol{q}')\gamma_\mu u(\boldsymbol{q})\overline{w}(\boldsymbol{p}')\gamma^\mu w(\boldsymbol{p})+2\overline{u}(\boldsymbol{q}')\gamma_\mu u(\boldsymbol{q})\overline{w}(\boldsymbol{p}')\gamma^\mu\gamma_\alpha\not{q}w(\boldsymbol{p})h^\alpha\\
&\quad\quad-2\overline{u}(\boldsymbol{q}')\gamma_\mu\gamma_\alpha\not{p}u(\boldsymbol{q})\overline{w}(\boldsymbol{p}')\gamma^\mu w(\boldsymbol{p})h^\alpha-\overline{u}(\boldsymbol{q}')\gamma_\mu\gamma_\alpha\gamma^\nu u(\boldsymbol{q})\overline{w}(\boldsymbol{p}')\gamma^\mu\gamma_\beta\gamma_\nu w(\boldsymbol{p})h^\alpha h^\beta.
\end{aligned}$$

In the denominators we take into account that $\tilde{q}^2=m_e^2c^4$, $\tilde{p}^2=m_p^2c^4$ and

$$\begin{aligned}
(\tilde{q}-\tilde{h})^2-m_e^2c^4&=\tilde{h}^2-2(\tilde{q}\cdot\tilde{h}),\\
(\tilde{p}+\tilde{h})^2-m_p^2c^4&=\tilde{h}^2+2(\tilde{p}\cdot\tilde{h}).
\end{aligned}$$

In the nonrelativistic approximation (B.79), we then obtain

$$\begin{aligned}
S_4^{(f)}\approx{}&\frac{e^4c^4\delta_{\sigma\sigma'}\delta_{\tau\tau'}}{(2\pi)^4(2\pi\hbar)^2}\\
&\times[4(\tilde{q}\cdot\tilde{p})\overline{u}(\boldsymbol{q}')\gamma_\mu u(\boldsymbol{q})\overline{w}(\boldsymbol{p}')\gamma^\mu w(\boldsymbol{p})b(\boldsymbol{p},\boldsymbol{q},\boldsymbol{k})\\
&+2\overline{u}(\boldsymbol{q}')\gamma_\mu u(\boldsymbol{q})\overline{w}(\boldsymbol{p}')\gamma^\mu\gamma_\alpha\not{q}w(\boldsymbol{p})b^\alpha(\boldsymbol{p},\boldsymbol{q},\boldsymbol{k})
\end{aligned}$$

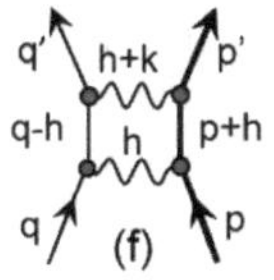

Figure 4.6: Ladder diagram for the electron–proton scattering. The same as in Figure 4.1 (f).

29 Here we omit spin indices of the functions u and w, since in our approximations the spin dependence will be lost (i. e., reduced to factors $\delta_{\sigma\sigma'}\delta_{\tau\tau'}$) anyway.

$$- 2\overline{u}(\boldsymbol{q}')\gamma_\mu\gamma_\alpha \not{p} u(\boldsymbol{q})\overline{w}(\boldsymbol{p}')\gamma^\mu w(\boldsymbol{p}) b^\alpha(\boldsymbol{p},\boldsymbol{q},\boldsymbol{k})$$
$$- \overline{u}(\boldsymbol{q}')\gamma_\mu\gamma_\alpha\gamma^\nu u(\boldsymbol{q})\overline{w}(\boldsymbol{p}')\gamma^\mu\gamma_\beta\gamma_\nu b^{\alpha\beta}(\boldsymbol{p},\boldsymbol{q},\boldsymbol{k})], \tag{4.49}$$

where

$$b(\boldsymbol{p},\boldsymbol{q},\boldsymbol{k}) \equiv \int \frac{d^4h}{[\tilde{h}^2 - 2(\tilde{q}\cdot\tilde{h})][\tilde{h}^2 + 2(\tilde{p}\cdot\tilde{h})][\tilde{h}^2 - \lambda^2c^4][(\tilde{h}+\tilde{k})^2 - \lambda^2c^4]}, \tag{4.50}$$
$$b^\alpha(\boldsymbol{p},\boldsymbol{q},\boldsymbol{k}) \equiv \int \frac{d^4h h^\alpha}{[\tilde{h}^2 - 2(\tilde{q}\cdot\tilde{h})][\tilde{h}^2 + 2(\tilde{p}\cdot\tilde{h})][\tilde{h}^2 - \lambda^2c^4][(\tilde{h}+\tilde{k})^2 - \lambda^2c^4]},$$
$$b^{\alpha\beta}(\boldsymbol{p},\boldsymbol{q},\boldsymbol{k}) \equiv \int \frac{d^4h h^\alpha h^\beta}{[\tilde{h}^2 - 2(\tilde{q}\cdot\tilde{h})][\tilde{h}^2 + 2(\tilde{p}\cdot\tilde{h})][\tilde{h}^2 - \lambda^2c^4][(\tilde{h}+\tilde{k})^2 - \lambda^2c^4]}.$$

In the framework of our approximate approach, we are only interested in dominant infrared-divergent terms in the above integrals. They come from those regions of the four-dimensional integration space of $\tilde{h}$, where the integrand's denominators vanish in the limit $\lambda \to 0$, i. e., near $\tilde{h} \approx 0$ and $\tilde{h} \approx -\tilde{k}$. Using these approximations in the numerators, we can express two other integrals in terms of (4.50), as follows:

$$b^\alpha(\boldsymbol{p},\boldsymbol{q},\boldsymbol{k}) \approx -k^\alpha b(\boldsymbol{p},\boldsymbol{q},\boldsymbol{k}),$$
$$b^{\alpha\beta}(\boldsymbol{p},\boldsymbol{q},\boldsymbol{k}) \approx k^\alpha k^\beta b(\boldsymbol{p},\boldsymbol{q},\boldsymbol{k}).$$

Next we substitute this result into (4.49) and use definitions (B.64)–(B.65) of the functions $\mathcal{U}^\mu$ and $\mathcal{W}^\mu$ to obtain

$$\begin{aligned} s_4^{(f)} &= \frac{e^4c^4\delta_{\sigma\sigma'}\delta_{\tau\tau'}}{(2\pi)^4(2\pi\hbar)^2} b(\boldsymbol{p},\boldsymbol{q},\boldsymbol{k}) \\ &\quad \times [4(\tilde{q}\cdot\tilde{p})\overline{u}(\boldsymbol{q}')\gamma^\nu u(\boldsymbol{q})\overline{w}(\boldsymbol{p}')\gamma_\nu w(\boldsymbol{p}) - 2\overline{u}(\boldsymbol{q}')\gamma_\mu u(\boldsymbol{q})\overline{w}(\boldsymbol{p}')\gamma^\mu \not{k}\not{q} w(\boldsymbol{p}) \\ &\quad + 2\overline{u}(\boldsymbol{q}')\gamma_\mu \not{k}\not{p} u(\boldsymbol{q})\overline{w}(\boldsymbol{p}')\gamma^\mu w(\boldsymbol{p}) - \overline{u}(\boldsymbol{q}')\gamma_\mu \not{k}\gamma^\nu u(\boldsymbol{q})\overline{w}(\boldsymbol{p}')\gamma^\mu \not{k}\gamma_\nu w(\boldsymbol{p})] \\ &= \frac{e^4c^4\delta_{\sigma\sigma'}\delta_{\tau\tau'}}{(2\pi)^4(2\pi\hbar)^2} b(\boldsymbol{p},\boldsymbol{q},\boldsymbol{k}) \\ &\quad \times [4(\tilde{q}\cdot\tilde{p})(\tilde{\mathcal{U}}\cdot\tilde{\mathcal{W}}) - 2\mathcal{U}_\mu\overline{w}(\boldsymbol{p}')\gamma^\mu \not{k}\not{q} w(\boldsymbol{p}) \\ &\quad + 2\overline{u}(\boldsymbol{q}')\gamma_\mu \not{k}\not{p} u(\boldsymbol{q})\mathcal{W}^\mu - \overline{u}(\boldsymbol{q}')\gamma_\mu \not{k}\gamma^\nu u(\boldsymbol{q})\overline{w}(\boldsymbol{p}')\gamma^\mu \not{k}\gamma_\nu w(\boldsymbol{p})]. \end{aligned}$$

Then we use equations (B.21)–(B.22) and (B.93)–(B.94) to simplify separate parts of this expression, as follows:

$$\begin{aligned} &\overline{w}(\boldsymbol{p}')\gamma^\mu \not{k}\not{q} w(\boldsymbol{p}) \\ &\quad = \overline{w}(\boldsymbol{p}')\gamma^\mu \not{p}\not{q} w(\boldsymbol{p}) - \overline{w}(\boldsymbol{p}')\gamma^\mu \not{p}'\not{q} w(\boldsymbol{p}) \\ &\quad = -\overline{w}(\boldsymbol{p}')\gamma^\mu \not{q}\not{p} w(\boldsymbol{p}) + 2\overline{w}(\boldsymbol{p}')\gamma^\mu(\tilde{p}\cdot\tilde{q})w(\boldsymbol{p}) + \overline{w}(\boldsymbol{p}')\not{p}'\gamma^\mu \not{q} w(\boldsymbol{p}) - 2\overline{w}(\boldsymbol{p}')(p')^\mu \not{q} w(\boldsymbol{p}) \\ &\quad = -\overline{w}(\boldsymbol{p}')\gamma^\mu \not{q} m_pc^2 w(\boldsymbol{p}) + 2\mathcal{W}^\mu(\tilde{p}\cdot\tilde{q}) + \overline{w}(\boldsymbol{p}')m_pc^2\gamma^\mu \not{q} w(\boldsymbol{p}) - 2(p')^\mu(\tilde{q}\cdot\tilde{\mathcal{W}}) \\ &\quad = 2\mathcal{W}^\mu(\tilde{p}\cdot\tilde{q}) - 2(p')^\mu(\tilde{q}\cdot\tilde{\mathcal{W}}), \end{aligned} \tag{4.51}$$

$$\begin{aligned}
\overline{u}(\boldsymbol{q}')\gamma_\mu \not{k}\not{p} u(\boldsymbol{q}) &= \overline{u}(\boldsymbol{q}')\gamma_\mu \not{q}'\not{p} u(\boldsymbol{q}) - \overline{u}(\boldsymbol{q}')\gamma_\mu \not{q}\not{p} u(\boldsymbol{q}) \\
&= -\overline{u}(\boldsymbol{q}')\not{q}'\gamma_\mu \not{p} u(\boldsymbol{q}) + 2\overline{u}(\boldsymbol{q}')(q')_\mu \not{p} u(\boldsymbol{q}) + \overline{u}(\boldsymbol{q}')\gamma_\mu \not{p}\not{q} u(\boldsymbol{q}) - 2\overline{u}(\boldsymbol{q}')\gamma_\mu(\tilde{q}\cdot\tilde{p})u(\boldsymbol{q}) \\
&= -\overline{u}(\boldsymbol{q}')m_e c^2 \gamma_\mu \not{p} u(\boldsymbol{q}) + 2(q')_\mu(\tilde{p}\cdot\tilde{\mathcal{U}}) + \overline{u}(\boldsymbol{q}')\gamma_\mu \not{p} m_e c^2 u(\boldsymbol{q}) - 2\mathcal{U}_\mu(\tilde{q}\cdot\tilde{p}) \\
&= 2(q')_\mu(\tilde{p}\cdot\tilde{\mathcal{U}}) - 2\mathcal{U}_\mu(\tilde{q}\cdot\tilde{p}),
\end{aligned}$$

$$\begin{aligned}
\overline{u}(\boldsymbol{q}')\gamma_\mu \not{k}\gamma^\nu u(\boldsymbol{q}) &= \overline{u}(\boldsymbol{q}')\gamma_\mu \not{q}'\gamma^\nu u(\boldsymbol{q}) - \overline{u}(\boldsymbol{q}')\gamma_\mu \not{q}\gamma^\nu u(\boldsymbol{q}) \\
&= -\overline{u}(\boldsymbol{q}')\not{q}'\gamma_\mu\gamma^\nu u(\boldsymbol{q}) + 2\overline{u}(\boldsymbol{q}')q'_\mu\gamma^\nu u(\boldsymbol{q}) + \overline{u}(\boldsymbol{q}')\gamma_\mu\gamma^\nu \not{q} u(\boldsymbol{q}) - 2\overline{u}(\boldsymbol{q}')\gamma_\mu q^\nu u(\boldsymbol{q}) \\
&= 2q'_\mu \mathcal{U}^\nu - 2q^\nu \mathcal{U}_\mu,
\end{aligned} \tag{4.52}$$

$$\overline{w}(\boldsymbol{p}')\gamma^\mu \not{k}\gamma_\nu w(\boldsymbol{p}) = \overline{w}(\boldsymbol{p}')\gamma^\mu(\not{p}-\not{p}')\gamma_\nu w(\boldsymbol{p}) = 2p_\nu \mathcal{W}^\mu - 2(p')^\mu \mathcal{W}_\nu. \tag{4.53}$$

Applying equations (F.37), (F.38) and nonrelativistic approximations $\tilde{q}' \approx \tilde{q} \approx (m_e c^2, 0,0,0)$, $\tilde{p}' \approx \tilde{p} \approx (m_p c^2, 0,0,0)$, $\tilde{\mathcal{U}} \approx \tilde{\mathcal{W}} \approx (1,0,0,0)$, we obtain

$$\begin{aligned}
s_4^{(f)} &= \frac{e^4c^4\delta_{\sigma\sigma'}\delta_{\tau\tau'}}{(2\pi)^4(2\pi\hbar)^2} b(\boldsymbol{p},\boldsymbol{q},\boldsymbol{k}) \\
&\quad \times [4(\tilde{q}\cdot\tilde{p})(\tilde{\mathcal{U}}\cdot\tilde{\mathcal{W}}) - 2\mathcal{U}_\mu(2\mathcal{W}^\mu(\tilde{p}\cdot\tilde{q}) - 2(p')^\mu(\tilde{q}\cdot\tilde{\mathcal{W}})) \\
&\quad + 2(2(q')_\mu(\tilde{p}\cdot\tilde{\mathcal{U}}) - 2\mathcal{U}_\mu(\tilde{q}\cdot\tilde{p}))\mathcal{W}^\mu - (2(q')_\mu\mathcal{U}^\nu - 2q^\nu\mathcal{U}_\mu)(2p_\nu\mathcal{W}^\mu - 2(p')^\mu\mathcal{W}_\nu)] \\
&= \frac{4e^4c^4\delta_{\sigma\sigma'}\delta_{\tau\tau'}}{(2\pi)^4(2\pi\hbar)^2} b(\boldsymbol{p},\boldsymbol{q},\boldsymbol{k})[(\tilde{q}\cdot\tilde{p})(\tilde{\mathcal{U}}\cdot\tilde{\mathcal{W}}) \\
&\quad - (\tilde{\mathcal{U}}\cdot\tilde{\mathcal{W}})(\tilde{p}\cdot\tilde{q}) + (\tilde{p}'\cdot\tilde{\mathcal{U}})(\tilde{q}\cdot\tilde{\mathcal{W}}) + (\tilde{q}'\cdot\tilde{\mathcal{W}})(\tilde{p}\cdot\tilde{\mathcal{U}}) - (\tilde{\mathcal{U}}\cdot\tilde{\mathcal{W}})(\tilde{q}\cdot\tilde{p}) \\
&\quad - (\tilde{p}\cdot\tilde{\mathcal{U}})(\tilde{q}'\cdot\tilde{\mathcal{W}}) + (\tilde{p}\cdot\tilde{q})(\tilde{\mathcal{U}}\cdot\tilde{\mathcal{W}}) + (\tilde{p}'\cdot\tilde{q}')(\tilde{\mathcal{U}}\cdot\tilde{\mathcal{W}}) - (\tilde{p}'\cdot\tilde{\mathcal{U}})(\tilde{q}\cdot\tilde{\mathcal{W}})] \\
&= \frac{4e^4c^4\delta_{\sigma\sigma'}\delta_{\tau\tau'}}{(2\pi)^4(2\pi\hbar)^2} b(\boldsymbol{p},\boldsymbol{q},\boldsymbol{k})(\tilde{p}'\cdot\tilde{q}')(\tilde{\mathcal{U}}\cdot\tilde{\mathcal{W}}) \\
&\approx \frac{4e^4m_pm_ec^8\delta_{\sigma\sigma'}\delta_{\tau\tau'}}{(2\pi)^4(2\pi\hbar)^2} b(\boldsymbol{p},\boldsymbol{q},\boldsymbol{k}).
\end{aligned} \tag{4.54}$$

The function $b(\boldsymbol{p},\boldsymbol{q},\boldsymbol{k})$ is calculated in (F.46). We have

$$b(\boldsymbol{p},\boldsymbol{q},\boldsymbol{k}) = \frac{\pi^2}{ic^3\tilde{k}^2}\ln\left(\frac{\tilde{k}^2}{\lambda^2c^4}\right)\int_0^1 \frac{dy}{(\tilde{p}+\tilde{q})^2y^2 - 2\tilde{p}\cdot(\tilde{p}+\tilde{q})y + \tilde{p}^2}. \tag{4.55}$$

The integral with respect to y is

$$\begin{aligned}
&\int_0^1 \frac{dy}{(\tilde{p}+\tilde{q})^2y^2 - 2\tilde{p}\cdot(\tilde{p}+\tilde{q})y + \tilde{p}^2} \\
&\quad = 2\mathfrak{B}^{-1}\arctan\left(\frac{2(\tilde{p}+\tilde{q})^2y - 2\tilde{p}\cdot(\tilde{p}+\tilde{q})}{\mathfrak{B}}\right)\Bigg|_{y=0}^{y=1} \\
&\quad = 2\mathfrak{B}^{-1}\left[\arctan\left(\frac{2\tilde{q}^2 + 2(\tilde{p}\cdot\tilde{q})}{\mathfrak{B}}\right) + \arctan\left(\frac{2\tilde{p}^2 + 2(\tilde{p}\cdot\tilde{q})}{\mathfrak{B}}\right)\right]
\end{aligned}$$

$$\approx \frac{1}{im_p c^3 q}\left[\arctan\left(\frac{m_e c}{iq}\right) + \arctan\left(\frac{m_p c}{iq}\right)\right], \tag{4.56}$$

where we used the inequality $m_p \gg m_e$ and denoted

$$\begin{aligned}
\mathfrak{B} &\equiv \sqrt{4(\tilde{p}+\tilde{q})^2\tilde{p}^2 - 4(\tilde{p}^2 + (\tilde{p}\cdot\tilde{q}))^2} = 2\sqrt{\tilde{p}^2\tilde{q}^2 - (\tilde{p}\cdot\tilde{q})^2} \\
&= 2\sqrt{m_p^2 m_e^2 c^8 - (\tilde{p}\cdot\tilde{q})^2} \\
&\approx 2\sqrt{m_p^2 m_e^2 c^8 - \left[\left(m_p c^2 + \frac{p^2}{2m_p}\right)\left(m_e c^2 + \frac{q^2}{2m_e}\right) - c^2(\boldsymbol{p}\cdot\boldsymbol{q})\right]^2} \\
&\approx 2\sqrt{-m_p^2 c^6 q^2} = 2im_p c^3 q.
\end{aligned}$$

Putting together results (4.54)–(4.56) and using $\tilde{k}^2 \approx -c^2k^2$, we finally obtain

$$s_4^{(f)} \approx \frac{\alpha^2 m_e c^2 \delta_{\sigma\sigma'}\delta_{\tau\tau'}}{\pi^2 q k^2}\left[\arctan\left(\frac{m_e c}{iq}\right) + \arctan\left(\frac{m_p c}{iq}\right)\right]\ln\left(-\frac{k^2}{\lambda^2 c^2}\right). \tag{4.57}$$

We will not transform this result any further, as we expect it to cancel partially with the cross ladder diagram in the next subsection.

4.3.4 Cross ladder diagram

Our calculation of the cross ladder diagram in Figure 4.7 is similar to the preceding subsection, i. e.,

$$\begin{aligned}
s_4^{(g)} &\approx \frac{e^4 c^4 \delta_{\sigma\sigma'}\delta_{\tau\tau'}}{(2\pi)^4(2\pi\hbar)^2}\int d^4h \frac{\overline{u}(\boldsymbol{q}')\gamma_\mu(\not{q} - \not{h} + m_e c^2)\gamma^\nu u(\boldsymbol{q})}{(\tilde{q}-\tilde{h})^2 - m_e^2 c^4} \\
&\quad\times \frac{\overline{w}(\boldsymbol{p}')\gamma_\nu(\not{p}' - \not{h} + m_p c^2)\gamma^\mu w(\boldsymbol{p})}{[(\tilde{p}'-\tilde{h})^2 - m_p^2 c^4][\tilde{h}^2 - \lambda^2 c^4][(\tilde{h}+\tilde{k})^2 - \lambda^2 c^4]}.
\end{aligned}$$

In the numerator, the application of (B.21)–(B.22) and (B.93)–(B.94) yields

$$\begin{aligned}
&[\overline{u}(\boldsymbol{q}')\gamma_\mu(\not{q} - \not{h} + m_e c^2)\gamma^\nu u(\boldsymbol{q})][\overline{w}(\boldsymbol{p}')\gamma_\nu(\not{p}' - \not{h} + m_p c^2)\gamma^\mu w(\boldsymbol{p})] \\
&\quad= [\overline{u}(\boldsymbol{q}')\gamma_\mu(\not{q} + m_e c^2)\gamma^\nu u(\boldsymbol{q}) - \overline{u}(\boldsymbol{q}')\gamma_\mu\not{h}\gamma^\nu u(\boldsymbol{q})] \\
&\quad\quad\times [\overline{w}(\boldsymbol{p}')\gamma_\nu(\not{p}' + m_p c^2)\gamma^\mu w(\boldsymbol{p}) - \overline{w}(\boldsymbol{p}')\gamma_\nu\not{h}\gamma_\mu w(\boldsymbol{p})]
\end{aligned}$$

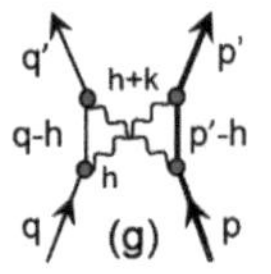

Figure 4.7: Cross ladder diagram for the electron–proton scattering. The same as in Figure 4.1 (g).

$$= [2\bar{u}(\boldsymbol{q}')\gamma_\mu q^\nu u(\boldsymbol{q}) - \bar{u}(\boldsymbol{q}')\gamma_\mu \not{h}\gamma^\nu u(\boldsymbol{q})][2\overline{w}(\boldsymbol{p}')(p')_\nu \gamma^\mu w(\boldsymbol{p}) - \overline{w}(\boldsymbol{p}')\gamma_\nu \not{h}\gamma^\mu w(\boldsymbol{p})]$$
$$= 4(\tilde{q}\cdot\tilde{p}')(\tilde{U}\cdot\tilde{W}) - 2U_\mu \overline{w}(\boldsymbol{p}')\not{q}\gamma_\alpha\gamma^\mu w(\boldsymbol{p})h^\alpha$$
$$- 2\bar{u}(\boldsymbol{q}')\gamma_\mu\gamma_\alpha \not{p}' u(\boldsymbol{q})W^\mu h^\alpha + \bar{u}(\boldsymbol{q}')\gamma_\mu\gamma_\alpha\gamma^\nu u(\boldsymbol{q})\overline{w}(\boldsymbol{p}')\gamma_\nu\gamma_\beta\gamma^\mu w(\boldsymbol{p})h^\alpha h^\beta,$$

so

$$\begin{aligned} s_4^{(g)} = {} & \frac{e^4c^4\delta_{\sigma\sigma'}\delta_{\tau\tau'}}{(2\pi)^4(2\pi\hbar)^2} \\ & \times [4(\tilde{q}\cdot\tilde{p}')(\tilde{U}\cdot\tilde{W})b(-\boldsymbol{p}',\boldsymbol{q},\boldsymbol{k}) - 2U_\mu \overline{w}(\boldsymbol{p}')\not{q}\gamma_\alpha\gamma^\mu w(\boldsymbol{p})b^\alpha(-\boldsymbol{p}',\boldsymbol{q},\boldsymbol{k}) \\ & - 2\bar{u}(\boldsymbol{q}')\gamma_\mu\gamma_\alpha\not{p}' u(\boldsymbol{q})W^\mu b^\alpha(-\boldsymbol{p}',\boldsymbol{q},\boldsymbol{k}) \\ & + \bar{u}(\boldsymbol{q}')\gamma_\mu\gamma_\alpha\gamma^\nu u(\boldsymbol{q})\overline{w}(\boldsymbol{p}')\gamma_\nu\gamma_\beta\gamma^\mu w(\boldsymbol{p})b^{\alpha\beta}(-\boldsymbol{p}',\boldsymbol{q},\boldsymbol{k})]. \end{aligned} \tag{4.58}$$

Here we notice that the integral

$$b(-\boldsymbol{p}',\boldsymbol{q},\boldsymbol{k}) \equiv \int \frac{d^4h}{[\tilde{h}^2 - 2(\tilde{q}\cdot\tilde{h})][\tilde{h}^2 - 2(\tilde{p}'\cdot\tilde{h})][\tilde{h}^2 - \lambda^2c^4][(\tilde{h}+\tilde{k})^2 - \lambda^2c^4]}$$

can be obtained from (4.50) by substituting $\tilde{p} \to -\tilde{p}'$. Using the same assumptions as in Subsection 4.3.3, the other two integrals can be expressed in terms of $b(-\boldsymbol{p}',\boldsymbol{q},\boldsymbol{k})$ as follows:

$$\begin{aligned} b^\alpha(-\boldsymbol{p}',\boldsymbol{q},\boldsymbol{k}) &\equiv \int \frac{d^4hh^\alpha}{[\tilde{h}^2 - 2(\tilde{q}\cdot\tilde{h})][\tilde{h}^2 - 2(\tilde{p}'\cdot\tilde{h})][\tilde{h}^2 - \lambda^2c^4][(\tilde{h}+\tilde{k})^2 - \lambda^2c^4]} \\ &\approx -k^\alpha b(-\boldsymbol{p}',\boldsymbol{q},\boldsymbol{k}), \\ b^{\alpha\beta}(-\boldsymbol{p}',\boldsymbol{q},\boldsymbol{k}) &\equiv \int \frac{d^4hh^\alpha h^\beta}{[\tilde{h}^2 - 2(\tilde{q}\cdot\tilde{h})][\tilde{h}^2 - 2(\tilde{p}'\cdot\tilde{h})][\tilde{h}^2 - \lambda^2c^4][(\tilde{h}+\tilde{k})^2 - \lambda^2c^4]} \\ &\approx k^\alpha k^\beta b(-\boldsymbol{p}',\boldsymbol{q},\boldsymbol{k}). \end{aligned}$$

Then

$$\begin{aligned} s_4^{(g)} = {} & \frac{e^4c^4\delta_{\sigma\sigma'}\delta_{\tau\tau'}}{(2\pi)^4(2\pi\hbar)^2} b(-\boldsymbol{p}',\boldsymbol{q},\boldsymbol{k})[4(\tilde{q}\cdot\tilde{p}')(\tilde{\mathcal{U}}\cdot\tilde{\mathcal{W}}) + 2\mathcal{U}_\mu \overline{w}(\boldsymbol{p}')\not{q}\not{k}\gamma^\mu w(\boldsymbol{p}) \\ & + 2\bar{u}(\boldsymbol{q}')\gamma_\mu \not{k}\not{p}' u(\boldsymbol{q})\mathcal{W}^\mu + \bar{u}(\boldsymbol{q}')\gamma_\mu\not{k}\gamma^\nu u(\boldsymbol{q})\overline{w}(\boldsymbol{p}')\gamma_\nu\not{k}\gamma^\mu w(\boldsymbol{p})]. \end{aligned}$$

Using familiar methods, we obtain

$$\begin{aligned} \overline{w}(\boldsymbol{p}')\not{q}\not{k}\gamma^\mu w(\boldsymbol{p}) &= \overline{w}(\boldsymbol{p}')\not{q}\not{p}\gamma^\mu w(\boldsymbol{p}) - \overline{w}(\boldsymbol{p}')\not{q}\not{p}'\gamma^\mu w(\boldsymbol{p}) \\ &= 2\overline{w}(\boldsymbol{p}')\not{q}p^\mu w(\boldsymbol{p}) - 2\overline{w}(\boldsymbol{p}')(\tilde{q}\cdot\tilde{p}')\gamma^\mu w(\boldsymbol{p}) = 2(\tilde{q}\cdot\tilde{\mathcal{W}})p^\mu - 2(\tilde{q}\cdot\tilde{p}')\mathcal{W}^\mu, \end{aligned}$$

$$\begin{aligned} \bar{u}(\boldsymbol{q}')\gamma_\mu\not{k}\not{p}' u(\boldsymbol{q}) &= \bar{u}(\boldsymbol{q}')\gamma_\mu \not{q}'\not{p}' u(\boldsymbol{q}) - \bar{u}(\boldsymbol{q}')\gamma_\mu\not{q}\not{p}' u(\boldsymbol{q}) \\ &= 2\bar{u}(\boldsymbol{q}')q'_\mu \not{p}' u(\boldsymbol{q}) - 2\bar{u}(\boldsymbol{q}')\gamma_\mu(\tilde{q}\cdot\tilde{p}')u(\boldsymbol{q}) = 2q'_\mu(\tilde{p}'\cdot\tilde{\mathcal{U}}) - 2\mathcal{U}_\mu(\tilde{q}\cdot\tilde{p}'), \end{aligned}$$

$$\bar{u}(\boldsymbol{q}')\gamma_\mu \not{k}\gamma^\nu u(\boldsymbol{q}) = \bar{u}(\boldsymbol{q}')\gamma_\mu \not{q}'\gamma^\nu u(\boldsymbol{q}) - \bar{u}(\boldsymbol{q}')\gamma_\mu \not{q}\gamma^\nu u(\boldsymbol{q})$$
$$= 2\bar{u}(\boldsymbol{q}')q'_\mu \gamma^\nu u(\boldsymbol{q}) - 2\bar{u}(\boldsymbol{q}')\gamma_\mu q^\nu u(\boldsymbol{q}) = 2q'_\mu \mathcal{U}^\nu - 2\mathcal{U}_\mu q^\nu,$$

$$\overline{w}(\boldsymbol{p}')\gamma_\nu \not{k}\gamma^\mu w(\boldsymbol{p}) = \overline{w}(\boldsymbol{p}')\gamma_\nu \not{p}\gamma^\mu w(\boldsymbol{p}) - \overline{w}(\boldsymbol{p}')\gamma_\nu \not{p}'\gamma^\mu w(\boldsymbol{p})$$
$$= 2\overline{w}(\boldsymbol{p}')\gamma_\nu p^\mu w(\boldsymbol{p}) - 2\overline{w}(\boldsymbol{p}')p'_\nu \gamma^\mu w(\boldsymbol{p}) = 2\mathcal{W}_\nu p^\mu - 2p'_\nu \mathcal{W}^\mu$$

and

$$\begin{aligned} s_4^{(g)} &= \frac{e^4c^4\delta_{\sigma\sigma'}\delta_{\tau\tau'}}{(2\pi)^4(2\pi\hbar)^2} b(-\boldsymbol{p}',\boldsymbol{q},\boldsymbol{k}) \\ &\quad \times [4(\tilde{q}\cdot\tilde{p}')(\tilde{\mathcal{U}}\cdot\tilde{\mathcal{W}}) + 2\mathcal{U}_\mu(2(\tilde{q}\cdot\tilde{\mathcal{W}})p^\mu - 2(\tilde{q}\cdot\tilde{p}')\mathcal{W}^\mu) \\ &\qquad + 2(2q'_\mu(\tilde{p}'\cdot\tilde{\mathcal{U}}) - 2\mathcal{U}_\mu(\tilde{q}\cdot\tilde{p}'))\mathcal{W}^\mu + (2q'_\mu\mathcal{U}^\nu - 2\mathcal{U}_\mu q^\nu)(2\mathcal{W}_\nu p^\mu - 2p'_\nu\mathcal{W}^\mu)] \\ &= \frac{4e^4c^4\delta_{\sigma\sigma'}\delta_{\tau\tau'}}{(2\pi)^4(2\pi\hbar)^2} b(-\boldsymbol{p}',\boldsymbol{q},\boldsymbol{k})[(\tilde{q}\cdot\tilde{p}')(\tilde{\mathcal{U}}\cdot\tilde{\mathcal{W}}) + (\tilde{p}\cdot\tilde{\mathcal{U}})(\tilde{q}\cdot\tilde{\mathcal{W}}) \\ &\quad - (\tilde{q}\cdot\tilde{p}')(\tilde{\mathcal{U}}\cdot\tilde{\mathcal{W}}) + (\tilde{q}'\cdot\tilde{\mathcal{W}})(\tilde{p}'\cdot\tilde{\mathcal{U}}) - (\tilde{\mathcal{U}}\cdot\tilde{\mathcal{W}})(\tilde{q}\cdot\tilde{p}') \\ &\quad + (\tilde{q}'\cdot\tilde{p})(\tilde{\mathcal{U}}\cdot\tilde{\mathcal{W}}) - (\tilde{p}\cdot\tilde{\mathcal{U}})(\tilde{q}\cdot\tilde{\mathcal{W}}) - (\tilde{q}'\cdot\tilde{\mathcal{W}})(\tilde{p}'\cdot\tilde{\mathcal{U}}) + (\tilde{\mathcal{U}}\cdot\mathcal{W})(\tilde{q}\cdot\tilde{p}')] \\ &= \frac{4e^4c^4\delta_{\sigma\sigma'}\delta_{\tau\tau'}}{(2\pi)^4(2\pi\hbar)^2} b(-\boldsymbol{p}',\boldsymbol{q},\boldsymbol{k})(\tilde{q}'\cdot\tilde{p})(\tilde{\mathcal{U}}\cdot\tilde{\mathcal{W}}) \\ &\approx \frac{4e^4m_em_pc^8\delta_{\sigma\sigma'}\delta_{\tau\tau'}}{(2\pi)^4(2\pi\hbar)^2} b(-\boldsymbol{p}',\boldsymbol{q},\boldsymbol{k}). \end{aligned}$$

For the integral

$$b(-\boldsymbol{p}',\boldsymbol{q},\boldsymbol{k}) = \frac{\pi^2}{ic^3\tilde{k}^2}\ln\left(\frac{\tilde{k}^2}{\lambda^2c^4}\right)\int_0^1 \frac{dy}{(-\tilde{p}'+\tilde{q})^2y^2 + 2\tilde{p}'\cdot(-\tilde{p}'+\tilde{q})y + (\tilde{p}')^2}$$

we use the same method as in (4.56). This time in our nonrelativistic approximation

$$\mathfrak{B}' \equiv \sqrt{4(\tilde{q}-\tilde{p}')^2(\tilde{p}')^2 - 4((\tilde{p}')^2 - (\tilde{p}'\cdot\tilde{q}))^2} \approx \mathfrak{B} = 2im_pc^3q,$$

$$\begin{aligned} &\int_0^1 \frac{dy}{(-\tilde{p}'+\tilde{q})^2y^2 + 2\tilde{p}'\cdot(-\tilde{p}'+\tilde{q})y + \tilde{p}^2} \\ &\quad \approx 2\mathfrak{B}^{-1}\arctan\left(\frac{2(-\tilde{p}'+\tilde{q})^2y + 2\tilde{p}'\cdot(-\tilde{p}'+\tilde{q})}{\mathfrak{B}}\right)\Bigg|_{y=0}^{y=1} \\ &\quad = 2\mathfrak{B}^{-1}\left[\arctan\left(\frac{2\tilde{q}^2 - 2(\tilde{p}'\cdot\tilde{q})}{\mathfrak{B}}\right) + \arctan\left(\frac{2\tilde{p}^2 - 2(\tilde{p}'\cdot\tilde{q})}{\mathfrak{B}}\right)\right] \\ &\quad \approx \frac{1}{im_pc^3q}\left[-\arctan\left(\frac{m_ec}{iq}\right) + \arctan\left(\frac{m_pc}{iq}\right)\right] \end{aligned}$$

and

$$s_4^{(g)} \approx \frac{\alpha^2m_ec^2}{\pi^2qk^2}\left[-\arctan\left(\frac{m_ec}{iq}\right) + \arctan\left(\frac{m_pc}{iq}\right)\right]\ln\left(-\frac{k^2}{\lambda^2c^2}\right)\delta_{\sigma\sigma'}\delta_{\tau\tau'}.$$

By adding this result to (4.57) and using the approximation

$$\arctan \frac{m_p c}{iq} \approx -\pi/2,$$

we obtain the combined contribution from the ladder and crossed-ladder diagrams,

$$s_4^{(f)+(g)} \approx -\frac{\alpha^2 m_e c^2}{\pi q k^2} \ln\left(\frac{k^2}{\lambda^2 c^2}\right) \delta_{\sigma\sigma'} \delta_{\tau\tau'}. \tag{4.59}$$

4.3.5 Renormalizability

Finally, we can put together our results (4.46), (4.48), (4.59) and obtain the total electron–proton scattering amplitude in the fourth order of perturbation theory. We have

$$\begin{aligned}
&\langle \text{vac} | a_{\boldsymbol{q}\sigma} d_{\boldsymbol{p}\tau} S_4^c d^\dagger_{\boldsymbol{p}'\tau'} a^\dagger_{\boldsymbol{q}'\sigma'} | \text{vac} \rangle \\
&= (s_4^{(d)+(k)} + s_4^{(e)+(h)} + s_4^{(f)} + s_4^{(g)}) \delta^4(\tilde{q} + \tilde{p} - \tilde{q}' - \tilde{p}') \\
&\approx \delta^4(\tilde{q} - \tilde{q}' - \tilde{p}' + \tilde{p}) \delta_{\tau\tau'} \left[\frac{i\alpha^2}{15\pi^2 m_e^2 c} \delta_{\sigma\sigma'} + \frac{i\alpha^2}{3\pi^2 m_e^2 c} \ln\left(\frac{\lambda}{m_e}\right) \delta_{\sigma\sigma'} \right. \\
&\left. - \frac{m_e c^2 \alpha^2}{\pi q k^2} \ln\left(\frac{k^2}{\lambda^2 c^2}\right) \delta_{\sigma\sigma'} - \frac{\alpha^2 \chi^\dagger_{\sigma'} (\boldsymbol{\sigma}_{\text{el}} \cdot [\boldsymbol{k} \times \boldsymbol{q}]) \chi_\sigma}{4\pi^2 m_e^2 c k^2} \right].
\end{aligned} \tag{4.60}$$

As expected, this result does not depend on the cutoff parameter Λ. In other words, it does not contain ultraviolet divergences, so we managed to keep the promise of renormalization theory.

Unfortunately, the amplitude (4.60) still contains unpleasant infrared-divergent logarithms. The physical reason for such divergences is related to the zero photon mass. Each collision involving charged particles[30] is inevitably accompanied by the emission of a large (even infinite) number of low-energy (soft) photons. In many cases, these photons (due to their low energy) escape experimental detection, but in a strict theoretical description of the scattering they must be taken into account. The interested reader will find detailed discussions of such calculations in many textbooks [21, 10]. As we shall see in Volume 3, the infrared divergences will cancel out in the physical quantities (for example, the Lamb shift) that are of interest to us.

Thus, we conclude that our approach to renormalization has reached its goal: ultraviolet divergences in loop integrals have canceled out, and we are close to an accurate description of scattering. Can we improve the accuracy of our approach by extending renormalization to higher orders of perturbation theory? The answer is "yes,"

30 In particular, $e^- + p^+$ collisions considered here.

but then we will have to add higher-order divergent counterterms to the interaction operator, so as to ensure the validity of the no-self-scattering and charge renormalization conditions in each perturbation order. Surprisingly, all these high-order counterterms will have the same operator structure as the counterterms discussed above. In other words, the complete interaction operator of the renormalized QED will have the same form as our expression (4.44). Only the values of renormalization constants δm, $(Z_2 - 1)$, $(Z_3 - 1)$ and $(Z_1 - 1)$ will be more complicated than our second-order approximations $(\delta m)_2$, $(Z_2 - 1)_2$, $(Z_3 - 1)_2$, $(Z_1 - 1)_2$. This nontrivial fact is referred to as the *renormalizability* of QED.

This concludes our discussion of the traditional QED. In the third volume, we will draw attention to the fact that, despite indisputable successes, this theory can not yet be regarded as the ultimate unification of the principle of relativity and quantum mechanics. Our main issue is that the QED Hamiltonian (4.44) diverges in the ultraviolet limit $\Lambda \to \infty$. This means that an adequate description of the time evolution (dynamics) with such a Hamiltonian is impossible.

In Volume 3, we will modify the above QED formalism in order to define a new Λ-independent Hamiltonian of interacting charges and photons. Our new approach will be referred to as the *relativistic quantum dynamics* (RQD).

A Useful integrals

In calculations we often find the following integral:

$$\int \frac{d\boldsymbol{r}}{r} e^{\frac{i}{\hbar}\boldsymbol{p}\cdot\boldsymbol{r}} = \int_0^{\pi} \sin\theta d\theta \int_0^{2\pi} d\varphi \int_0^{\infty} r^2 dr \frac{e^{\frac{i}{\hbar}pr\cos\theta}}{r} = 2\pi \int_{-1}^{1} dz \int_0^{\infty} dr r e^{\frac{i}{\hbar}prz}$$

$$= 2\pi\hbar \int_0^{\infty} r dr \frac{e^{\frac{i}{\hbar}pr} - e^{-\frac{i}{\hbar}pr}}{ipr} = \frac{4\pi\hbar}{p} \int_0^{\infty} dr \sin\left(\frac{pr}{\hbar}\right) = \frac{4\pi\hbar^2}{p^2} \int_0^{\infty} d\rho \sin(\rho)$$

$$= -\frac{4\pi\hbar^2}{p^2}(\cos(\infty) - \cos(0)) = \frac{4\pi\hbar^2}{p^2}. \quad \text{(A.1)}$$

In this derivation we set $\cos(\infty) = 0$, because in applications of this formula, the plane wave $e^{\frac{i}{\hbar}\boldsymbol{p}\cdot\boldsymbol{r}}$ in the integrand is usually present multiplied by a smooth damping factor, which tends to zero for large values of $\boldsymbol{r}$. This attenuation is analogous to the slow (adiabatic) switching of the interaction described in Subsection **1**-7.1.4.

To calculate the similar integral

$$K = \int d\boldsymbol{x} d\boldsymbol{y} \frac{e^{\frac{i}{\hbar}(\boldsymbol{p}\cdot\boldsymbol{x} + \boldsymbol{q}\cdot\boldsymbol{y})}}{|\boldsymbol{x} - \boldsymbol{y}|},$$

we change integration variables as follows:

$$\boldsymbol{x} = \frac{1}{2}(\boldsymbol{z} + \boldsymbol{t}),$$
$$\boldsymbol{y} = \frac{1}{2}(\boldsymbol{z} - \boldsymbol{t}),$$
$$\boldsymbol{x} - \boldsymbol{y} = \boldsymbol{t},$$
$$\boldsymbol{x} + \boldsymbol{y} = \boldsymbol{z}.$$

The Jacobian of this transformation is

$$J \equiv \left|\det\left[\frac{\partial(\boldsymbol{x}, \boldsymbol{y})}{\partial(\boldsymbol{z}, \boldsymbol{t})}\right]\right| = 1/8.$$

Then, using formulas (**1**-A.1) and (A.1), we get

$$K = \frac{1}{8} \int d\boldsymbol{t} d\boldsymbol{z} \frac{e^{\frac{i}{2\hbar}(\boldsymbol{p}\cdot(\boldsymbol{z}+\boldsymbol{t}) + \boldsymbol{q}\cdot(\boldsymbol{z}-\boldsymbol{t}))}}{t} = \frac{1}{8} \int d\boldsymbol{t} d\boldsymbol{z} \frac{e^{\frac{i}{2\hbar}(\boldsymbol{z}\cdot(\boldsymbol{p}+\boldsymbol{q}) + \boldsymbol{t}\cdot(\boldsymbol{p}-\boldsymbol{q}))}}{t}$$

$$= (2\pi\hbar)^3 \delta(\boldsymbol{p} + \boldsymbol{q}) \int d\boldsymbol{t} \frac{e^{\frac{i}{2\hbar}\boldsymbol{t}\cdot(\boldsymbol{p}-\boldsymbol{q})}}{t} = \frac{(2\pi\hbar)^6}{2\pi^2\hbar} \frac{\delta(\boldsymbol{p} + \boldsymbol{q})}{(\boldsymbol{p} - \boldsymbol{q})^2}. \quad \text{(A.2)}$$

Other useful integrals are

$$\int \frac{d\boldsymbol{k}}{k^2} e^{\frac{i}{\hbar}\boldsymbol{k}\cdot\boldsymbol{r}} = \frac{(2\pi\hbar)^3}{4\pi\hbar^2 r}, \quad \text{(A.3)}$$

https://doi.org/10.1515/9783110493207-005

$$\int \frac{d\mathbf{k}\mathbf{k}}{k^2} e^{\frac{i}{\hbar}\mathbf{k}\cdot\mathbf{r}} = -i\hbar\frac{\partial}{\partial \mathbf{r}} \int \frac{d\mathbf{k}}{k^2} e^{\frac{i}{\hbar}\mathbf{k}\cdot\mathbf{r}} = -\frac{i(2\pi\hbar)^3}{4\pi\hbar}\frac{\partial}{\partial \mathbf{r}}\left(\frac{1}{r}\right) = \frac{i(2\pi\hbar)^3\mathbf{r}}{4\pi\hbar r^3}, \tag{A.4}$$

$$\int \frac{d\mathbf{k}(\mathbf{q}\cdot\mathbf{k})(\mathbf{p}\cdot\mathbf{k})}{k^4} e^{\frac{i}{\hbar}\mathbf{k}\cdot\mathbf{r}} = \frac{(2\pi\hbar)^3}{8\pi\hbar^2 r}\left[(\mathbf{q}\cdot\mathbf{p}) - \frac{(\mathbf{q}\cdot\mathbf{r})(\mathbf{p}\cdot\mathbf{r})}{r^2}\right], \tag{A.5}$$

$$\int \frac{d\mathbf{k}(\mathbf{q}\cdot\mathbf{k})(\mathbf{p}\cdot\mathbf{k})}{k^2} e^{\frac{i}{\hbar}\mathbf{k}\cdot\mathbf{r}} = \frac{(2\pi\hbar)^3}{4\pi r^3}\left[(\mathbf{q}\cdot\mathbf{p}) - 3\frac{(\mathbf{q}\cdot\mathbf{r})(\mathbf{p}\cdot\mathbf{r})}{r^2}\right] + \frac{(2\pi\hbar)^3}{3}(\mathbf{q}\cdot\mathbf{p})\delta(\mathbf{r}). \tag{A.6}$$

Their calculations can be found, for example, in § 83 in [1].

In the covariant time-ordered perturbation theory, we often encounter the delta function in the 4D energy–momentum space (p_0, p_x, p_y, p_z),

$$\delta^4(\tilde{p}) \equiv \delta(p_0)\delta(p_x)\delta(p_y)\delta(p_z) = \delta(p_0)\delta(\mathbf{p}). \tag{A.7}$$

It has the following integral representation:

$$\frac{1}{(2\pi\hbar)^4}\int e^{\frac{i}{\hbar}\tilde{p}\cdot\tilde{x}} d^4x = \delta^4(\tilde{p}). \tag{A.8}$$

The *step-function* $\theta(t)$ is defined as[1]

$$\theta(t) \equiv \begin{cases} 1, & \text{if } t > 0, \\ 0, & \text{if } t < 0. \end{cases} \tag{A.9}$$

It has the following integral representation:

$$\theta(t) = -\frac{1}{2\pi i}\int_{-\infty}^{\infty} ds \frac{e^{-ist}}{s + i\epsilon}, \tag{A.10}$$

where ϵ is a small positive number. This integral can be calculated in the complex plane along the contour shown in Figure A.1. For $t > 0$, the integration contour can be

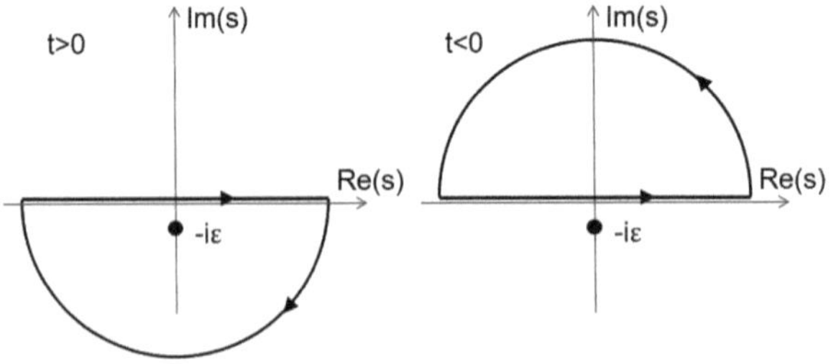

Figure A.1: To the calculation of the integral (A.10).

1 Sometimes, for symmetry, this definition is augmented by $\theta(0) = 1/2$.

closed in the lower half-plane, because the exponent is large and negative there, i. e., the integral over the large semicircle tends to 0. Then the integral in (A.10) is equal to the residue ($2\pi i$) at the pole $s = -i\epsilon$, taken with the opposite sign, because the contour is traversed clockwise. Thus, $\theta(t > 0) = 1$. For $t < 0$, we should close the contour in the upper half-plane. This contour does not contain poles, and the integral vanishes.

Lemma A.1 (Riemann–Lebesgue). *The Fourier integral of a smooth function $f(x)$ tends to zero as the frequency k tends to infinity, i. e.,*

$$\lim_{k\to\infty} \int_{-\infty}^{\infty} f(x)e^{ikx} = 0.$$

B Quantum fields of fermions

Usually QFT textbooks claim that (quantum) fields are the fundamental ingredients of nature, and the main task of QFT is to apply the laws of quantum mechanics to these systems with an infinite number of degrees of freedom. We do not adhere to this point of view. We believe that matter consists of particles, and quantum fields are just abstract mathematical creations, whose purpose is to simplify the construction of relativistically invariant and cluster-separable operators of interactions between particles. Therefore we placed our discussion of quantum fields in this appendix, and not in the main body of the book. Here we will talk about quantum fields for massive fermions with spin 1/2 (electrons, protons and their antiparticles). In the next appendix we will consider the photon quantum field.

B.1 Pauli matrices

Generators of the spin-1/2 representation of the rotation group (see Table **1**-I.1) are conveniently expressed through so-called *Pauli matrices* σ_i $(i = x, y, z)$. We have

$$S_i = \frac{\hbar}{2}\sigma_i, \tag{B.1}$$

where

$$\sigma_x \equiv \sigma_1 = \begin{bmatrix} 0 & 1 \\ 1 & 0 \end{bmatrix},$$

$$\sigma_y \equiv \sigma_2 = \begin{bmatrix} 0 & -i \\ i & 0 \end{bmatrix},$$

$$\sigma_z \equiv \sigma_3 = \begin{bmatrix} 1 & 0 \\ 0 & -1 \end{bmatrix}.$$

In calculations we will need the following properties of these matrices:

$$[\sigma_i, \sigma_j] = 2i \sum_{i=1}^{3} \epsilon_{ijk}\sigma_k,$$

$$\{\sigma_i, \sigma_j\} = 2\delta_{ij},$$

$$\sigma_i^2 = 1.$$

Sometimes it is useful to define the fourth Pauli matrix

$$\sigma_t \equiv \sigma_0 = \begin{bmatrix} 1 & 0 \\ 0 & 1 \end{bmatrix}.$$

https://doi.org/10.1515/9783110493207-006

For arbitrary numerical 3-vectors $\boldsymbol{a}$ and $\boldsymbol{b}$ we have

$$(\boldsymbol{\sigma}\cdot\boldsymbol{a})\boldsymbol{\sigma} = \boldsymbol{a}\sigma_0 + i[\boldsymbol{\sigma}\times\boldsymbol{a}], \tag{B.2}$$

$$\boldsymbol{\sigma}(\boldsymbol{\sigma}\cdot\boldsymbol{a}) = \boldsymbol{a}\sigma_0 - i[\boldsymbol{\sigma}\times\boldsymbol{a}], \tag{B.3}$$

$$(\boldsymbol{\sigma}\cdot\boldsymbol{a})(\boldsymbol{\sigma}\cdot\boldsymbol{b}) = (\boldsymbol{a}\cdot\boldsymbol{b})\sigma_0 + i\boldsymbol{\sigma}\cdot[\boldsymbol{a}\times\boldsymbol{b}]. \tag{B.4}$$

B.2 Dirac gamma matrices

Let us introduce the following 4 × 4 Dirac *gamma matrices*[1]:

$$\gamma^0 \equiv \begin{bmatrix} 1 & 0 & 0 & 0 \\ 0 & 1 & 0 & 0 \\ 0 & 0 & -1 & 0 \\ 0 & 0 & 0 & -1 \end{bmatrix} = \begin{bmatrix} \sigma_0 & 0 \\ 0 & -\sigma_0 \end{bmatrix} = \begin{bmatrix} 1 & 0 \\ 0 & -1 \end{bmatrix}, \tag{B.5}$$

$$\gamma^x \equiv \begin{bmatrix} 0 & 0 & 0 & 1 \\ 0 & 0 & 1 & 0 \\ 0 & -1 & 0 & 0 \\ -1 & 0 & 0 & 0 \end{bmatrix} = \begin{bmatrix} 0 & \sigma_x \\ -\sigma_x & 0 \end{bmatrix},$$

$$\gamma^y \equiv \begin{bmatrix} 0 & 0 & 0 & -i \\ 0 & 0 & i & 0 \\ 0 & i & 0 & 0 \\ -i & 0 & 0 & 0 \end{bmatrix} = \begin{bmatrix} 0 & \sigma_y \\ -\sigma_y & 0 \end{bmatrix},$$

$$\gamma^z \equiv \begin{bmatrix} 0 & 0 & 1 & 0 \\ 0 & 0 & 0 & -1 \\ -1 & 0 & 0 & 0 \\ 0 & 1 & 0 & 0 \end{bmatrix} = \begin{bmatrix} 0 & \sigma_z \\ -\sigma_z & 0 \end{bmatrix},$$

$$\boldsymbol{\gamma} \equiv \begin{bmatrix} 0 & \boldsymbol{\sigma} \\ -\boldsymbol{\sigma} & 0 \end{bmatrix}. \tag{B.6}$$

These matrices have the following properties[2]:

$$\gamma^0\boldsymbol{\gamma} = \boldsymbol{\gamma}^\dagger\gamma^0 = -\boldsymbol{\gamma}\gamma^0, \tag{B.7}$$

$$\gamma^\mu\gamma^\nu + \gamma^\nu\gamma^\mu = 2\eta^{\mu\nu}, \tag{B.8}$$

$$\gamma^0\gamma^0 = 1, \tag{B.9}$$

$$\gamma^i\gamma^i = -1, \tag{B.10}$$

1 On the right-hand sides, each 2×2 subblock is expressed through Pauli matrices from Appendix B.1.

2 The indices take values $\mu, \nu = 0, 1, 2, 3$; $i = 1, 2, 3$; $\eta_{\mu\nu}$ is the metric tensor (**1**-J.2).

$$\mathrm{Tr}(\gamma^\mu) = 0, \tag{B.11}$$
$$\mathrm{Tr}(\gamma^\mu\gamma^\nu) = 4\eta_{\mu\nu}. \tag{B.12}$$

Lorentz indices can be lowered by the usual rule from Appendix **1**-J.1,

$$\gamma_\mu = \eta_{\mu\nu}\gamma^\nu,$$
$$\gamma_0 = \gamma^0,$$
$$\gamma_i = -\gamma^i.$$

Then

$$\gamma_\mu\gamma^\mu = -\gamma^x\gamma^x - \gamma^y\gamma^y - \gamma^z\gamma^z + \gamma^0\gamma^0 = 4, \tag{B.13}$$
$$\gamma_\mu\gamma_\nu\gamma^\mu = -\gamma_\nu\gamma_\mu\gamma^\mu + 2\eta_{\mu\nu}\gamma^\mu = -4\gamma_\nu + 2\gamma_\nu = -2\gamma_\nu. \tag{B.14}$$

If A, B, C are arbitrary linear combinations of gamma matrices, then

$$\gamma_\mu A\gamma^\mu = -2A, \tag{B.15}$$
$$\gamma_\mu AB\gamma^\mu = 2(AB + BA), \tag{B.16}$$
$$\gamma_\mu ABC\gamma^\mu = -2CBA. \tag{B.17}$$

It will be convenient to introduce the *slash* notation for the pseudo-scalar product of γ^μ with a 4-vector $\tilde{k}$,

$$\not{k} \equiv \gamma^\mu k_\mu \equiv \gamma_0 k_0 - \boldsymbol{\gamma}\cdot\boldsymbol{k}. \tag{B.18}$$

This symbol has the following properties:

$$\not{k}^2 = \gamma^\mu k_\mu \gamma^\nu k_\nu = 1/2(\gamma^\mu\gamma^\nu + \gamma^\nu\gamma^\mu)k_\mu k_\nu = \eta^{\mu\nu}k_\mu k_\nu = \tilde{k}^2, \tag{B.19}$$
$$(\not{k} - mc^2)(\not{k} + mc^2) = \not{k}^2 - m^2c^4 = \tilde{k}^2 - m^2c^4, \tag{B.20}$$
$$\gamma_\mu\not{k} + \not{k}\gamma_\mu = 2k_\mu, \tag{B.21}$$
$$\not{p}\not{q} + \not{q}\not{p} = 2(\tilde{p}\cdot\tilde{q}). \tag{B.22}$$

B.3 Dirac representation of Lorentz group

Here we will construct the *Dirac representation* $\mathcal{D}(\Lambda)$ of the Lorentz group. Just as the pseudo-orthogonal representation from Appendix **1**-J, the Dirac representation is realized by 4×4 matrices.

In the Dirac representation, the generators of pure boosts and rotations are defined through commutators of the gamma matrices. We have

$$\mathfrak{K} \equiv \frac{i\hbar}{4c}[\gamma^0, \boldsymbol{\gamma}] = \frac{i\hbar}{2c}\begin{bmatrix} 0 & \boldsymbol{\sigma} \\ \boldsymbol{\sigma} & 0 \end{bmatrix}, \tag{B.23}$$

$$\mathfrak{J}_x \equiv \frac{i\hbar}{4}[\gamma^y, \gamma^z] = \frac{\hbar}{2}\begin{bmatrix} \sigma_x & 0 \\ 0 & \sigma_x \end{bmatrix}, \tag{B.24}$$

$$\mathfrak{J}_y \equiv \frac{i\hbar}{4}[\gamma^z, \gamma^x] = \frac{\hbar}{2}\begin{bmatrix} \sigma_y & 0 \\ 0 & \sigma_y \end{bmatrix}, \tag{B.25}$$

$$\mathfrak{J}_z \equiv \frac{i\hbar}{4}[\gamma^x, \gamma^y] = \frac{\hbar}{2}\begin{bmatrix} \sigma_z & 0 \\ 0 & \sigma_z \end{bmatrix}. \tag{B.26}$$

Using properties of the Pauli matrices from Appendix B.1, one can verify that commutators of these generators satisfy the usual relations of the Lorentz Lie algebra (**1**-3.50), (**1**-3.51) and (**1**-3.53). For example,

$$[\mathfrak{J}_x, \mathfrak{J}_y] = \frac{\hbar^2}{4}\begin{bmatrix} [\sigma_x, \sigma_y] & 0 \\ 0 & [\sigma_x, \sigma_y] \end{bmatrix} = \frac{i\hbar^2}{2}\begin{bmatrix} \sigma_z & 0 \\ 0 & \sigma_z \end{bmatrix} = i\hbar\mathfrak{J}_z,$$

$$\begin{aligned}[\mathfrak{J}_x, \mathfrak{K}_y] &= \frac{i\hbar^2}{4c}\left(\begin{bmatrix} \sigma_x & 0 \\ 0 & \sigma_x \end{bmatrix}\begin{bmatrix} 0 & \sigma_y \\ \sigma_y & 0 \end{bmatrix} - \begin{bmatrix} 0 & \sigma_y \\ \sigma_y & 0 \end{bmatrix}\begin{bmatrix} \sigma_x & 0 \\ 0 & \sigma_x \end{bmatrix}\right) \\ &= -\frac{\hbar^2}{2c}\begin{bmatrix} 0 & \sigma_z \\ \sigma_z & 0 \end{bmatrix} = i\hbar\mathfrak{K}_z,\end{aligned}$$

$$\begin{aligned}[\mathfrak{K}_x, \mathfrak{K}_y] &= -\frac{\hbar^2}{4c^2}\left(\begin{bmatrix} 0 & \sigma_x \\ \sigma_x & 0 \end{bmatrix}\begin{bmatrix} 0 & \sigma_y \\ \sigma_y & 0 \end{bmatrix} - \begin{bmatrix} 0 & \sigma_y \\ \sigma_y & 0 \end{bmatrix}\begin{bmatrix} 0 & \sigma_x \\ \sigma_x & 0 \end{bmatrix}\right) \\ &= -\frac{\hbar^2}{4c^2}\begin{bmatrix} [\sigma_x, \sigma_y] & 0 \\ 0 & [\sigma_x, \sigma_y] \end{bmatrix} = -\frac{i\hbar^2}{2c^2}\begin{bmatrix} \sigma_z & 0 \\ 0 & \sigma_z \end{bmatrix} = -\frac{i\hbar}{c^2}\mathfrak{J}_z.\end{aligned}$$

The representation of the Lorentz Lie algebra constructed above is connected with finite inertial transformations by exponential functions. For a Lorentz boost with rapidity $\boldsymbol{\theta}$ we have

$$\begin{aligned}\mathcal{D}(e^{c\mathcal{K}\cdot\boldsymbol{\theta}}) &\equiv \mathcal{D}(\boldsymbol{\theta}) = e^{-\frac{ic}{\hbar}\mathfrak{K}\cdot\boldsymbol{\theta}} = \exp\left(\frac{1}{2}\begin{bmatrix} 0 & \boldsymbol{\sigma}\cdot\boldsymbol{\theta} \\ \boldsymbol{\sigma}\cdot\boldsymbol{\theta} & 0 \end{bmatrix}\right) \\ &= 1 + \frac{1}{2}\begin{bmatrix} 0 & \boldsymbol{\sigma}\cdot\boldsymbol{\theta} \\ \boldsymbol{\sigma}\cdot\boldsymbol{\theta} & 0 \end{bmatrix} + \frac{1}{2!}\left(\frac{\theta}{2}\right)^2\begin{bmatrix} 1 & 0 \\ 0 & 1 \end{bmatrix} + O(\theta^3) \\ &= I\cosh\frac{\theta}{2} + \frac{2c}{i\hbar}\sinh\frac{\theta}{2}\left(\mathfrak{K}\cdot\frac{\boldsymbol{\theta}}{\theta}\right).\end{aligned} \tag{B.27}$$

Boost generators (B.23) are not Hermitian, so representatives of finite boosts (B.27) are not unitary.[3] We have

$$\mathcal{D}^\dagger(\boldsymbol{\theta}) = \mathcal{D}(\boldsymbol{\theta}). \tag{B.28}$$

3 See footnote 7 on page 59.

However, generators of rotations (B.24)–(B.26) are Hermitian, so representatives of finite rotations are unitary, i. e.,

$$\mathcal{D}^\dagger(\boldsymbol{\varphi}) = \mathcal{D}^{-1}(\boldsymbol{\varphi}) = \mathcal{D}(-\boldsymbol{\varphi}). \tag{B.29}$$

This is easily verified in the explicit matrix form

$$\begin{aligned}
\mathcal{D}(\boldsymbol{\varphi}) &\equiv \mathcal{D}(e^{\mathcal{J}\cdot\boldsymbol{\varphi}}) = e^{-\frac{i}{\hbar}\mathfrak{J}\cdot\boldsymbol{\varphi}} = \exp\left(-\frac{i}{2}\begin{bmatrix}\boldsymbol{\sigma}\cdot\boldsymbol{\varphi} & 0\\ 0 & \boldsymbol{\sigma}\cdot\boldsymbol{\varphi}\end{bmatrix}\right)\\
&= 1 - \frac{i}{2}\begin{bmatrix}\boldsymbol{\sigma}\cdot\boldsymbol{\varphi} & 0\\ 0 & \boldsymbol{\sigma}\cdot\boldsymbol{\varphi}\end{bmatrix} - \frac{1}{2!}\left(\frac{\varphi}{2}\right)^2\begin{bmatrix}1 & 0\\ 0 & 1\end{bmatrix} + O(\varphi^3)\\
&= I\cos\frac{\varphi}{2} + \frac{2}{i\hbar}\sin\frac{\varphi}{2}\left(\mathfrak{J}\cdot\frac{\boldsymbol{\varphi}}{\varphi}\right),\\
\mathcal{D}^\dagger(\boldsymbol{\varphi}) &= I\cos\frac{\varphi}{2} - \frac{2}{i\hbar}\sin\frac{\varphi}{2}\left(\mathfrak{J}\cdot\frac{\boldsymbol{\varphi}}{\varphi}\right) = \mathcal{D}(-\boldsymbol{\varphi}).
\end{aligned} \tag{B.30}$$

From equations (B.27) and (B.30) we obtain one more important property of gamma matrices, i. e.,

$$\mathcal{D}^{-1}(\Lambda)\gamma^\mu\mathcal{D}(\Lambda) = \sum_\nu \Lambda^\mu_{\ \nu}\gamma^\nu, \tag{B.31}$$

where Λ is any transformation from the Lorentz group and $\Lambda^\mu_{\ \nu}$ is a 4×4 matrix (**1**-J.15), realizing the pseudo-orthogonal representation of Λ. To see this, let us consider a particular case, where $\mu = 0$ and Λ is a boost along the x-axis with the rapidity θ. Then

$$\begin{aligned}
\mathcal{D}^{-1}(\boldsymbol{\theta})\gamma^0\mathcal{D}(\boldsymbol{\theta}) &= \left(I\cosh\frac{\theta}{2} - \frac{2c}{i\hbar}\mathfrak{K}_x\sinh\frac{\theta}{2}\right)\gamma^0\left(I\cosh\frac{\theta}{2} + \frac{2c}{i\hbar}\mathfrak{K}_x\sinh\frac{\theta}{2}\right)\\
&= \left(\cosh\frac{\theta}{2}\begin{bmatrix}1 & 0\\ 0 & 1\end{bmatrix} - \sinh\frac{\theta}{2}\begin{bmatrix}0 & \sigma_x\\ \sigma_x & 0\end{bmatrix}\right)\begin{bmatrix}1 & 0\\ 0 & -1\end{bmatrix}\\
&\quad\times\left(\cosh\frac{\theta}{2}\begin{bmatrix}1 & 0\\ 0 & 1\end{bmatrix} + \sinh\frac{\theta}{2}\begin{bmatrix}0 & \sigma_x\\ \sigma_x & 0\end{bmatrix}\right)\\
&= \cosh^2\frac{\theta}{2}\begin{bmatrix}1 & 0\\ 0 & -1\end{bmatrix} + 2\sinh\frac{\theta}{2}\cosh\frac{\theta}{2}\begin{bmatrix}0 & \sigma_x\\ -\sigma_x & 0\end{bmatrix} + \sinh^2\frac{\theta}{2}\begin{bmatrix}1 & 0\\ 0 & -1\end{bmatrix}\\
&= \gamma^0\cosh\theta + \gamma^x\sinh\theta = (\tilde{\boldsymbol{\theta}}_x)^0_{\ \nu}\gamma^\nu,
\end{aligned}$$

which is in agreement with our formula (**1**-J.11) for the boost matrix.

We can also verify that for pure boosts

$$\begin{aligned}
\gamma^0\mathcal{D}(\boldsymbol{\theta})\gamma^0 &= \gamma^0 e^{-\frac{ic}{\hbar}\mathfrak{K}\cdot\boldsymbol{\theta}}\gamma^0\\
&= 1 + \frac{1}{2}\gamma^0\begin{bmatrix}0 & \boldsymbol{\sigma}\cdot\boldsymbol{\theta}\\ \boldsymbol{\sigma}\cdot\boldsymbol{\theta} & 0\end{bmatrix}\gamma^0 + \frac{1}{2!}\left(\frac{\theta}{2}\right)^2\begin{bmatrix}1 & 0\\ 0 & 1\end{bmatrix} + O(\theta^3)
\end{aligned}$$

$$= 1 - \frac{1}{2}\begin{bmatrix} 0 & \boldsymbol{\sigma}\cdot\boldsymbol{\theta} \\ \boldsymbol{\sigma}\cdot\boldsymbol{\theta} & 0 \end{bmatrix} + \frac{1}{2!}\left(\frac{\theta}{2}\right)^2 \begin{bmatrix} 1 & 0 \\ 0 & 1 \end{bmatrix} + O(\theta^3)$$

$$= \mathcal{D}^{-1}(\boldsymbol{\theta}).$$

A similar calculation for rotations should convince us that for a general Lorentz transformation Λ

$$\gamma^0 \mathcal{D}(\Lambda)\gamma^0 = \mathcal{D}^{-1}(\Lambda). \tag{B.32}$$

Here we present one more useful result:

$$\mathcal{D}(\Lambda)\gamma^0\mathcal{D}(\Lambda) = \mathcal{D}(\Lambda)\gamma^0\mathcal{D}(\Lambda)\gamma^0\gamma^0 = \mathcal{D}(\Lambda)\mathcal{D}^{-1}(\Lambda)\gamma^0 = \gamma^0. \tag{B.33}$$

B.4 Construction of Dirac field

According to **Step 1** from Subsection 3.1.2, in order to build relativistic interaction operators, we have to associate with each particle species some finite-dimensional representation $D_{\alpha\beta}(\Lambda)$ of the Lorentz group. We also should define an operator function (= quantum field) $\phi_a(\tilde{x})$ on the Minkowski space–time $\mathcal{M}$, such that transformation properties of this function are governed by the representation D. In this appendix we will construct a quantum field for electrons/positrons. We postulate that the desired *Dirac field* has four components, which transform by the Dirac representation $\mathcal{D}(\Lambda)$ from Appendix B.3. In its explicit form the Dirac quantum field operator is[4]

$$\begin{aligned}\psi_a(\tilde{x}) &\equiv \psi_a(t, \boldsymbol{x}) \\ &= \int \frac{d\boldsymbol{p}}{(2\pi\hbar)^{3/2}} \sqrt{\frac{m_e c^2}{\omega_{\boldsymbol{p}}}} \sum_{s_z} (e^{-\frac{i}{\hbar}\tilde{p}\cdot\tilde{x}} u_a(\boldsymbol{p}, s_z) a_{\boldsymbol{p}s_z} + e^{\frac{i}{\hbar}\tilde{p}\cdot\tilde{x}} v_a(\boldsymbol{p}, s_z) b^\dagger_{\boldsymbol{p}s_z}).\end{aligned} \tag{B.34}$$

Here $a_{\boldsymbol{p}s_z}$ is the electron annihilation operator and $b^\dagger_{\boldsymbol{p}s_z}$ is the positron creation operator. For brevity we denoted by

$$\tilde{p} \equiv (\omega_{\boldsymbol{p}}, cp_x, cp_y, cp_z),$$
$$\tilde{x} \equiv (t, x/c, y/c, z/c)$$

the energy–momentum 4-vector and the Minkowski coordinate 4-vector, respectively. As in Appendix **1**-J.1, the pseudo-scalar product of two 4-vectors is denoted by the dot: $\tilde{p}\cdot\tilde{x} \equiv p_\mu x^\mu \equiv \omega_{\boldsymbol{p}} t - \boldsymbol{p}\cdot\boldsymbol{x}$, so that exponents in (B.34) have the forms $\exp[\pm\frac{i}{\hbar}(\omega_{\boldsymbol{p}} t - \boldsymbol{p}\cdot\boldsymbol{x})]$, where $\omega_{\boldsymbol{p}} \equiv \sqrt{m_e^2 c^4 + p^2 c^2}$. Note that according to equations (1.45) and (1.46)

$$\psi_a(t, \boldsymbol{x}) = e^{\frac{i}{\hbar}H_0 t}\psi_a(0, \boldsymbol{x})e^{-\frac{i}{\hbar}H_0 t}. \tag{B.35}$$

4 This form (up to an arbitrary normalization factor) can be established unambiguously [21] from properties (I)–(III) in **Step 1** of Subsection 3.1.2. The Dirac index takes values $a = 1, 2, 3, 4$.

So, the t-dependence required by equation (1.60) for regular operators is obvious in our definition (B.34). Numerical factors $u_a(\boldsymbol{p}, s_z)$ and $v_a(\boldsymbol{p}, s_z)$ will be discussed in Appendix B.5.

It is convenient to represent the Dirac field by a four-component column of operator functions, i. e.,

$$\psi(\tilde{x}) = \begin{bmatrix} \psi_1(\tilde{x}) \\ \psi_2(\tilde{x}) \\ \psi_3(\tilde{x}) \\ \psi_4(\tilde{x}) \end{bmatrix}.$$

We will also need a Hermitian-*conjugated* field

$$\psi_a^\dagger(\tilde{x}) = \int \frac{d\boldsymbol{p}}{(2\pi\hbar)^{3/2}} \sqrt{\frac{m_e c^2}{\omega_{\boldsymbol{p}}}} \sum_{s_z} (e^{\frac{i}{\hbar}\tilde{p}\cdot\tilde{x}} u_a^\dagger(\boldsymbol{p}, s_z) a_{\boldsymbol{p}s_z}^\dagger + e^{-\frac{i}{\hbar}\tilde{p}\cdot\tilde{x}} v_a^\dagger(\boldsymbol{p}, s_z) b_{\boldsymbol{p}s_z}),$$

which can be represented by a row as follows:

$$\psi^\dagger = [\psi_1^*, \psi_2^*, \psi_3^*, \psi_4^*].$$

The *Dirac-conjugated* field

$$\overline{\psi}_a(\tilde{x}) \equiv \sum_b \psi_b^\dagger(\tilde{x}) \gamma_{ba}^0 \tag{B.36}$$

is also represented by a row, i. e.,

$$\begin{aligned} \overline{\psi} \equiv \psi^\dagger \gamma^0 &= [\psi_1^*, \psi_2^*, \psi_3^*, \psi_4^*] \begin{bmatrix} 1 & 0 & 0 & 0 \\ 0 & 1 & 0 & 0 \\ 0 & 0 & -1 & 0 \\ 0 & 0 & 0 & -1 \end{bmatrix} \\ &= [\psi_1^*, \psi_2^*, -\psi_3^*, -\psi_4^*]. \end{aligned}$$

The proton/antiproton quantum field is built in a similar way. We have

$$\Psi(\tilde{x}) = \int \frac{d\boldsymbol{p}}{(2\pi\hbar)^{3/2}} \sqrt{\frac{m_p c^2}{\Omega_{\boldsymbol{p}}}} \sum_{s_z} (e^{-\frac{i}{\hbar}\tilde{P}\cdot\tilde{x}} w(\boldsymbol{p}, s_z) d_{\boldsymbol{p}s_z} + e^{\frac{i}{\hbar}\tilde{P}\cdot\tilde{x}} y(\boldsymbol{p}, s_z) f_{\boldsymbol{p}s_z}^\dagger), \tag{B.37}$$

where $\Omega_{\boldsymbol{p}} = \sqrt{m_p^2 c^4 + p^2 c^2}$, $\tilde{P} \cdot \tilde{x} \equiv \Omega_{\boldsymbol{p}} t - \boldsymbol{p} \cdot \boldsymbol{x}$ and the coefficient functions $w(\boldsymbol{p}, s_z)$ and $y(\boldsymbol{p}, s_z)$ can be obtained from $u(\boldsymbol{p}, s_z)$ and $v(\boldsymbol{p}, s_z)$ simply by replacing the electron mass m_e with the proton mass m_p.

B.5 Properties of functions *u* and *v*

The key components of the quantum field (B.34) are the numerical functions $u_a(\mathbf{p}, s_z)$ and $v_a(\mathbf{p}, s_z)$.[5] We can represent them as 4 × 2 matrices with the Dirac index $a = 1, 2, 3, 4$ numbering the rows and the spin projection index $s_z = -1/2, 1/2$ numbering the columns. Let us first postulate the form of these matrices at zero momentum.[6] We have

$$u(\mathbf{0}) = \begin{bmatrix} 0 & 1 \\ 1 & 0 \\ 0 & 0 \\ 0 & 0 \end{bmatrix}, \quad v(\mathbf{0}) = \begin{bmatrix} 0 & 0 \\ 0 & 0 \\ 1 & 0 \\ 0 & 1 \end{bmatrix}. \tag{B.38}$$

Our formulas will be more compact if we introduce the two-component quantities

$$\chi_{1/2} = \begin{bmatrix} 1 \\ 0 \end{bmatrix}, \quad \chi_{-1/2} = \begin{bmatrix} 0 \\ 1 \end{bmatrix}, \quad \chi^{\dagger}_{1/2} = (1, 0), \quad \chi^{\dagger}_{-1/2} = (0, 1). \tag{B.39}$$

Then we can write

$$u(\mathbf{0}, s_z) = \begin{bmatrix} \chi_{s_z} \\ 0 \end{bmatrix}, \quad v(\mathbf{0}, s_z) = \begin{bmatrix} 0 \\ \chi_{-s_z} \end{bmatrix}.$$

First check that matrix $u(\mathbf{0})$ has the following property:

$$\sum_b \mathcal{D}_{ab}(\boldsymbol{\varphi}) u_b(\mathbf{0}, s_z) = \sum_\tau u_a(\mathbf{0}, \tau) \mathscr{D}^{1/2}_{\tau s_z}(\boldsymbol{\varphi}), \tag{B.40}$$

where $\mathcal{D}$ is the Dirac representation of the Lorentz group from Appendix B.3, $\mathscr{D}^{1/2}$ is the 2D unitary irreducible representation of the rotation group from Appendix **1**-I.5 and $\boldsymbol{\varphi}$ is any rotation. Denoting $\mathfrak{J}_k$ the generators of rotations (B.24)–(B.26) in the representation $\mathcal{D}(\boldsymbol{\varphi})$ and S_k the generators of rotations (B.1) in the representation $\mathscr{D}^{1/2}(\boldsymbol{\varphi})$, we can rewrite equation (B.40) in the equivalent differential form

$$\sum_b (\mathfrak{J}_k)_{ab} u_b(\mathbf{0}, s_z) = \sum_\tau u_a(\mathbf{0}, \tau)(S_k)_{\tau s_z}.$$

Let us verify that this equality is valid, for example, for rotations about the x-axis. By acting with the 4 × 4 matrix (B.24) on the index b in $u_b(\mathbf{0}, s_z)$, we obtain

$$\mathfrak{J}_x u(\mathbf{0}) = \frac{\hbar}{2} \begin{bmatrix} 0 & 1 & 0 & 0 \\ 1 & 0 & 0 & 0 \\ 0 & 0 & 0 & 1 \\ 0 & 0 & 1 & 0 \end{bmatrix} \begin{bmatrix} 0 & 1 \\ 1 & 0 \\ 0 & 0 \\ 0 & 0 \end{bmatrix} = \frac{\hbar}{2} \begin{bmatrix} 1 & 0 \\ 0 & 1 \\ 0 & 0 \\ 0 & 0 \end{bmatrix}.$$

---

5 They are sometimes incorrectly called wave functions of free particles, although they have nothing to do with wave functions. See Subsection 8.1.5 in Volume 3.

6 This choice is not accidental, since the zero vector $\boldsymbol{\kappa} = (0, 0, 0)$ is the standard momentum in the method of induced representations of the Poincaré group for massive particles; see Section **1**-5.1.

This is the same as applying the 2×2 spin matrix

$$S_x = \frac{\hbar}{2}\begin{bmatrix} 0 & 1 \\ 1 & 0 \end{bmatrix}$$

to the index τ in $u_a(\mathbf{0}, \tau)$, which gives

$$u(\mathbf{0})S_x = \frac{\hbar}{2}\begin{bmatrix} 0 & 1 \\ 1 & 0 \\ 0 & 0 \\ 0 & 0 \end{bmatrix}\begin{bmatrix} 0 & 1 \\ 1 & 0 \end{bmatrix} = \frac{\hbar}{2}\begin{bmatrix} 1 & 0 \\ 0 & 1 \\ 0 & 0 \\ 0 & 0 \end{bmatrix}.$$

This proves equation (B.40).

The corresponding equation for the Dirac-conjugated factor $\overline{u}$ is obtained as follows. First we apply the Hermitian conjugation to (B.40). Then we multiply by the matrix γ^0 from the right and take into account equations (B.9), (B.29) and (B.32), to obtain

$$u^\dagger(\mathbf{0}, s_z)\gamma^0\gamma^0\mathcal{D}^\dagger(\boldsymbol{\varphi})\gamma^0 = \sum_\tau u^\dagger(\mathbf{0}, \tau)\gamma^0\mathscr{D}^{1/2}_{\tau s_z}(\boldsymbol{\varphi}),$$

$$\overline{u}(\mathbf{0}, s_z)\mathcal{D}(\boldsymbol{\varphi}) = \sum_\tau \overline{u}(\mathbf{0}, \tau)\mathscr{D}^{1/2}_{\tau s_z}(\boldsymbol{\varphi}). \tag{B.41}$$

So far we have discussed the functions u and v at zero momentum. Their values $u_a(\boldsymbol{p}, s_z)$ and $v_a(\boldsymbol{p}, s_z)$ at arbitrary momenta $\boldsymbol{p}$ are defined by application of representation matrices (B.27) of the standard boosts $\boldsymbol{\theta_p}$ defined in Subsection **1**-5.1.3. We have

$$u_a(\boldsymbol{p}, s_z) \equiv \sum_b \mathcal{D}_{ab}(\boldsymbol{\theta_p})u_b(\mathbf{0}, s_z), \tag{B.42}$$

$$v_a(\boldsymbol{p}, s_z) \equiv \sum_b \mathcal{D}_{ab}(\boldsymbol{\theta_p})v_b(\mathbf{0}, s_z). \tag{B.43}$$

By taking Hermitian conjugates of (B.42)–(B.43) and multiplying them by γ^0 from the right, we obtain the following coefficient functions of the Dirac-conjugated field:

$$\begin{aligned}\overline{u}(\boldsymbol{p}, s_z) &\equiv u^\dagger(\boldsymbol{p}, s_z)\gamma^0 = u^\dagger(\mathbf{0}, s_z)\mathcal{D}^\dagger(\boldsymbol{\theta_p})\gamma^0 = u^\dagger(\mathbf{0}, s_z)\gamma^0\gamma^0\mathcal{D}(\boldsymbol{\theta_p})\gamma^0 \\ &= u^\dagger(\mathbf{0}, s_z)\gamma^0\mathcal{D}^{-1}(\boldsymbol{\theta_p}) = \overline{u}(\mathbf{0}, s_z)\mathcal{D}^{-1}(\boldsymbol{\theta_p}), \\ \overline{v}(\boldsymbol{p}, s_z) &= \overline{v}(\mathbf{0}, s_z)\mathcal{D}^{-1}(\boldsymbol{\theta_p}).\end{aligned} \tag{B.44}$$

B.6 Explicit formulas for *u* and *v*

Next we are going to find explicit expressions for the factors u, v, $\overline{u}$ and $\overline{v}$ as functions of the momentum. Using equations (B.27), (B.23) and denoting $w = c\tanh\theta_{\boldsymbol{p}}$ – the speed of the standard boost $\theta_{\boldsymbol{p}}$ – we obtain

$$\theta_{\boldsymbol{p}} = \tanh^{-1}(w/c),$$

$$\tanh(\theta_{\boldsymbol{p}}/2) = \frac{\tanh\theta_{\boldsymbol{p}}}{1+\sqrt{1-\tanh^2\theta_{\boldsymbol{p}}}} = \frac{w/c}{1+\sqrt{1-w^2/c^2}} = \frac{pc}{\omega_{\boldsymbol{p}}+mc^2},$$

$$\cosh(\theta_{\boldsymbol{p}}/2) = \frac{1}{\sqrt{1-\tanh^2(\theta_{\boldsymbol{p}}/2)}} = \sqrt{\frac{\omega_{\boldsymbol{p}}+m_ec^2}{2m_ec^2}},$$

$$\sinh(\theta_{\boldsymbol{p}}/2) = \tanh(\theta_{\boldsymbol{p}}/2)\cosh(\theta_{\boldsymbol{p}}/2),$$

$$\begin{aligned}
\mathcal{D}(\boldsymbol{\theta}_{\boldsymbol{p}}) = e^{-\frac{ic}{\hbar}(\boldsymbol{\mathfrak{K}}\cdot\frac{\boldsymbol{p}}{p})\theta_{\boldsymbol{p}}} &= I\cosh(\theta_{\boldsymbol{p}}/2) + \frac{2c}{i\hbar}\frac{\boldsymbol{\mathfrak{K}}\cdot\boldsymbol{p}}{p}\sinh(\theta_{\boldsymbol{p}}/2) \\
&= \cosh(\theta_{\boldsymbol{p}}/2)\begin{bmatrix} 1 & 0 \\ 0 & 1 \end{bmatrix} + \sinh(\theta_{\boldsymbol{p}}/2)\begin{bmatrix} 0 & \frac{\boldsymbol{\sigma}\cdot\boldsymbol{p}}{p} \\ \frac{\boldsymbol{\sigma}\cdot\boldsymbol{p}}{p} & 0 \end{bmatrix} \\
&= \cosh(\theta_{\boldsymbol{p}}/2)\left(1+\tanh(\theta_{\boldsymbol{p}}/2)\begin{bmatrix} 0 & \frac{\boldsymbol{\sigma}\cdot\boldsymbol{p}}{p} \\ \frac{\boldsymbol{\sigma}\cdot\boldsymbol{p}}{p} & 0 \end{bmatrix}\right) \\
&= \sqrt{\frac{\omega_{\boldsymbol{p}}+m_ec^2}{2m_ec^2}}\left(1+\frac{pc}{\omega_{\boldsymbol{p}}+m_ec^2}\begin{bmatrix} 0 & \frac{\boldsymbol{\sigma}\cdot\boldsymbol{p}}{p} \\ \frac{\boldsymbol{\sigma}\cdot\boldsymbol{p}}{p} & 0 \end{bmatrix}\right) \\
&= \sqrt{\frac{\omega_{\boldsymbol{p}}+m_ec^2}{2m_ec^2}}\begin{bmatrix} 1 & \frac{\boldsymbol{\sigma}\cdot\boldsymbol{p}c}{\omega_{\boldsymbol{p}}+m_ec^2} \\ \frac{\boldsymbol{\sigma}\cdot\boldsymbol{p}c}{\omega_{\boldsymbol{p}}+m_ec^2} & 1 \end{bmatrix}.
\end{aligned}$$

Substituting this result in (B.42), we have

$$\begin{aligned}
u(\boldsymbol{p},s_z) &= \sqrt{\frac{\omega_{\boldsymbol{p}}+m_ec^2}{2m_ec^2}}\begin{bmatrix} 1 & \frac{\boldsymbol{\sigma}\cdot\boldsymbol{p}c}{\omega_{\boldsymbol{p}}+m_ec^2} \\ \frac{\boldsymbol{\sigma}\cdot\boldsymbol{p}c}{\omega_{\boldsymbol{p}}+m_ec^2} & 1 \end{bmatrix}\begin{bmatrix} \chi_{s_z} \\ 0 \end{bmatrix} \\
&= \begin{bmatrix} \sqrt{\omega_{\boldsymbol{p}}+m_ec^2} \\ \sqrt{\omega_{\boldsymbol{p}}-m_ec^2}(\boldsymbol{\sigma}\cdot\frac{\boldsymbol{p}}{p}) \end{bmatrix}\frac{\chi_{s_z}}{\sqrt{2m_ec^2}}.
\end{aligned} \tag{B.45}$$

Similar calculations yield explicit expressions for v, $\overline{u}$ and $\overline{v}$, i. e.,

$$v(\boldsymbol{p},s_z) = \begin{bmatrix} \sqrt{\omega_{\boldsymbol{p}}-m_ec^2}(\boldsymbol{\sigma}\cdot\frac{\boldsymbol{p}}{p}) \\ \sqrt{\omega_{\boldsymbol{p}}+m_ec^2} \end{bmatrix}\frac{\chi_{-s_z}}{\sqrt{2m_ec^2}}, \tag{B.46}$$

$$\overline{u}(\boldsymbol{p},s_z) = \frac{\chi_{s_z}^\dagger}{\sqrt{2m_ec^2}}\left[\sqrt{\omega_{\boldsymbol{p}}+m_ec^2}, -\sqrt{\omega_{\boldsymbol{p}}-m_ec^2}\left(\boldsymbol{\sigma}\cdot\frac{\boldsymbol{p}}{p}\right)\right], \tag{B.47}$$

$$\overline{v}(\boldsymbol{p},s_z) = \frac{\chi_{-s_z}^\dagger}{\sqrt{2m_ec^2}}\left[\sqrt{\omega_{\boldsymbol{p}}-m_ec^2}\left(\boldsymbol{\sigma}\cdot\frac{\boldsymbol{p}}{p}\right), \sqrt{\omega_{\boldsymbol{p}}+m_ec^2}\right]. \tag{B.48}$$

These functions are normalized to unity in the sense that (here we used (B.4))

$$\overline{u}(\boldsymbol{p},s_z)u(\boldsymbol{p},s_z') = \chi_{s_z}^\dagger\left[\sqrt{\omega_{\boldsymbol{p}}+m_ec^2}, -\sqrt{\omega_{\boldsymbol{p}}-m_ec^2}\left(\frac{\boldsymbol{p}}{p}\cdot\boldsymbol{\sigma}\right)\right]$$

$$
\begin{aligned}
&\times \begin{bmatrix} \sqrt{\omega_{\boldsymbol{p}} + m_e c^2} \\ \sqrt{\omega_{\boldsymbol{p}} - m_e c^2}(\frac{\boldsymbol{p}}{p} \cdot \boldsymbol{\sigma}) \end{bmatrix} \chi_{s_z'} \frac{1}{2m_e c^2} \\
&= \chi_{s_z}^\dagger \left(\omega_{\boldsymbol{p}} + m_e c^2 - (\omega_{\boldsymbol{p}} - m_e c^2) \frac{(\boldsymbol{p} \cdot \boldsymbol{\sigma})(\boldsymbol{p} \cdot \boldsymbol{\sigma})}{p^2} \right) \chi_{s_z'} \frac{1}{2m_e c^2} \\
&= \chi_{s_z}^\dagger \chi_{s_z'} = \delta_{s_z s_z'}. \qquad (B.49)
\end{aligned}
$$

We also need to calculate the sum $\sum_{s_z=-1/2}^{1/2} u(\boldsymbol{p}, s_z) u^\dagger(\boldsymbol{p}, s_z)$. At zero momentum, we can use the explicit representation (B.38)

$$
\begin{aligned}
\sum_{s_z-1/2}^{1/2} u(\mathbf{0}, s_z) u^\dagger(\mathbf{0}, s_z) &= \begin{bmatrix} 0 & 0 & 0 & 0 \\ 0 & 1 & 0 & 0 \\ 0 & 0 & 0 & 0 \\ 0 & 0 & 0 & 0 \end{bmatrix} + \begin{bmatrix} 1 & 0 & 0 & 0 \\ 0 & 0 & 0 & 0 \\ 0 & 0 & 0 & 0 \\ 0 & 0 & 0 & 0 \end{bmatrix} \\
&= \begin{bmatrix} 1 & 0 & 0 & 0 \\ 0 & 1 & 0 & 0 \\ 0 & 0 & 0 & 0 \\ 0 & 0 & 0 & 0 \end{bmatrix} = \frac{1}{2}(1 + \gamma^0).
\end{aligned}
$$

For arbitrary momenta, we apply (B.42), (B.44), use the Hermiticity of the matrix $\mathcal{D}(\boldsymbol{\theta_p})$ and properties (B.9), (B.31)–(B.33) to obtain

$$
\begin{aligned}
\sum_{s_z=-1/2}^{1/2} u(\boldsymbol{p}, s_z) u^\dagger(\boldsymbol{p}, s_z) &= \mathcal{D}(\boldsymbol{\theta_p}) \left(\sum_{s_z=-1/2}^{1/2} u(\mathbf{0}, s_z) u^\dagger(\mathbf{0}, s_z) \right) \mathcal{D}^\dagger(\boldsymbol{\theta_p}) \\
&= \frac{1}{2} \mathcal{D}(\boldsymbol{\theta_p})(1 + \gamma^0) \mathcal{D}(\boldsymbol{\theta_p}) = \frac{1}{2}(\mathcal{D}(\boldsymbol{\theta_p}) \mathcal{D}(\boldsymbol{\theta_p}) + \gamma^0) \\
&= \frac{1}{2}(\mathcal{D}(\boldsymbol{\theta_p}) \gamma^0 \gamma^0 \mathcal{D}(\boldsymbol{\theta_p}) \gamma^0 \gamma^0 + \gamma^0) = \frac{1}{2}(\mathcal{D}(\boldsymbol{\theta_p}) \gamma^0 \mathcal{D}^{-1}(\boldsymbol{\theta_p}) + 1) \gamma^0 \\
&= \frac{1}{2}\left(\gamma^0 \cosh \theta_{\boldsymbol{p}} - \boldsymbol{\gamma} \cdot \frac{\boldsymbol{\theta_p}}{\theta_{\boldsymbol{p}}} \sinh \theta_{\boldsymbol{p}} + 1 \right) \gamma^0 \\
&= \frac{1}{2m_e c^2} (\gamma^0 \omega_{\boldsymbol{p}} - \boldsymbol{\gamma} \cdot \boldsymbol{p} c + m_e c^2) \gamma^0 \\
&= \frac{1}{2m_e c^2} (\not{p} + m_e c^2) \gamma^0. \qquad (B.50)
\end{aligned}
$$

Similarly, we obtain

$$
\sum_{s_z=-1/2}^{1/2} v(\boldsymbol{p}, s_z) v^\dagger(\boldsymbol{p}, s_z) = \frac{1}{2m_e c^2} (\not{p} - m_e c^2) \gamma^0. \qquad (B.51)
$$

B.7 Useful notation

To simplify our calculations we will introduce the following combinations of particle operators:

$$A_a(\boldsymbol{p}) \equiv \sqrt{\frac{m_e c^2}{\omega_{\boldsymbol{p}}}} \sum_{s_z} u_a(\boldsymbol{p}, s_z) a_{\boldsymbol{p}s_z}, \tag{B.52}$$

$$\overline{A}_a^\dagger(\boldsymbol{p}) = \sqrt{\frac{m_e c^2}{\omega_{\boldsymbol{p}}}} \sum_{s_z} \overline{u}_a(\boldsymbol{p}, s_z) a_{\boldsymbol{p}s_z}^\dagger, \tag{B.53}$$

$$B_a^\dagger(\boldsymbol{p}) \equiv \sqrt{\frac{m_e c^2}{\omega_{\boldsymbol{p}}}} \sum_{s_z} v_a(\boldsymbol{p}, s_z) b_{\boldsymbol{p}s_z}^\dagger, \tag{B.54}$$

$$\overline{B}_a(\boldsymbol{p}) \equiv \sqrt{\frac{m_e c^2}{\omega_{\boldsymbol{p}}}} \sum_{s_z} \overline{v}_a(\boldsymbol{p}, s_z) b_{\boldsymbol{p}s_z}, \tag{B.55}$$

$$D_a(\boldsymbol{p}) \equiv \sqrt{\frac{m_p c^2}{\Omega_{\boldsymbol{p}}}} \sum_{s_z} w_a(\boldsymbol{p}, s_z) d_{\boldsymbol{p}s_z}, \tag{B.56}$$

$$\overline{D}_a^\dagger(\boldsymbol{p}) \equiv \sqrt{\frac{m_p c^2}{\Omega_{\boldsymbol{p}}}} \sum_{s_z} \overline{w}_a(\boldsymbol{p}, s_z) d_{\boldsymbol{p}s_z}^\dagger, \tag{B.57}$$

$$F_a^\dagger(\boldsymbol{p}) \equiv \sqrt{\frac{m_p c^2}{\Omega_{\boldsymbol{p}}}} \sum_{s_z} y_a(\boldsymbol{p}, s_z) f_{\boldsymbol{p}s_z}^\dagger, \tag{B.58}$$

$$\overline{F}_a(\boldsymbol{p}) \equiv \sqrt{\frac{m_p c^2}{\Omega_{\boldsymbol{p}}}} \sum_{s_z} \overline{y}_a(\boldsymbol{p}, s_z) f_{\boldsymbol{p}s_z}. \tag{B.59}$$

In this notation the electron–positron and proton–antiproton fields look a bit more compact, i. e.,

$$\psi_a(\tilde{x}) = \frac{1}{(2\pi\hbar)^{3/2}} \int d\boldsymbol{p}[e^{-\frac{i}{\hbar}\tilde{p}\cdot\tilde{x}} A_a(\boldsymbol{p}) + e^{\frac{i}{\hbar}\tilde{p}\cdot\tilde{x}} B_a^\dagger(\boldsymbol{p})], \tag{B.60}$$

$$\overline{\psi}_a(\tilde{x}) = \frac{1}{(2\pi\hbar)^{3/2}} \int d\boldsymbol{p}[e^{\frac{i}{\hbar}\tilde{p}\cdot\tilde{x}} \overline{A}_a^\dagger(\boldsymbol{p}) + e^{-\frac{i}{\hbar}\tilde{p}\cdot\tilde{x}} \overline{B}_a(\boldsymbol{p})], \tag{B.61}$$

$$\Psi_a(\tilde{x}) = \frac{1}{(2\pi\hbar)^{3/2}} \int d\boldsymbol{p}[e^{-\frac{i}{\hbar}\tilde{P}\cdot\tilde{x}} D_a(\boldsymbol{p}) + e^{\frac{i}{\hbar}\tilde{P}\cdot\tilde{x}} F_a^\dagger(\boldsymbol{p})], \tag{B.62}$$

$$\overline{\Psi}_a(\tilde{x}) = \frac{1}{(2\pi\hbar)^{3/2}} \int d\boldsymbol{p}[e^{\frac{i}{\hbar}\tilde{P}\cdot\tilde{x}} \overline{D}_a^\dagger(\boldsymbol{p}) + e^{-\frac{i}{\hbar}\tilde{P}\cdot\tilde{x}} \overline{F}_a(\boldsymbol{p})]. \tag{B.63}$$

In QED calculations, we often find products like $\overline{u}\gamma^\mu u$ and $\overline{w}\gamma^\mu w$. It will be convenient to introduce special symbols for them:

$$\mathcal{U}^\mu(\boldsymbol{p}'s_z', \boldsymbol{p}s_z) \equiv \overline{u}(\boldsymbol{p}', s_z')\gamma^\mu u(\boldsymbol{p}, s_z), \tag{B.64}$$

$$\mathcal{W}^\mu(\boldsymbol{p}'s_z', \boldsymbol{p}s_z) \equiv \overline{w}(\boldsymbol{p}', s_z')\gamma^\mu w(\boldsymbol{p}, s_z). \tag{B.65}$$

B.8 Poincaré transformations of fields

Operators (B.52)–(B.59) have simple transformation properties with respect to the noninteracting representation of boosts in the Fock space. For example, using (1.42), (1-5.18), (B.40) and (B.42), we obtain

$$\begin{aligned}
e^{-\frac{ic}{\hbar}\boldsymbol{K}_0\cdot\boldsymbol{\theta}}A(\boldsymbol{p})e^{\frac{ic}{\hbar}\boldsymbol{K}_0\cdot\boldsymbol{\theta}} &= \sqrt{\frac{m_ec^2}{\omega_{\boldsymbol{p}}}}\sum_{s_z}u(\boldsymbol{p},s_z)e^{-\frac{ic}{\hbar}\boldsymbol{K}_0\cdot\boldsymbol{\theta}}a_{\boldsymbol{p}s_z}e^{\frac{ic}{\hbar}\boldsymbol{K}_0\cdot\boldsymbol{\theta}} \\
&= \frac{\sqrt{m_ec^2\omega_{\boldsymbol{\theta p}}}}{\omega_{\boldsymbol{p}}}\sum_{s_z}u(\boldsymbol{p},s_z)\sum_{s'_z}\mathscr{D}^{1/2}_{s_zs'_z}(-\boldsymbol{\varphi}_W(\boldsymbol{p},\boldsymbol{\theta}))a_{(\boldsymbol{\theta p})s'_z} \\
&= \frac{\sqrt{m_ec^2\omega_{\boldsymbol{\theta p}}}}{\omega_{\boldsymbol{p}}}\mathcal{D}(\boldsymbol{\theta}_{\boldsymbol{p}})\sum_{s_z}u(\boldsymbol{0},s_z)\sum_{s'_z}\mathscr{D}^{1/2}_{s_zs'_z}(-\boldsymbol{\varphi}_W(\boldsymbol{p},\boldsymbol{\theta}))a_{(\boldsymbol{\theta p})s'_z} \\
&= \frac{\sqrt{m_ec^2\omega_{\boldsymbol{\theta p}}}}{\omega_{\boldsymbol{p}}}\mathcal{D}(\boldsymbol{\theta}_{\boldsymbol{p}})\mathcal{D}(-\boldsymbol{\varphi}_W(\boldsymbol{p},\boldsymbol{\theta}))\sum_{s_z}u(\boldsymbol{0},s_z)a_{(\boldsymbol{\theta p})s_z} \\
&= \frac{\sqrt{m_ec^2\omega_{\boldsymbol{\theta p}}}}{\omega_{\boldsymbol{p}}}\mathcal{D}(\boldsymbol{\theta}_{\boldsymbol{p}})\mathcal{D}(\boldsymbol{\theta}_{\boldsymbol{p}}^{-1}\circ\boldsymbol{\theta}^{-1}\circ\boldsymbol{\theta}_{\boldsymbol{\theta p}})\sum_{s_z}u(\boldsymbol{0},s_z)a_{(\boldsymbol{\theta p})s_z} \\
&= \frac{\sqrt{m_ec^2\omega_{\boldsymbol{\theta p}}}}{\omega_{\boldsymbol{p}}}\mathcal{D}(\boldsymbol{\theta}^{-1})\mathcal{D}(\boldsymbol{\theta}_{\boldsymbol{\theta p}})\sum_{s_z}u(\boldsymbol{0},s_z)a_{(\boldsymbol{\theta p})s_z} \\
&= \frac{\sqrt{m_ec^2\omega_{\boldsymbol{\theta p}}}}{\omega_{\boldsymbol{p}}}\mathcal{D}(\boldsymbol{\theta}^{-1})\sum_{s_z}u(\boldsymbol{\theta p},s_z)a_{(\boldsymbol{\theta p})s_z} \\
&= \frac{\omega_{\boldsymbol{\theta p}}}{\omega_{\boldsymbol{p}}}\mathcal{D}(\boldsymbol{\theta}^{-1})A(\boldsymbol{\theta p}).
\end{aligned} \tag{B.66}$$

From here, with the help of Hermitian conjugation and (B.28) it follows that

$$e^{-\frac{ic}{\hbar}\boldsymbol{K}_0\cdot\boldsymbol{\theta}}A^\dagger(\boldsymbol{p})e^{\frac{ic}{\hbar}\boldsymbol{K}_0\cdot\boldsymbol{\theta}} = \frac{\omega_{\boldsymbol{\theta p}}}{\omega_{\boldsymbol{p}}}\mathcal{D}^\dagger(\boldsymbol{\theta}^{-1})A^\dagger(\boldsymbol{\theta p}) = \frac{\omega_{\boldsymbol{\theta p}}}{\omega_{\boldsymbol{p}}}\mathcal{D}(\boldsymbol{\theta}^{-1})A^\dagger(\boldsymbol{\theta p}), \tag{B.67}$$

$$e^{-\frac{ic}{\hbar}\boldsymbol{K}_0\cdot\boldsymbol{\theta}}B^\dagger(\boldsymbol{p})e^{\frac{ic}{\hbar}\boldsymbol{K}_0\cdot\boldsymbol{\theta}} = \frac{\omega_{\boldsymbol{\theta p}}}{\omega_{\boldsymbol{p}}}\mathcal{D}(\boldsymbol{\theta}^{-1})B^\dagger(\boldsymbol{\theta p}). \tag{B.68}$$

Let us now verify that the quantum field $\psi_a(\tilde{x})$ has the required covariant transformation law (3.1)

$$U_0(\Lambda;\tilde{a})\psi_a(\tilde{x})U_0^{-1}(\Lambda;\tilde{a}) = \sum_j \mathcal{D}_{ab}(\Lambda^{-1})\psi_b(\Lambda(\tilde{x}+\tilde{a})). \tag{B.69}$$

Transformations with respect to translations are obtained from equations (1.45), (1.46) for the creation and annihilation operators. We have

$$
\begin{aligned}
& e^{-\frac{i}{\hbar}\mathbf{P}_0\cdot\mathbf{r}} e^{\frac{i}{\hbar}H_0 t}\psi_a(\tilde{x}) e^{-\frac{i}{\hbar}H_0 t} e^{\frac{i}{\hbar}\mathbf{P}_0\cdot\mathbf{r}} \\
&= \int \frac{d\mathbf{p}}{(2\pi\hbar)^{3/2}} (e^{-\frac{i}{\hbar}\tilde{p}\cdot\tilde{x}} e^{-\frac{i}{\hbar}\mathbf{P}_0\cdot\mathbf{r}} e^{\frac{i}{\hbar}H_0 t} A_a(\mathbf{p}) e^{\frac{i}{\hbar}\mathbf{P}_0\cdot\mathbf{r}} e^{-\frac{i}{\hbar}H_0 t} \\
&+ e^{\frac{i}{\hbar}\tilde{p}\cdot\tilde{x}} e^{-\frac{i}{\hbar}\mathbf{P}_0\cdot\mathbf{r}} e^{\frac{i}{\hbar}H_0 t} B_a^\dagger(\mathbf{p}) e^{\frac{i}{\hbar}\mathbf{P}_0\cdot\mathbf{r}} e^{-\frac{i}{\hbar}H_0 t}) \\
&= \int \frac{d\mathbf{p}}{(2\pi\hbar)^{3/2}} (e^{-\frac{i}{\hbar}\tilde{p}\cdot(\tilde{x}+\tilde{a})} A_a(\mathbf{p}) + e^{\frac{i}{\hbar}\tilde{p}\cdot(\tilde{x}+\tilde{a})} B_a^\dagger(\mathbf{p})) = \psi_a(\tilde{x}+\tilde{a}),
\end{aligned}
$$

where $\tilde{a} = (t, \mathbf{r}/c)$ is an arbitrary translation 4-vector. For boosts we use equations (B.66), (B.68), (**1**-5.31) and (**1**-J.6) to obtain

$$
\begin{aligned}
e^{-\frac{ic}{\hbar}\mathbf{K}_0\cdot\boldsymbol{\theta}}\psi(\tilde{x}) e^{\frac{ic}{\hbar}\mathbf{K}_0\cdot\boldsymbol{\theta}} &= \int \frac{d\mathbf{p}}{(2\pi\hbar)^{3/2}} (e^{-\frac{i}{\hbar}\tilde{p}\cdot\tilde{x}} e^{-\frac{ic}{\hbar}\mathbf{K}_0\cdot\boldsymbol{\theta}} A(\mathbf{p}) e^{\frac{ic}{\hbar}\mathbf{K}_0\cdot\boldsymbol{\theta}} + e^{\frac{i}{\hbar}\tilde{p}\cdot\tilde{x}} e^{-\frac{ic}{\hbar}\mathbf{K}_0\cdot\boldsymbol{\theta}} B^\dagger(\mathbf{p}) e^{\frac{ic}{\hbar}\mathbf{K}_0\cdot\boldsymbol{\theta}}) \\
&= \mathcal{D}(\boldsymbol{\theta}^{-1}) \int \frac{d\mathbf{p}}{(2\pi\hbar)^{3/2}} \frac{\omega_{\boldsymbol{\theta}\mathbf{p}}}{\omega_{\mathbf{p}}} (e^{-\frac{i}{\hbar}\tilde{p}\cdot\tilde{x}} A(\boldsymbol{\theta}\mathbf{p}) + e^{\frac{i}{\hbar}\tilde{p}\cdot\tilde{x}} B^\dagger(\boldsymbol{\theta}\mathbf{p})) \\
&= \mathcal{D}(\boldsymbol{\theta}^{-1}) \int \frac{d\mathbf{q}}{(2\pi\hbar)^{3/2}} (e^{-\frac{i}{\hbar}\tilde{\boldsymbol{\theta}}^{-1}\tilde{q}\cdot\tilde{x}} A(\mathbf{q}) + e^{\frac{i}{\hbar}\tilde{\boldsymbol{\theta}}^{-1}\tilde{q}\cdot\tilde{x}} B^\dagger(\mathbf{q})) \\
&= \mathcal{D}(\boldsymbol{\theta}^{-1}) \int \frac{d\mathbf{q}}{(2\pi\hbar)^{3/2}} (e^{-\frac{i}{\hbar}\tilde{q}\cdot\tilde{\boldsymbol{\theta}}\tilde{x}} A(\mathbf{q}) + e^{\frac{i}{\hbar}\tilde{q}\cdot\tilde{\boldsymbol{\theta}}\tilde{x}} B^\dagger(\mathbf{q})) \\
&= \mathcal{D}(\boldsymbol{\theta}^{-1})\psi(\tilde{\boldsymbol{\theta}}\tilde{x}),
\end{aligned} \tag{B.70}
$$

where $\tilde{\boldsymbol{\theta}}$ is the 4×4 boost matrix (**1**-J.10). The verification of (B.69) in the case of rotations is left as an exercise for the reader.

B.9 Approximation $(v/c)^2$

We will often make QED calculations in weakly relativistic or nonrelativistic approximations, when particle velocities are much lower than the speed of light and their momenta are smaller than $m_e c$. In these cases, with reasonable accuracy, one can expand all quantities as series in powers of (v/c) and leave only quadratic $\propto (v/c)^2$ and lower-order terms. Then with the help of (2.21) we obtain

$$
\begin{aligned}
\sqrt{\omega_{\mathbf{p}} + m_e c^2} &\approx \sqrt{m_e c^2 + \frac{p^2}{2m_e} + m_e c^2} = \sqrt{2m_e c^2 + \frac{p^2}{2m_e}} \\
&= \sqrt{2m_e c^2}\sqrt{1 + \frac{p^2}{4m_e^2 c^2}} \approx \sqrt{2m_e c^2}\left(1 + \frac{p^2}{8m_e^2 c^2}\right),
\end{aligned} \tag{B.71}
$$

$$
\sqrt{\omega_{\mathbf{p}} - m_e c^2} \approx \sqrt{m_e c^2 + \frac{p^2}{2m_e} - m_e c^2} = \frac{p}{\sqrt{2m_e}}, \tag{B.72}
$$

$$
(\omega_{\mathbf{q}+\mathbf{k}} - \omega_{\mathbf{q}})^2 - c^2k^2 \approx -c^2k^2, \tag{B.73}
$$

$$
\frac{m_p m_e c^4}{\sqrt{\Omega_{\mathbf{p}-\mathbf{k}}\Omega_{\mathbf{p}}\omega_{\mathbf{q}+\mathbf{k}}\omega_{\mathbf{q}}}} \approx \frac{1}{\sqrt{1 + \frac{(\mathbf{p}-\mathbf{k})^2}{2m_p^2c^2}}} \frac{1}{\sqrt{1 + \frac{p^2}{2m_p^2c^2}}} \frac{1}{\sqrt{1 + \frac{(\mathbf{q}+\mathbf{k})^2}{2m_e^2c^2}}} \frac{1}{\sqrt{1 + \frac{q^2}{2m_e^2c^2}}}
$$

$$\approx 1 - \frac{(\boldsymbol{p}-\boldsymbol{k})^2}{4m_p^2c^2} - \frac{p^2}{4m_p^2c^2} - \frac{(\boldsymbol{q}+\boldsymbol{k})^2}{4m_e^2c^2} - \frac{q^2}{4m_e^2c^2}$$
$$= 1 - \frac{p^2}{2m_p^2c^2} + \frac{\boldsymbol{p}\cdot\boldsymbol{k}}{2m_p^2c^2} - \frac{k^2}{4m_p^2c^2} - \frac{q^2}{2m_e^2c^2} - \frac{\boldsymbol{q}\cdot\boldsymbol{k}}{2m_e^2c^2} - \frac{k^2}{4m_e^2c^2}. \tag{B.74}$$

In order to get the $(v/c)^2$ approximation for equations (B.64) and (B.65), we use formulas (B.71), (B.72), (B.45)–(B.48) and (B.2)–(B.4). We have

$$\begin{aligned}
\mathcal{U}^0(\boldsymbol{p}s_z, \boldsymbol{p}'s_z') &\equiv \overline{u}(\boldsymbol{p}, s_z)\gamma^0 u(\boldsymbol{p}', s_z') \\
&= \chi_{s_z}^\dagger \left[\sqrt{\omega_{\boldsymbol{p}} + m_ec^2}, \sqrt{\omega_{\boldsymbol{p}} - m_ec^2}\left(\frac{\boldsymbol{p}}{p}\cdot\boldsymbol{\sigma}\right)\right] \left[\begin{array}{c} \sqrt{\omega_{\boldsymbol{p}'} + m_ec^2} \\ \sqrt{\omega_{\boldsymbol{p}'} - m_ec^2}(\frac{\boldsymbol{p}'}{p'}\cdot\boldsymbol{\sigma}) \end{array}\right] \frac{\chi_{s_z'}}{2m_ec^2} \\
&= \chi_{s_z}^\dagger \Big(\sqrt{\omega_{\boldsymbol{p}} + m_ec^2}\sqrt{\omega_{\boldsymbol{p}'} + m_ec^2} \\
&\quad + \sqrt{\omega_{\boldsymbol{p}} - m_ec^2}\sqrt{\omega_{\boldsymbol{p}'} - m_ec^2}\frac{(\boldsymbol{p}\cdot\boldsymbol{\sigma})(\boldsymbol{p}'\cdot\boldsymbol{\sigma})}{pp'}\Big)\frac{\chi_{s_z'}}{2m_ec^2} \\
&\approx \chi_{s_z}^\dagger \left(\left(1 + \frac{p^2}{8m_e^2c^2}\right)\left(1 + \frac{(p')^2}{8m_e^2c^2}\right) + \frac{pp'}{4m_e^2c^2}\frac{(\boldsymbol{p}\cdot\boldsymbol{\sigma})(\boldsymbol{p}'\cdot\boldsymbol{\sigma})}{pp'}\right)\chi_{s_z'} \\
&= \chi_{s_z}^\dagger \left(1 + \frac{(\boldsymbol{p}+\boldsymbol{p}')^2 + 2i\boldsymbol{\sigma}\cdot[\boldsymbol{p}\times\boldsymbol{p}']}{8m_e^2c^2}\right)\chi_{s_z'},
\end{aligned} \tag{B.75}$$

$$\mathcal{W}^0(\boldsymbol{p}s_z, \boldsymbol{p}'s_z') \equiv \overline{w}(\boldsymbol{p}, s_z)\gamma^0 w(\boldsymbol{p}', s_z') \approx \chi_{s_z}^\dagger \left(1 + \frac{(\boldsymbol{p}+\boldsymbol{p}')^2 + 2i\boldsymbol{\sigma}\cdot[\boldsymbol{p}\times\boldsymbol{p}']}{8m_p^2c^2}\right)\chi_{s_z'}, \tag{B.76}$$

$$\begin{aligned}
\boldsymbol{\mathcal{U}}(\boldsymbol{p}s_z, \boldsymbol{p}'s_z') &\equiv \overline{u}(\boldsymbol{p}, s_z)\boldsymbol{\gamma} u(\boldsymbol{p}', s_z') \\
&= \chi_{s_z}^\dagger \left[\sqrt{\omega_{\boldsymbol{p}} + m_ec^2}, -\sqrt{\omega_{\boldsymbol{p}} - m_ec^2}\left(\frac{\boldsymbol{p}}{p}\cdot\boldsymbol{\sigma}\right)\right] \left[\begin{array}{cc} 0 & \boldsymbol{\sigma} \\ -\boldsymbol{\sigma} & 0 \end{array}\right] \\
&\quad \times \left[\begin{array}{c} \sqrt{\omega_{\boldsymbol{p}'} + m_ec^2} \\ \sqrt{\omega_{\boldsymbol{p}'} - m_ec^2}(\frac{\boldsymbol{p}'}{p'}\cdot\boldsymbol{\sigma}) \end{array}\right] \frac{\chi_{s_z'}}{2m_ec^2} \\
&= \chi_{s_z}^\dagger \left[\sqrt{\omega_{\boldsymbol{p}} + m_ec^2}, -\sqrt{\omega_{\boldsymbol{p}} - m_ec^2}\left(\frac{\boldsymbol{p}}{p}\cdot\boldsymbol{\sigma}\right)\right] \left[\begin{array}{c} \sqrt{\omega_{\boldsymbol{p}'} - m_ec^2}\frac{\boldsymbol{\sigma}(\boldsymbol{p}'\cdot\boldsymbol{\sigma})}{p'} \\ -\boldsymbol{\sigma}\sqrt{\omega_{\boldsymbol{p}'} + m_ec^2} \end{array}\right] \frac{\chi_{s_z'}}{2m_ec^2} \\
&= \chi_{s_z}^\dagger \Big(\sqrt{\omega_{\boldsymbol{p}} + m_ec^2}\sqrt{\omega_{\boldsymbol{p}'} - m_ec^2}\frac{\boldsymbol{\sigma}(\boldsymbol{p}'\cdot\boldsymbol{\sigma})}{p'} \\
&\quad + \sqrt{\omega_{\boldsymbol{p}} - m_ec^2}\sqrt{\omega_{\boldsymbol{p}'} + m_ec^2}\frac{(\boldsymbol{p}\cdot\boldsymbol{\sigma})\boldsymbol{\sigma}}{p}\Big)\frac{\chi_{s_z'}}{2m_ec^2} \\
&\approx \chi_{s_z}^\dagger \left(\sqrt{2m_ec^2}\frac{p'}{\sqrt{2m_e}}\frac{\boldsymbol{\sigma}(\boldsymbol{p}'\cdot\boldsymbol{\sigma})}{p'} + \sqrt{2m_ec^2}\frac{p}{\sqrt{2m_e}}\frac{(\boldsymbol{p}\cdot\boldsymbol{\sigma})\boldsymbol{\sigma}}{p}\right)\frac{\chi_{s_z'}}{2m_ec^2}
\end{aligned}$$

$$= \chi_{s_z}^\dagger((\boldsymbol{\sigma}\cdot\boldsymbol{p})\boldsymbol{\sigma} + \boldsymbol{\sigma}(\boldsymbol{\sigma}\cdot\boldsymbol{p}'))\frac{\chi_{s_z'}}{2m_e c}$$

$$= \chi_{s_z}^\dagger(\boldsymbol{p}+\boldsymbol{p}' + i[\boldsymbol{\sigma}\times(\boldsymbol{p}-\boldsymbol{p}')])\frac{\chi_{s_z'}}{2m_e c}, \tag{B.77}$$

$$\mathcal{W}(\boldsymbol{p}s_z,\boldsymbol{p}'s_z') \approx \chi_{s_z}^\dagger(\boldsymbol{p}+\boldsymbol{p}' + i[\boldsymbol{\sigma}\times(\boldsymbol{p}-\boldsymbol{p}')])\frac{\chi_{s_z'}}{2m_p c}. \tag{B.78}$$

In the extreme nonrelativistic limit ($c \to \infty$), all formulas simplify even more. Then we have

$$\lim_{c\to\infty}\omega_{\boldsymbol{p}} = m_e c^2,$$

$$\lim_{c\to\infty}\Omega_{\boldsymbol{p}} = m_p c^2,$$

$$\lim_{c\to\infty}\frac{m_p m_e c^4}{\sqrt{\Omega_{\boldsymbol{p}'}\omega_{\boldsymbol{q}'}\Omega_{\boldsymbol{p}}\omega_{\boldsymbol{q}}}} = 1, \tag{B.79}$$

$$\lim_{c\to\infty}\mathcal{U}_0(\boldsymbol{p}s_z,\boldsymbol{p}'s_z') = \chi_{s_z}^\dagger\chi_{s_z'} = \delta_{s_z s_z'}, \tag{B.80}$$

$$\lim_{c\to\infty}\mathcal{W}_0(\boldsymbol{p}s_z,\boldsymbol{p}'s_z') = \delta_{s_z s_z'}, \tag{B.81}$$

$$\lim_{c\to\infty}\mathcal{U}(\boldsymbol{p}s_z,\boldsymbol{p}'s_z') = 0, \tag{B.82}$$

$$\lim_{c\to\infty}\mathcal{W}(\boldsymbol{p}s_z,\boldsymbol{p}'s_z') = 0. \tag{B.83}$$

B.10 Anticommutation relations

In order to verify anticommutators (3.2), we use (1.24)–(1.25), (B.50)–(B.51) and obtain

$$\{\psi_a(0,\boldsymbol{x}),\psi_b^\dagger(0,\boldsymbol{y})\}$$

$$= \int\frac{d\boldsymbol{p}}{(2\pi\hbar)^{3/2}}\sqrt{\frac{m_e c^2}{\omega_{\boldsymbol{p}}}}\frac{d\boldsymbol{p}'}{(2\pi\hbar)^{3/2}}\sqrt{\frac{m_e c^2}{\omega_{\boldsymbol{p}'}}}$$

$$\times\sum_{s_z s_z'=-1/2}^{1/2}\{(e^{-\frac{i}{\hbar}\boldsymbol{p}\cdot\boldsymbol{x}}u_a(\boldsymbol{p},s_z)a_{\boldsymbol{p}s_z} + e^{\frac{i}{\hbar}\boldsymbol{p}\cdot\boldsymbol{x}}v_a(\boldsymbol{p},s_z)b_{\boldsymbol{p}s_z}^\dagger),$$

$$(e^{\frac{i}{\hbar}\boldsymbol{p}'\cdot\boldsymbol{y}}u_b^\dagger(\boldsymbol{p}',s_z')a_{\boldsymbol{p}'s_z'}^\dagger + e^{-\frac{i}{\hbar}\boldsymbol{p}'\cdot\boldsymbol{y}}v_b^\dagger(\boldsymbol{p}',s_z')b_{\boldsymbol{p}'s_z'})\}$$

$$= \int\frac{d\boldsymbol{p}d\boldsymbol{p}'}{(2\pi\hbar)^3}\frac{m_e c^2}{\sqrt{\omega_{\boldsymbol{p}}\omega_{\boldsymbol{p}'}}}\sum_{s_z s_z=-1/2}^{1/2}(e^{-\frac{i}{\hbar}\boldsymbol{p}\cdot\boldsymbol{x}+\frac{i}{\hbar}\boldsymbol{p}'\cdot\boldsymbol{y}}u_a(\boldsymbol{p},s_z)u_b^\dagger(\boldsymbol{p}',s_z')\{a_{\boldsymbol{p}s_z},a_{\boldsymbol{p}'s_z'}^\dagger\}$$

$$+ e^{\frac{i}{\hbar}\boldsymbol{p}\cdot\boldsymbol{x}-\frac{i}{\hbar}\boldsymbol{p}'\cdot\boldsymbol{y}}v_a(\boldsymbol{p},s_z)v_b^\dagger(\boldsymbol{p}',s_z')\{b_{\boldsymbol{p}s_z}^\dagger,b_{\boldsymbol{p}'s_z'}\})$$

$$= \int\frac{d\boldsymbol{p}d\boldsymbol{p}' m_e c^2}{(2\pi\hbar)^3\omega_{\boldsymbol{p}}}\sum_{s_z s_z'=-1/2}^{1/2}(e^{-\frac{i}{\hbar}\boldsymbol{p}\cdot(\boldsymbol{x}-\boldsymbol{y})}u_a(\boldsymbol{p},s_z)u_b^\dagger(\boldsymbol{p}',s_z')\delta(\boldsymbol{p}-\boldsymbol{p}')\delta_{s_z s_z'}$$

$$+ e^{\frac{i}{\hbar}\boldsymbol{p}\cdot(\boldsymbol{x}-\boldsymbol{y})}v_a(\boldsymbol{p},s_z)v_b^\dagger(\boldsymbol{p}',s_z')\delta(\boldsymbol{p}-\boldsymbol{p}')\delta_{s_z s_z'})$$

$$
\begin{aligned}
&= \int \frac{d\boldsymbol{p} m_e c^2}{(2\pi\hbar)^3 \omega_{\boldsymbol{p}}} \sum_{s_z=-1/2}^{1/2} (e^{-\frac{i}{\hbar}\boldsymbol{p}\cdot(\boldsymbol{x}-\boldsymbol{y})} u_a(\boldsymbol{p}, s_z) u_b^\dagger(\boldsymbol{p}, s_z) + e^{\frac{i}{\hbar}\boldsymbol{p}\cdot(\boldsymbol{x}-\boldsymbol{y})} v_a(\boldsymbol{p}, s_z) v_b^\dagger(\boldsymbol{p}, s_z)) \\
&= \int \frac{d\boldsymbol{p} m_e c^2}{(2\pi\hbar)^3 \omega_{\boldsymbol{p}}} e^{-\frac{i}{\hbar}\boldsymbol{p}\cdot(\boldsymbol{x}-\boldsymbol{y})} \sum_{s_z=-1/2}^{1/2} (u_a(\boldsymbol{p}, s_z) u_b^\dagger(\boldsymbol{p}, s_z) + v_a(-\boldsymbol{p}, s_z) v_b^\dagger(-\boldsymbol{p}, s_z)) \\
&= \int \frac{d\boldsymbol{p} m_e c^2}{(2\pi\hbar)^3 \omega_{\boldsymbol{p}}} \cdot \frac{e^{-\frac{i}{\hbar}\boldsymbol{p}\cdot(\boldsymbol{x}-\boldsymbol{y})}}{2 m_e c^2} [(\gamma^0 \omega_{\boldsymbol{p}} - \boldsymbol{\gamma}\cdot\boldsymbol{p} + m_e c^2)\gamma^0 + (\gamma^0 \omega_{\boldsymbol{p}} + \boldsymbol{\gamma}\cdot\boldsymbol{p} - m_e c^2)\gamma^0] \\
&= \int \frac{d\boldsymbol{p} m_e c^2}{(2\pi\hbar)^3 \omega_{\boldsymbol{p}}} e^{-\frac{i}{\hbar}\boldsymbol{p}\cdot(\boldsymbol{x}-\boldsymbol{y})} \frac{\omega_{\boldsymbol{p}}}{m_e c^2} (\gamma^0 \gamma^0)_{ab} \\
&= \delta(\boldsymbol{x}-\boldsymbol{y})\delta_{ab}.
\end{aligned}
\tag{B.84}
$$

We will also find useful the following anticommutators:

$$
\{\psi_a(0,\boldsymbol{x}), \psi_b(0,\boldsymbol{y})\} = \{\psi_a^\dagger(0,\boldsymbol{x}), \psi_b^\dagger(0,\boldsymbol{y})\} = 0,
\tag{B.85}
$$

$$
\begin{aligned}
\{A_a(\boldsymbol{p}), \overline{A}_b^\dagger(\boldsymbol{p}')\} &= \frac{m_e c^2}{\omega_{\boldsymbol{p}}} \sum_{s_z s_z'} u_a(\boldsymbol{p}, s_z) \overline{u}_b(\boldsymbol{p}', s_z') \{a_{\boldsymbol{p}s_z}, a_{\boldsymbol{p}'s_z'}^\dagger\} \\
&= \frac{m_e c^2}{\omega_{\boldsymbol{p}}} \Big(\sum_{s_z} u_a(\boldsymbol{p}, s_z) u_b^\dagger(\boldsymbol{p}, s_z) \Big) \gamma^0 \delta(\boldsymbol{p}-\boldsymbol{p}') \\
&= \frac{(\gamma^0 \omega_{\boldsymbol{p}} - \boldsymbol{\gamma}\cdot\boldsymbol{p}c + m_e c^2)_{ab}}{2\omega_{\boldsymbol{p}}} \delta(\boldsymbol{p}-\boldsymbol{p}'),
\end{aligned}
\tag{B.86}
$$

$$
\{\overline{B}_a(\boldsymbol{p}), B_b^\dagger(\boldsymbol{p}')\} = \frac{(\gamma^0 \omega_{\boldsymbol{p}} - \boldsymbol{\gamma}\cdot\boldsymbol{p}c - m_e c^2)_{ab}}{2\omega_{\boldsymbol{p}}} \delta(\boldsymbol{p}-\boldsymbol{p}').
\tag{B.87}
$$

B.11 Dirac equation

The electron–positron quantum field (B.34) can be written as a sum of two terms, i. e.,

$$
\begin{aligned}
\psi_a(\tilde{x}) &= \psi_a^-(\tilde{x}) + \psi_a^+(\tilde{x}), \\
\psi_a^-(\tilde{x}) &\equiv \sum_{s_z} \int \frac{d\boldsymbol{p}}{(2\pi\hbar)^{3/2}} \sqrt{\frac{m_e c^2}{\omega_{\boldsymbol{p}}}} e^{-\frac{i}{\hbar}\tilde{p}\cdot\tilde{x}} u_a(\boldsymbol{p}, s_z) a_{\boldsymbol{p}s_z}, \\
\psi_a^+(\tilde{x}) &\equiv \sum_{s_z} \int \frac{d\boldsymbol{p}}{(2\pi\hbar)^{3/2}} \sqrt{\frac{m_e c^2}{\omega_{\boldsymbol{p}}}} e^{\frac{i}{\hbar}\tilde{p}\cdot\tilde{x}} v_a(\boldsymbol{p}, s_z) b_{\boldsymbol{p}s_z}^\dagger.
\end{aligned}
$$

Let us now apply the operator in parentheses to the field component $\psi^-(\tilde{x})$. Then we have

$$
\begin{aligned}
&\left(c\gamma^\mu \partial_\mu - \frac{imc^2}{\hbar} \right) \psi^-(\tilde{x}) \\
&= \left(-\gamma^0 \frac{\partial}{\partial t} - c\boldsymbol{\gamma}\cdot\frac{\partial}{\partial \boldsymbol{x}} - \frac{i m_e c^2}{\hbar} \right) \sum_{s_z} \int \frac{d\boldsymbol{p}}{(2\pi\hbar)^{3/2}} \sqrt{\frac{m_e c^2}{\omega_{\boldsymbol{p}}}} e^{\frac{i}{\hbar}\boldsymbol{p}\cdot\boldsymbol{x} - \frac{i}{\hbar}\omega_{\boldsymbol{p}} t} u(\boldsymbol{p}, s_z) a_{\boldsymbol{p}s_z}
\end{aligned}
$$

$$= \frac{i}{\hbar}\sum_{s_z}\int \frac{d\boldsymbol{p}}{(2\pi\hbar)^{3/2}}\sqrt{\frac{m_ec^2}{\omega_{\boldsymbol{p}}}}(\gamma^0\omega_{\boldsymbol{p}} - c\boldsymbol{\gamma}\cdot\boldsymbol{p} - m_ec^2)u(\boldsymbol{p},s_z)e^{\frac{i}{\hbar}\boldsymbol{p}\cdot\boldsymbol{x}-\frac{i}{\hbar}\omega_{\boldsymbol{p}}t}a_{\boldsymbol{p}s_z}.$$

For the product on the right-hand side we use the equality

$$pc = \sqrt{\omega_{\boldsymbol{p}} + m_ec^2}\sqrt{\omega_{\boldsymbol{p}} - m_ec^2},$$

explicit expressions for gamma matrices (B.5)–(B.6) and equation (B.45) to obtain

$$\begin{aligned}
&(\gamma^0\omega_{\boldsymbol{p}} - c\boldsymbol{\gamma}\cdot\boldsymbol{p} - m_ec^2)u(\boldsymbol{p},s_z)\\
&= \omega_{\boldsymbol{p}}\begin{bmatrix}\sqrt{\omega_{\boldsymbol{p}} + m_ec^2}\\ -\sqrt{\omega_{\boldsymbol{p}} - m_ec^2}(\boldsymbol{\sigma}\cdot\frac{\boldsymbol{p}}{p})\end{bmatrix}\frac{\chi_{s_z}}{\sqrt{2m_ec^2}} - \begin{bmatrix}\sqrt{\omega_{\boldsymbol{p}} - m_ec^2}pc\\ -\sqrt{\omega_{\boldsymbol{p}} + m_ec^2}(\boldsymbol{\sigma}\cdot\boldsymbol{p})c\end{bmatrix}\frac{\chi_{s_z}}{\sqrt{2m_ec^2}}\\
&\quad - m_ec^2\begin{bmatrix}\sqrt{\omega_{\boldsymbol{p}} + m_ec^2}\\ \sqrt{\omega_{\boldsymbol{p}} - m_ec^2}(\boldsymbol{\sigma}\cdot\frac{\boldsymbol{p}}{p})\end{bmatrix}\frac{\chi_{s_z}}{\sqrt{2m_ec^2}}\\
&= \begin{bmatrix}\sqrt{\omega_{\boldsymbol{p}} + m_ec^2}(\omega_{\boldsymbol{p}} - (\omega_{\boldsymbol{p}} - m_ec^2) - m_ec^2)\\ [-(\omega_{\boldsymbol{p}} + m_ec^2)\sqrt{\omega_{\boldsymbol{p}} - m_ec^2} + \sqrt{\omega_{\boldsymbol{p}} + m_ec^2}pc](\boldsymbol{\sigma}\cdot\frac{\boldsymbol{p}}{p})\end{bmatrix}\frac{\chi_{s_z}}{\sqrt{2m_ec^2}} = 0.
\end{aligned} \tag{B.88}$$

This leads to the *Dirac equation* for the field component $\psi^-(\tilde{x})$,

$$\left(\gamma^0\frac{\partial}{\partial t} + c\boldsymbol{\gamma}\cdot\frac{\partial}{\partial\boldsymbol{x}} - \frac{im_ec^2}{\hbar}\right)\psi^-(\tilde{x}) = 0. \tag{B.89}$$

The same equation is valid for the component $\psi^+(\tilde{x})$ and for the full field operator

$$\left(\gamma^0\frac{\partial}{\partial t} + c\boldsymbol{\gamma}\cdot\frac{\partial}{\partial\boldsymbol{x}} - \frac{im_ec^2}{\hbar}\right)\psi(x) = 0. \tag{B.90}$$

Taking the Hermitian conjugate of (B.88) and multiplying it on γ^0 from the right, we obtain

$$\begin{aligned}
0 &= u^\dagger(\boldsymbol{p},s_z)((\gamma^0)^\dagger\omega_{\boldsymbol{p}} - c(\boldsymbol{\gamma})^\dagger\cdot\boldsymbol{p} - m_ec^2)\gamma^0\\
&= u^\dagger(\boldsymbol{p},s_z)(\gamma^0\omega_{\boldsymbol{p}} + c\boldsymbol{\gamma}\cdot\boldsymbol{p} - m_ec^2)\gamma^0\\
&= u^\dagger(\boldsymbol{p},s_z)\gamma^0\gamma^0(\gamma^0\omega_{\boldsymbol{p}} + c\boldsymbol{\gamma}\cdot\boldsymbol{p} - m_ec^2)\gamma^0\\
&= \overline{u}(\boldsymbol{p},s_z)\gamma^0(\gamma^0\omega_{\boldsymbol{p}} + c\boldsymbol{\gamma}\cdot\boldsymbol{p} - m_ec^2)\gamma^0\\
&= \overline{u}(\boldsymbol{p},s_z)(\gamma^0\omega_{\boldsymbol{p}} - c\boldsymbol{\gamma}\cdot\boldsymbol{p} - m_ec^2).
\end{aligned} \tag{B.91}$$

This implies the following Dirac equation for the conjugate field

$$\frac{\partial\overline{\psi}(\tilde{x})}{\partial t}\gamma^0 + c\frac{\partial\overline{\psi}(\tilde{x})}{\partial\boldsymbol{x}}\cdot\boldsymbol{\gamma} + \frac{im_ec^2}{\hbar}\overline{\psi}(\tilde{x}) = 0. \tag{B.92}$$

In the "slash" notation (B.18), Dirac equations (B.88) and (B.91) take more compact forms, i. e.,

$$(\not{p} - m_e c^2)u(\boldsymbol{p}, s_z) = 0, \quad \text{(B.93)}$$

$$\overline{u}(\boldsymbol{p}, s_z)(\not{p} - m_e c^2) = 0. \quad \text{(B.94)}$$

If we denote $\tilde{k} \equiv \tilde{p}' - \tilde{p}$, then $\not{k} \equiv \not{p}' - \not{p}$, and (B.93)–(B.94) imply

$$\begin{aligned}\mathcal{U}^{\mu}(\boldsymbol{p}s_z, \boldsymbol{p}'s_z')k_\mu &= \overline{u}(\boldsymbol{p}, s_z)\not{k}u(\boldsymbol{p}', s_z') \\ &= \overline{u}(\boldsymbol{p}, s_z)[\,\not{p}'\,u(\boldsymbol{p}', s_z')] - [\overline{u}(\boldsymbol{p}, s_z)\,\not{p}\,]u(\boldsymbol{p}', s_z') \\ &= (m_e c^2 - m_e c^2)\overline{u}(\boldsymbol{p}, s_z)u(\boldsymbol{p}', s_z') = 0, \quad \text{(B.95)} \\ \mathcal{W}^{\mu}(\boldsymbol{p}s_z, \boldsymbol{p}'s_z')k_\mu &= 0. \quad \text{(B.96)}\end{aligned}$$

It must be emphasized that in our approach the Dirac equation is an unremarkable property of the fields of fermions with spin 1/2. This equation makes it possible to simplify many calculations, but it does not play the fundamental role attributed to it in many textbooks. In any case, the Dirac equation cannot be regarded as a "relativistic analog of the Schrödinger equation for electrons". The correct relativistic electron wave functions were constructed in Chapter **1**-5. The relativistic analog of the Schrödinger equation for an interacting electron–proton system will be formulated in Chapter 3 of the third volume.

B.12 Fermion propagator

Let us calculate the *electron propagator*, a matrix element that often shows up in the covariant time-ordered perturbation theory,

$$\mathfrak{D}_{ab}(\tilde{x}_1, \tilde{x}_2) \equiv \langle \text{vac}|T[\psi_a(\tilde{x}_1)\overline{\psi}_b(\tilde{x}_2)]|\text{vac}\rangle. \quad \text{(B.97)}$$

Note that for anticommuting fermion fields the definition of time ordering includes a change of sign (compare with equation (**1**-7.16) for bosonic operators), i. e.,

$$T[\psi_a(\tilde{x}_1)\overline{\psi}_b(\tilde{x}_2)] = \begin{cases}\psi_a(\tilde{x}_1)\overline{\psi}_b(\tilde{x}_2), & \text{if } t_1 > t_2, \\ -\overline{\psi}_b(\tilde{x}_2)\psi_a(\tilde{x}_1), & \text{if } t_1 < t_2.\end{cases} \quad \text{(B.98)}$$

If $t_1 > t_2$, we can drop the time ordering sign in (B.97) and use (B.50) to obtain

$$\begin{aligned}\mathfrak{D}_{ab}(\tilde{x}_1, \tilde{x}_2) &= \langle \text{vac}|\psi_a(\tilde{x}_1)\overline{\psi}_b(\tilde{x}_2)|\text{vac}\rangle \propto \langle \text{vac}|(a + b^\dagger)(a^\dagger + b)|\text{vac}\rangle \propto \langle \text{vac}|aa^\dagger|\text{vac}\rangle \\ &= \langle \text{vac}|\left(\int \frac{d\boldsymbol{p}}{(2\pi\hbar)^{3/2}}\sqrt{\frac{m_e c^2}{\omega_{\boldsymbol{p}}}}\sum_{s_z} e^{-\frac{i}{\hbar}\tilde{p}\cdot\tilde{x}_1}u_a(\boldsymbol{p}, s_z)a_{\boldsymbol{p}s_z}\right) \\ &\quad \times \left(\int \frac{d\boldsymbol{q}}{(2\pi\hbar)^{3/2}}\sqrt{\frac{m_e c^2}{\omega_{\boldsymbol{q}}}}\sum_{\tau} e^{\frac{i}{\hbar}\tilde{q}\cdot\tilde{x}_2}u_b^\dagger(\boldsymbol{q}, \tau)a_{\boldsymbol{q}\tau}^\dagger\right)\gamma^0|\text{vac}\rangle\end{aligned}$$

$$= \int \frac{d\mathbf{p}d\mathbf{q}}{(2\pi\hbar)^3} \frac{m_e c^2}{\sqrt{\omega_{\mathbf{p}}\omega_{\mathbf{q}}}} \sum_{s_z\tau} e^{-\frac{i}{\hbar}\tilde{p}\cdot\tilde{x}_1} u_a(\mathbf{p}, s_z) e^{\frac{i}{\hbar}\tilde{q}\cdot\tilde{x}_2} u_b^\dagger(\mathbf{q}, \tau)\gamma^0 \delta(\mathbf{p}-\mathbf{q})\delta_{s_z\tau}$$

$$= \int \frac{d\mathbf{p}}{(2\pi\hbar)^3} \frac{m_e c^2}{\omega_{\mathbf{p}}} e^{\frac{i}{\hbar}\tilde{p}\cdot(\tilde{x}_2-\tilde{x}_1)} \sum_{s_z} u_a(\mathbf{p}, s_z) u_b^\dagger(\mathbf{p}, s_z)\gamma^0$$

$$= \int \frac{d\mathbf{p}}{(2\pi\hbar)^3} e^{\frac{i}{\hbar}(\omega_{\mathbf{p}}(t_2-t_1)-\mathbf{p}\cdot(\mathbf{x}_2-\mathbf{x}_1))} \frac{(\gamma^0\omega_{\mathbf{p}} - \boldsymbol{\gamma}\cdot\mathbf{p}c + m_e c^2)_{ab}}{2\omega_{\mathbf{p}}}.$$

In the case where $t_1 < t_2$, we apply (B.51) to obtain

$$\begin{aligned}
\mathfrak{D}_{ab}(\tilde{x}_1, \tilde{x}_2) &= -\langle \text{vac}|\overline{\psi}_b(\tilde{x}_2)\psi_a(\tilde{x}_1)|\text{vac}\rangle \\
&\propto -\langle \text{vac}|(a^\dagger + b)(a + b^\dagger)|\text{vac}\rangle \propto -\langle \text{vac}|bb^\dagger|\text{vac}\rangle \\
&= -\langle \text{vac}| \int \frac{d\mathbf{p}}{(2\pi\hbar)^3} e^{-\frac{i}{\hbar}\tilde{p}\cdot\tilde{x}_2} B_b(\mathbf{p}) \int d\mathbf{q} e^{\frac{i}{\hbar}\tilde{q}\cdot\tilde{x}_1} B_a^\dagger(\mathbf{q})|\text{vac}\rangle \\
&= -\int \frac{d\mathbf{p}}{(2\pi\hbar)^3} \frac{m_e c^2}{\omega_{\mathbf{p}}} e^{\frac{i}{\hbar}\tilde{p}\cdot(\tilde{x}_1-\tilde{x}_2)} \sum_{s_z} v_a(\mathbf{p}, s_z) v_b^\dagger(\mathbf{p}, s_z)\gamma^0 \\
&= -\int \frac{d\mathbf{p}}{(2\pi\hbar)^3} e^{\frac{i}{\hbar}(\omega_{\mathbf{p}}(t_1-t_2)-\mathbf{p}\cdot(\mathbf{x}_1-\mathbf{x}_2))} \frac{(\gamma^0\omega_{\mathbf{p}} - \boldsymbol{\gamma}\cdot\mathbf{p}c - m_e c^2)_{ab}}{2\omega_{\mathbf{p}}}.
\end{aligned}$$

The sum of these two expressions is

$$\begin{aligned}
\mathfrak{D}_{ab}(\tilde{x}_1, \tilde{x}_2) &= \theta(t_1 - t_2) \int \frac{d\mathbf{p}}{(2\pi\hbar)^3} e^{\frac{i}{\hbar}(\omega_{\mathbf{p}}(t_2-t_1)-\mathbf{p}\cdot(\mathbf{x}_2-\mathbf{x}_1))} \frac{\mathcal{P}_{ab}(\mathbf{p}, \omega_{\mathbf{p}})}{2\omega_{\mathbf{p}}} \\
&+ \theta(t_2 - t_1) \int \frac{d\mathbf{p}}{(2\pi\hbar)^3} e^{\frac{i}{\hbar}(\omega_{\mathbf{p}}(t_1-t_2)-\mathbf{p}\cdot(\mathbf{x}_1-\mathbf{x}_2))} \frac{\mathcal{P}_{ab}(-\mathbf{p}, -\omega_{\mathbf{p}})}{2\omega_{\mathbf{p}}},
\end{aligned} \tag{B.99}$$

where we denoted

$$\mathcal{P}_{ab}(\mathbf{p}, \omega_{\mathbf{p}}) = (\gamma^0\omega_{\mathbf{p}} - \boldsymbol{\gamma}\cdot\mathbf{p}c + m_e c^2)_{ab}$$

and $\theta(t)$ is the step function defined in (A.9). Our next goal is to rewrite equation (B.99) so that integration is carried out with respect to four independent variables (p_0, p_x, p_y, p_z). Using the integral representation (A.10) of the step function, we obtain

$$\begin{aligned}
\mathfrak{D}_{ab}(\tilde{x}_1, \tilde{x}_2) &= -\frac{1}{2\pi i} \int \frac{d\mathbf{p}\mathcal{P}_{ab}(\mathbf{p}, \omega_{\mathbf{p}})}{(2\pi\hbar)^3 2\omega_{\mathbf{p}}} \int ds \frac{e^{-is(t_1-t_2)}}{s + i\epsilon} e^{-\frac{i}{\hbar}(\omega_{\mathbf{p}}(t_1-t_2)-\mathbf{p}\cdot(\mathbf{x}_1-\mathbf{x}_2))} \\
&\quad - \frac{1}{2\pi i} \int \frac{d\mathbf{p}\mathcal{P}_{ab}(-\mathbf{p}, -\omega_{\mathbf{p}})}{(2\pi\hbar)^3 2\omega_{\mathbf{p}}} \int ds \frac{e^{is(t_1-t_2)}}{s + i\epsilon} e^{\frac{i}{\hbar}(\omega_{\mathbf{p}}(t_1-t_2)-\mathbf{p}\cdot(\mathbf{x}_1-\mathbf{x}_2))} \\
&= -\frac{1}{2\pi i} \int \frac{d\mathbf{p}}{(2\pi\hbar)^3} \int ds \frac{1}{(s + i\epsilon)2\omega_{\mathbf{p}}} [e^{-\frac{i}{\hbar}((\omega_{\mathbf{p}}+\hbar s)(t_1-t_2)-\mathbf{p}\cdot(\mathbf{x}_1-\mathbf{x}_2))} \mathcal{P}_{ab}(\mathbf{p}, \omega_{\mathbf{p}}) \\
&\quad + e^{\frac{i}{\hbar}((\omega_{\mathbf{p}}+\hbar s)(t_1-t_2)-\mathbf{p}\cdot(\mathbf{x}_1-\mathbf{x}_2))} \mathcal{P}_{ab}(-\mathbf{p}, -\omega_{\mathbf{p}})]
\end{aligned}$$

$$= -\frac{1}{2\pi i}\int\frac{d\mathbf{p}}{2(2\pi\hbar)^3}\int\frac{dp_0}{(p_0-\omega_{\mathbf{p}}+i\epsilon)\omega_{\mathbf{p}}}$$
$$\times\left[e^{-\frac{i}{\hbar}(p_0(t_1-t_2)-\mathbf{p}\cdot(\mathbf{x}_1-\mathbf{x}_2))}\mathcal{P}_{ab}(\mathbf{p},\omega_{\mathbf{p}})+e^{\frac{i}{\hbar}(p_0(t_1-t_2)-\mathbf{p}\cdot(\mathbf{x}_1-\mathbf{x}_2))}\mathcal{P}_{ab}(-\mathbf{p},-\omega_{\mathbf{p}})\right].$$

The integral with respect to p_0 can be calculated in the complex plane, closing the contour in its lower part. Then the integral will be equal to the residue at the pole $p_0=\omega_{\mathbf{p}}-i\epsilon$. This value will not change if we replace $\omega_{\mathbf{p}}$ with p_0 in the arguments of the function $\mathcal{P}_{ab}$. Hence

$$\begin{aligned}
\mathfrak{D}_{ab}(\tilde{x}_1,\tilde{x}_2) &= -\frac{1}{2\pi i}\int\frac{d\mathbf{p}}{(2\pi\hbar)^3}\int dp_0\frac{1}{(p_0-\omega_{\mathbf{p}}+i\epsilon)2\omega_{\mathbf{p}}}\\
&\quad\times\left[e^{-\frac{i}{\hbar}(p_0(t_1-t_2)-\mathbf{p}\cdot(\mathbf{x}_1-\mathbf{x}_2))}\mathcal{P}_{ab}(\mathbf{p},p_0)+e^{-\frac{i}{\hbar}(-p_0(t_1-t_2)+\mathbf{p}\cdot(\mathbf{x}_1-\mathbf{x}_2))}\mathcal{P}_{ab}(-\mathbf{p},-p_0)\right]\\
&= -\frac{1}{2\pi i}\int\frac{d\mathbf{p}}{(2\pi\hbar)^3}\int dp_0e^{-\frac{i}{\hbar}(p_0(t_1-t_2)-\mathbf{p}\cdot(\mathbf{x}_1-\mathbf{x}_2))}\frac{1}{2\omega_{\mathbf{p}}}\\
&\quad\times\left[\frac{\mathcal{P}_{ab}(\mathbf{p},p_0)}{p_0-\omega_{\mathbf{p}}+i\epsilon}+\frac{\mathcal{P}_{ab}(\mathbf{p},p_0)}{-p_0-\omega_{\mathbf{p}}+i\epsilon}\right]\\
&= \frac{1}{2\pi i}\int\frac{d\mathbf{p}}{(2\pi\hbar)^3}\int dp_0e^{-\frac{i}{\hbar}(p_0(t_1-t_2)-\mathbf{p}\cdot(\mathbf{x}_1-\mathbf{x}_2))}\frac{\mathcal{P}_{ab}(\mathbf{p},p_0)}{p_0^2-\omega_{\mathbf{p}}^2+i\epsilon}\\
&= \frac{1}{2\pi i(2\pi\hbar)^3}\int d^4pe^{-\frac{i}{\hbar}\tilde{p}\cdot(\tilde{x}_1-\tilde{x}_2)}\frac{\mathcal{P}_{ab}(\mathbf{p},p_0)}{p_0^2-c^2\mathbf{p}^2-m_e^2c^4+i\epsilon}\\
&= \frac{1}{2\pi i(2\pi\hbar)^3}\int d^4pe^{-\frac{i}{\hbar}\tilde{p}\cdot(\tilde{x}_1-\tilde{x}_2)}\frac{(\not{p}+m_ec^2)_{ab}}{\tilde{p}^2-m_e^2c^4+i\epsilon}.
\end{aligned}\tag{B.100}$$

C Quantum field of photons

C.1 Construction of photon quantum field

Here we are going to construct a quantum field based on the creation ($c^{\dagger}_{\boldsymbol{p}\tau}$) and annihilation ($c_{\boldsymbol{p}\tau}$) operators of photons. Our goal is to satisfy conditions (I)–(V) listed in **Step 1** of Subsection 3.1.2. In particular, we require that the Poincaré transformation (3.1) of the photon field $\mathcal{A}_\mu(\tilde{x})$ is associated with the four-dimensional pseudo-orthogonal representation of the Lorentz group from Appendix **1**-J. So, our intention is to obtain

$$U_0(\Lambda;\tilde{a})\mathcal{A}_\mu(\tilde{x})U_0^{-1}(\Lambda;\tilde{a}) = \sum_\nu (\Lambda^{-1})_\mu^{\ \nu}\mathcal{A}_\nu(\Lambda(\tilde{x}+\tilde{a})), \tag{C.1}$$

where the indices μ and ν run through the values 0, 1, 2, 3. We will try to define the photon four-component quantum field by analogy with the Dirac field (B.34),

$$\begin{aligned}\mathcal{A}_\mu(\tilde{x}) &\equiv \mathcal{A}_\mu(t,\boldsymbol{x})\\ &= \frac{\hbar c}{(2\pi\hbar)^{3/2}}\int \frac{d\boldsymbol{p}}{\sqrt{2pc}}\sum_\tau [e^{-\frac{i}{\hbar}\tilde{p}\cdot\tilde{x}}e_\mu(\boldsymbol{p},\tau)c_{\boldsymbol{p}\tau} + e^{\frac{i}{\hbar}\tilde{p}\cdot\tilde{x}}e^*_\mu(\boldsymbol{p},\tau)c^\dagger_{\boldsymbol{p}\tau}],\end{aligned} \tag{C.2}$$

where $\tilde{p}\cdot\tilde{x} \equiv cpt - \boldsymbol{p}\cdot\boldsymbol{x}$. Following the recipe from Appendix B.5, we first postulate the following value of the coefficient function $e_\mu(\boldsymbol{\kappa},\tau)$ at the standard momentum $\boldsymbol{\kappa} \equiv (0,0,1)$, which was selected in (**1**-5.61):

$$e_\mu(\boldsymbol{\kappa},\tau) \equiv \frac{1}{\sqrt{2}}\begin{bmatrix}0\\1\\i\tau\\0\end{bmatrix}. \tag{C.3}$$

For all other photon momenta $\boldsymbol{p}$ we define[1]

$$\tilde{e}(\boldsymbol{p},\tau) = \lambda_{\boldsymbol{p}}\tilde{e}(\boldsymbol{\kappa},\tau), \tag{C.4}$$

$$\tilde{e}^\dagger(\boldsymbol{p},\tau) = \tilde{e}^\dagger(\boldsymbol{\kappa},\tau)\lambda_{\boldsymbol{p}}^{-1}, \tag{C.5}$$

where $\lambda_{\boldsymbol{p}}$ is a 4×4 matrix of the Lorentz transformation (**1**-5.67) that transforms the standard momentum $\boldsymbol{\kappa}$ into the momentum $\boldsymbol{p}$, i. e.,

$$\lambda_{\boldsymbol{p}} \equiv \boldsymbol{\theta}_{\boldsymbol{p}} \circ \boldsymbol{\varphi}_{\boldsymbol{p}}. \tag{C.6}$$

---

1 Note the analogy with the massive case (B.42)–(B.44). We mark the coefficient function $\tilde{e}(\boldsymbol{p},\tau)$ by the tilde to underscore its four-component nature, although, strictly speaking, it is not a 4-vector.

https://doi.org/10.1515/9783110493207-007

Here $\boldsymbol{\varphi}_{\boldsymbol{p}}$ is a pure rotation that takes $\boldsymbol{\kappa}$ to the vector $\boldsymbol{p}/p$, and $\boldsymbol{\theta}_{\boldsymbol{p}}$ is a boost along the direction $\boldsymbol{p}/p$, which takes the latter vector to $\boldsymbol{p}$. For the vector $\boldsymbol{\varphi}_{\boldsymbol{p}}$, it is not difficult to obtain the following explicit formulas:

$$\frac{\boldsymbol{\varphi}_{\boldsymbol{p}}}{\varphi_{\boldsymbol{p}}} = \frac{(-p_y, p_x, 0)}{\sqrt{p_x^2 + p_y^2}}, \tag{C.7}$$

$$\cos\varphi_{\boldsymbol{p}} = \frac{p_z}{p}, \tag{C.8}$$

$$\sin\varphi_{\boldsymbol{p}} = \frac{\sqrt{p_x^2 + p_y^2}}{p}. \tag{C.9}$$

Indeed, the substitution in the definition (**1**-D.22) gives[2]

$$\boldsymbol{\varphi}_{\boldsymbol{p}}\boldsymbol{\kappa} = \boldsymbol{\kappa}\frac{p_z}{p} - \left[\boldsymbol{\kappa} \times \frac{(-p_y, p_x, 0)}{\sqrt{p_x^2 + p_y^2}}\right]\frac{\sqrt{p_x^2 + p_y^2}}{p} = \frac{\mathbf{i}p_x + \mathbf{j}p_y + \mathbf{k}p_z}{p} = \frac{\boldsymbol{p}}{p}.$$

C.2 Properties of function $e_\mu(p, \tau)$

Let us rewrite definition (C.6) in the form

$$\lambda_{\boldsymbol{p}} = \boldsymbol{\theta}_{\boldsymbol{p}} \circ \boldsymbol{\varphi}_{\boldsymbol{p}} = \boldsymbol{\varphi}_{\boldsymbol{p}} \circ \boldsymbol{\varphi}_{\boldsymbol{p}}^{-1} \circ \boldsymbol{\theta}_{\boldsymbol{p}} \circ \boldsymbol{\varphi}_{\boldsymbol{p}} = \boldsymbol{\varphi}_{\boldsymbol{p}} \circ \boldsymbol{\theta}_{\varphi_{\boldsymbol{p}}^{-1}\boldsymbol{p}} = \boldsymbol{\varphi}_{\boldsymbol{p}} \circ \boldsymbol{\theta}_{p\boldsymbol{\kappa}}, \tag{C.10}$$

where $\boldsymbol{\theta}_{p\boldsymbol{\kappa}}$ is a boost along the z-axis, taking $\boldsymbol{\kappa}$ to $p\boldsymbol{\kappa}$. Obviously, this boost does not change the 4-vector (C.3), i. e.,

$$\boldsymbol{\theta}_{p\boldsymbol{\kappa}}\tilde{e}(\boldsymbol{\kappa}, \tau) = \tilde{e}(\boldsymbol{\kappa}, \tau).$$

The zero component of this 4-vector does not change under rotations $\boldsymbol{\varphi}_{\boldsymbol{p}}$ as well. Hence, we conclude that for all $\boldsymbol{p}$ and $\boldsymbol{x}$

$$e_0(\boldsymbol{p}, \tau) = \lambda_{\boldsymbol{p}} e_0(\boldsymbol{\kappa}, \tau) = \boldsymbol{\varphi}_{\boldsymbol{p}} \circ \boldsymbol{\theta}_{p\boldsymbol{\kappa}} e_0(\boldsymbol{\kappa}, \tau) = \boldsymbol{\varphi}_{\boldsymbol{p}} e_0(\boldsymbol{\kappa}, \tau) = e_0(\boldsymbol{\kappa}, \tau) = 0, \tag{C.11}$$

$$\mathcal{A}_0(\boldsymbol{x}, t) = 0. \tag{C.12}$$

Denoting $\boldsymbol{e}(\boldsymbol{p}, \tau)$ the 3-vector part of the quantity $e_\mu(\boldsymbol{p}, \tau)$ and taking into account the rotational invariance of the scalar product, we obtain

$$\begin{aligned} \boldsymbol{p} &= \lambda_{\boldsymbol{p}}\boldsymbol{\kappa} = (\boldsymbol{\varphi}_{\boldsymbol{p}} \circ \boldsymbol{\theta}_{p\boldsymbol{\kappa}})\boldsymbol{\kappa} = p\boldsymbol{\varphi}_{\boldsymbol{p}}\boldsymbol{\kappa}, \\ \boldsymbol{e}(\boldsymbol{p}, \tau) &= \lambda_{\boldsymbol{p}}\boldsymbol{e}(\boldsymbol{\kappa}, \tau) = \boldsymbol{\varphi}_{\boldsymbol{p}} \circ \boldsymbol{\theta}_{p\boldsymbol{\kappa}}\boldsymbol{e}(\boldsymbol{\kappa}, \tau) = \boldsymbol{\varphi}_{\boldsymbol{p}}\boldsymbol{e}(\boldsymbol{\kappa}, \tau), \\ \boldsymbol{p} \cdot \boldsymbol{e}(\boldsymbol{p}, \tau) &= p\boldsymbol{\varphi}_{\boldsymbol{p}}\boldsymbol{\kappa} \cdot \boldsymbol{\varphi}_{\boldsymbol{p}}\boldsymbol{e}(\boldsymbol{\kappa}, \tau) = p\boldsymbol{\kappa} \cdot \boldsymbol{e}(\boldsymbol{\kappa}, \tau) = 0. \end{aligned} \tag{C.13}$$

2 We change the sign of φ in (**1**-D.22), because this is an active rotation.

Taking into account (C.11), we then obtain

$$p^\mu e_\mu(\boldsymbol{p},\tau) = -\boldsymbol{p}\cdot\boldsymbol{e}(\boldsymbol{p},\tau) = 0. \tag{C.14}$$

In the third volume we will also need an explicit expression for $e_z(\boldsymbol{p},\tau)$. Inserting (C.7)–(C.9) into (C.13) and (**1**-D.22), we obtain

$$e_z(\boldsymbol{p},\tau) = (\boldsymbol{\varphi}_{\boldsymbol{p}}\boldsymbol{e}(\boldsymbol{\kappa},\tau))_z = -\left[\boldsymbol{e}(\boldsymbol{\kappa},\tau)\times\frac{\boldsymbol{\varphi}_{\boldsymbol{p}}}{\varphi_{\boldsymbol{p}}}\right]_z \sin(\varphi_{\boldsymbol{p}}) = \frac{-p_x - i\tau p_y}{\sqrt{2}p} \tag{C.15}$$

for all momenta $\boldsymbol{p}$.

C.3 Useful commutator

It will be convenient to introduce a special notation for the following combination of photon annihilation operators:

$$\mathcal{C}_{ab}(\boldsymbol{p}) \equiv \frac{\hbar c}{\sqrt{2pc}}\gamma^\mu_{ab}\sum_\tau e_\mu(\boldsymbol{p},\tau)c_{\boldsymbol{p},\tau}. \tag{C.16}$$

In this notation, the pseudo-scalar product $\gamma^\mu\mathcal{A}_\mu$ of the photon field with gamma matrices takes the form

$$\gamma^\mu_{ab}\mathcal{A}_\mu(\tilde{x}) = \int\frac{d\boldsymbol{p}}{(2\pi\hbar)^{3/2}}\left(e^{-\frac{i}{\hbar}\tilde{p}\cdot\tilde{x}}\mathcal{C}_{ab}(\boldsymbol{p}) + e^{\frac{i}{\hbar}\tilde{p}\cdot\tilde{x}}\mathcal{C}^\dagger_{ab}(\boldsymbol{p})\right). \tag{C.17}$$

In Subsection 3.2.1 we will need the commutator of $\mathcal{C}$-operators

$$\begin{aligned}
[\mathcal{C}^\dagger_{ab}(\boldsymbol{p}),\mathcal{C}_{cd}(\boldsymbol{p}')] &= \frac{\hbar^2 c}{2\sqrt{pp'}}\sum_{\tau\tau'}\gamma^\mu_{ab}\gamma^\nu_{cd}e^\dagger_\mu(\boldsymbol{p},\tau)e_\nu(\boldsymbol{p}',\tau')[c^\dagger_{\boldsymbol{p}\tau},c_{\boldsymbol{p}'\tau'}] \\
&= -\frac{\hbar^2 c}{2p}\sum_{\tau\tau'}\gamma^\mu_{ab}\gamma^\nu_{cd}e^\dagger_\mu(\boldsymbol{p},\tau)e_\nu(\boldsymbol{p}',\tau')\delta(\boldsymbol{p}-\boldsymbol{p}')\delta_{\tau\tau'} \\
&= -\frac{\hbar^2 c}{2p}\sum_{\tau}\gamma^\mu_{ab}\gamma^\nu_{cd}e^\dagger_\mu(\boldsymbol{p},\tau)e_\nu(\boldsymbol{p},\tau)\delta(\boldsymbol{p}-\boldsymbol{p}') \\
&= -\frac{\hbar^2 c}{2p}\gamma^\mu_{ab}\gamma^\nu_{cd}h_{\mu\nu}(\boldsymbol{p})\delta(\boldsymbol{p}-\boldsymbol{p}'),
\end{aligned} \tag{C.18}$$

where the sum

$$h_{\mu\nu}(\boldsymbol{p}) = h_{\nu\mu}(\boldsymbol{p}) \equiv \sum_\tau e_\mu(\boldsymbol{p},\tau)e^\dagger_\nu(\boldsymbol{p},\tau) \tag{C.19}$$

often appears in calculations. First we compute this sum at the standard momentum $\boldsymbol{\kappa} = (0,0,1)$ with the help of (C.3),

$$h_{\mu\nu}(\boldsymbol{\kappa}) = \sum_\tau e_\mu(\boldsymbol{\kappa},\tau)e^\dagger_\nu(\boldsymbol{\kappa},\tau)$$

$$= \frac{1}{2}\begin{bmatrix} 0 \\ 1 \\ i \\ 0 \end{bmatrix} [0 \quad 1 \quad -i \quad 0] + \frac{1}{2}\begin{bmatrix} 0 \\ 1 \\ -i \\ 0 \end{bmatrix} [0 \quad 1 \quad i \quad 0]$$

$$= \frac{1}{2}\begin{bmatrix} 0 & 0 & 0 & 0 \\ 0 & 1 & -i & 0 \\ 0 & i & 1 & 0 \\ 0 & 0 & 0 & 0 \end{bmatrix} + \frac{1}{2}\begin{bmatrix} 0 & 0 & 0 & 0 \\ 0 & 1 & i & 0 \\ 0 & -i & 1 & 0 \\ 0 & 0 & 0 & 0 \end{bmatrix}$$

$$= \begin{bmatrix} 0 & 0 & 0 & 0 \\ 0 & 1 & 0 & 0 \\ 0 & 0 & 1 & 0 \\ 0 & 0 & 0 & 0 \end{bmatrix}.$$

This result can be also written in terms of the components of the standard vector $\boldsymbol{\kappa} = (0, 0, 1)$, i. e.,

$$h_{0\mu}(\boldsymbol{\kappa}) = h_{\mu 0}(\boldsymbol{\kappa}) = 0,$$
$$h_{ij}(\boldsymbol{\kappa}) = \delta_{ij} - \frac{\kappa_i \kappa_j}{\kappa^2},$$

where $i, j = 1, 2, 3$; $\mu = 0, 1, 2, 3$. For arbitrary values of $\boldsymbol{p}$ we now use equations (C.4), (C.5), (C.6) and (C.10) and obtain

$$\begin{aligned} h_{\mu\nu}(\boldsymbol{p}) &\equiv \sum_{\tau} e_\mu(\boldsymbol{p}, \tau) e_\nu^\dagger(\boldsymbol{p}, \tau) \\ &= \sum_{\tau} (\boldsymbol{\varphi}_{\boldsymbol{p}} \circ \boldsymbol{\theta}_{\boldsymbol{p}\boldsymbol{\kappa}} e_\mu(\boldsymbol{\kappa}, \tau))(e_\nu^\dagger(\boldsymbol{\kappa}, \tau) \boldsymbol{\theta}_{\boldsymbol{p}\boldsymbol{\kappa}}^{-1} \circ \boldsymbol{\varphi}_{\boldsymbol{p}}^{-1}) \\ &= \boldsymbol{\varphi}_{\boldsymbol{p}} \left(\sum_{\tau} e_\mu(\boldsymbol{\kappa}, \tau) e_\nu^\dagger(\boldsymbol{\kappa}, \tau) \right) \boldsymbol{\varphi}_{\boldsymbol{p}}^{-1} \\ &= \boldsymbol{\varphi}_{\boldsymbol{p}} \begin{bmatrix} 0 & 0 & 0 & 0 \\ 0 & 1 & 0 & 0 \\ 0 & 0 & 1 & 0 \\ 0 & 0 & 0 & 0 \end{bmatrix} \boldsymbol{\varphi}_{\boldsymbol{p}}^{-1}. \end{aligned}$$

This implies that $h_{0\mu}(\boldsymbol{p}) = h_{\mu 0}(\boldsymbol{p}) = 0$, that the 3×3 submatrix is equal to

$$h_{ij}(\boldsymbol{p}) = \boldsymbol{\varphi}_{\boldsymbol{p}} \left[\delta_{ij} - \frac{\kappa_i \kappa_j}{\kappa^2} \right] \boldsymbol{\varphi}_{\boldsymbol{p}}^{-1} = \delta_{ij} - \frac{p_i p_j}{p^2} \tag{C.20}$$

and that the final formula for $h_{\mu\nu}(\boldsymbol{p})$ is

$$h_{\mu\nu}(\boldsymbol{p}) = \begin{bmatrix} 0 & 0 & 0 & 0 \\ 0 & 1 - \frac{p_x^2}{p^2} & -\frac{p_x p_y}{p^2} & -\frac{p_x p_z}{p^2} \\ 0 & -\frac{p_x p_y}{p^2} & 1 - \frac{p_y^2}{p^2} & -\frac{p_y p_z}{p^2} \\ 0 & -\frac{p_x p_z}{p^2} & -\frac{p_z p_y}{p^2} & 1 - \frac{p_z^2}{p^2} \end{bmatrix}. \tag{C.21}$$

C.4 Commutator of photon fields

The photon quantum field (C.2) commutes with itself at space-like intervals, as required in (3.4). Indeed, assuming that $\boldsymbol{x} \neq \boldsymbol{y}$, we obtain

$$
\begin{aligned}
&[\mathcal{A}_\mu(0,\boldsymbol{x}), \mathcal{A}_\nu^\dagger(0,\boldsymbol{y})] \\
&= \frac{\hbar^2 c}{2(2\pi\hbar)^3} \int \frac{d\boldsymbol{p}d\boldsymbol{p}'}{\sqrt{pp'}} \sum_{\tau\tau'} [(e^{-\frac{i}{\hbar}\boldsymbol{p}\cdot\boldsymbol{x}} e_\mu(\boldsymbol{p},\tau) c_{\boldsymbol{p}\tau} + e^{\frac{i}{\hbar}\boldsymbol{p}\cdot\boldsymbol{x}} e_\mu^*(\boldsymbol{p},\tau) c_{\boldsymbol{p}\tau}^\dagger), \\
&\quad (e^{\frac{i}{\hbar}\boldsymbol{p}'\cdot\boldsymbol{y}} e_\nu^*(\boldsymbol{p}',\tau') c_{\boldsymbol{p}'\tau'}^\dagger + e^{-\frac{i}{\hbar}\boldsymbol{p}'\cdot\boldsymbol{y}} e_\nu(\boldsymbol{p}',\tau') c_{\boldsymbol{p}'\tau'})] \\
&= \frac{\hbar^2 c}{2(2\pi\hbar)^3} \int \frac{d\boldsymbol{p}d\boldsymbol{p}'}{\sqrt{pp'}} \sum_{\tau\tau'} (e^{-\frac{i}{\hbar}\boldsymbol{p}\cdot\boldsymbol{x}} e^{\frac{i}{\hbar}\boldsymbol{p}'\cdot\boldsymbol{y}} e_\mu(\boldsymbol{p},\tau) e_\nu^\dagger(\boldsymbol{p}',\tau') [c_{\boldsymbol{p}\tau}, c_{\boldsymbol{p}'\tau'}^\dagger] \\
&\quad + e^{\frac{i}{\hbar}\boldsymbol{p}\cdot\boldsymbol{x}} e^{-\frac{i}{\hbar}\boldsymbol{p}'\cdot\boldsymbol{y}} e_\mu^*(\boldsymbol{p},\tau) e_\nu^{*\dagger}(\boldsymbol{p}',\tau') [c_{\boldsymbol{p}\tau}^\dagger, c_{\boldsymbol{p}'\tau'}]) \\
&= \frac{\hbar^2 c}{2(2\pi\hbar)^3} \int \frac{d\boldsymbol{p}d\boldsymbol{p}'}{p} \delta(\boldsymbol{p}-\boldsymbol{p}') \sum_{\tau\tau'} \delta_{\tau\tau'} \\
&\quad \times (e^{-\frac{i}{\hbar}\boldsymbol{p}\cdot(\boldsymbol{x}-\boldsymbol{y})} e_\mu(\boldsymbol{p},\tau) e_\nu^\dagger(\boldsymbol{p}',\tau') - e^{\frac{i}{\hbar}\boldsymbol{p}\cdot(\boldsymbol{x}-\boldsymbol{y})} e_\mu^*(\boldsymbol{p},\tau) e_\nu^{*\dagger}(\boldsymbol{p}',\tau')) \\
&= \frac{\hbar^2 c}{2(2\pi\hbar)^3} \int \frac{d\boldsymbol{p}}{p} \sum_{\tau} (e^{-\frac{i}{\hbar}\boldsymbol{p}\cdot(\boldsymbol{x}-\boldsymbol{y})} e_\mu(\boldsymbol{p},\tau) e_\nu^\dagger(\boldsymbol{p},\tau) - e^{\frac{i}{\hbar}\boldsymbol{p}\cdot(\boldsymbol{x}-\boldsymbol{y})} e_\mu^*(\boldsymbol{p},\tau) e_\nu^{*\dagger}(\boldsymbol{p},\tau)) \\
&= \frac{\hbar^2 c}{2(2\pi\hbar)^3} \int \frac{d\boldsymbol{p}}{p} (e^{-\frac{i}{\hbar}\boldsymbol{p}\cdot(\boldsymbol{x}-\boldsymbol{y})} - e^{\frac{i}{\hbar}\boldsymbol{p}\cdot(\boldsymbol{x}-\boldsymbol{y})}) h_{\mu\nu}(\boldsymbol{p}) \\
&= -\frac{i\hbar^2 c}{(2\pi\hbar)^3} \int \frac{d\boldsymbol{p}}{p} \sin(\hbar^{-1}\boldsymbol{p}\cdot(\boldsymbol{x}-\boldsymbol{y})) h_{\mu\nu}(\boldsymbol{p}) \qquad \text{(C.22)} \\
&= 0,
\end{aligned}
$$

because the integrand in (C.22) is an odd function of $\boldsymbol{p}$.

C.5 Photon propagator

Next we should calculate the *photon propagator*, defined by

$$\langle \text{vac}|T[\mathcal{A}_\mu(\tilde{x}_1)\mathcal{A}_\nu(\tilde{x}_2)]|\text{vac}\rangle.$$

We can drop the time ordering sign and consider the two cases $t_1 > t_2$ and $t_1 < t_2$ separately. For $t_1 > t_2$, we have

$$
\begin{aligned}
&\langle \text{vac}|\mathcal{A}_\mu(\tilde{x}_1)\mathcal{A}_\nu(\tilde{x}_2)|\text{vac}\rangle \\
&= \theta(t_1 - t_2)\langle \text{vac}| \frac{\hbar^2 c}{2(2\pi\hbar)^3} \int \frac{d\boldsymbol{p}d\boldsymbol{p}'}{\sqrt{pp'}} \sum_{\tau\tau'} (e^{-\frac{i}{\hbar}\tilde{p}\cdot\tilde{x}_1} e_\mu(\boldsymbol{p},\tau) c_{\boldsymbol{p}\tau} + e^{\frac{i}{\hbar}\tilde{p}\cdot\tilde{x}_1} e_\mu^*(\boldsymbol{p},\tau) c_{\boldsymbol{p}\tau}^\dagger) \\
&\quad \times (e^{-\frac{i}{\hbar}\tilde{p}'\cdot\tilde{x}_2} e_\nu(\boldsymbol{p}',\tau') c_{\boldsymbol{p}'\tau'} + e^{\frac{i}{\hbar}\tilde{p}'\cdot\tilde{x}_2} e_\nu^*(\boldsymbol{p}',\tau') c_{\boldsymbol{p}'\tau'}^\dagger)|\text{vac}\rangle
\end{aligned}
$$

$$= \theta(t_1 - t_2)\langle \mathrm{vac}|\frac{\hbar^2 c}{2(2\pi\hbar)^3}\int \frac{d\boldsymbol{p}d\boldsymbol{p}'}{\sqrt{pp'}}\sum_{\tau\tau'} e^{-\frac{i}{\hbar}\tilde{p}\cdot\tilde{x}_1} e^{\frac{i}{\hbar}\tilde{p}'\cdot\tilde{x}_2} e_\mu(\boldsymbol{p},\tau)e_\nu^*(\boldsymbol{p}',\tau')c_{\boldsymbol{p}\tau}c^\dagger_{\boldsymbol{p}'\tau'}|\mathrm{vac}\rangle$$
$$= \theta(t_1 - t_2)\frac{\hbar^2 c}{2(2\pi\hbar)^3}\int \frac{d\boldsymbol{p}d\boldsymbol{p}'}{\sqrt{pp'}}\sum_{\tau\tau'} e^{-\frac{i}{\hbar}\tilde{p}\cdot(\tilde{x}_1-\tilde{x}_2)} e_\mu(\boldsymbol{p},\tau)e_\nu^*(\boldsymbol{p}',\tau')\delta(\boldsymbol{p}-\boldsymbol{p}')\delta_{\tau\tau'}$$
$$= \theta(t_1 - t_2)\frac{\hbar^2 c}{2(2\pi\hbar)^3}\int \frac{d\boldsymbol{p}}{p}\sum_{\tau} e^{-\frac{i}{\hbar}\tilde{p}\cdot(\tilde{x}_1-\tilde{x}_2)} e_\mu(\boldsymbol{p},\tau)e_\nu^*(\boldsymbol{p},\tau)$$
$$= \theta(t_1 - t_2)\frac{\hbar^2 c}{2(2\pi\hbar)^3}\int \frac{d\boldsymbol{p}}{p} e^{-\frac{i}{\hbar}\tilde{p}\cdot(\tilde{x}_1-\tilde{x}_2)} h_{\mu\nu}(\boldsymbol{p}).$$

The propagator's value for $t_1 < t_2$ is obtained by swapping the indices $1 \rightleftharpoons 2$ and $\mu \rightleftharpoons \nu$, so that

$$\langle \mathrm{vac}|\mathcal{A}_\nu(\tilde{x}_2)\mathcal{A}_\mu(\tilde{x}_1)|\mathrm{vac}\rangle = \theta(t_2 - t_1)\frac{\hbar^2 c}{2(2\pi\hbar)^3}\int \frac{d\boldsymbol{p}}{p} e^{-\frac{i}{\hbar}\tilde{p}\cdot(\tilde{x}_2-\tilde{x}_1)} h_{\mu\nu}(\boldsymbol{p}).$$

Next we use the integral representation (A.10) of the step function $\theta(t)$, to write down the full expression

$$\begin{aligned}
&\langle \mathrm{vac}|T[\mathcal{A}_\mu(\tilde{x}_1)\mathcal{A}_\nu(\tilde{x}_2)]|\mathrm{vac}\rangle \\
&= \hbar^2 c \int \frac{d\boldsymbol{p}}{2(2\pi\hbar)^3 p} h_{\mu\nu}(\boldsymbol{p})[e^{-\frac{i}{\hbar}\tilde{p}\cdot(\tilde{x}_1-\tilde{x}_2)}\theta(t_1-t_2) + e^{-\frac{i}{\hbar}\tilde{p}\cdot(\tilde{x}_2-\tilde{x}_1)}\theta(t_2-t_1)] \\
&= -\frac{\hbar^2 c}{2\pi i}\int_{-\infty}^{\infty} ds \int \frac{d\boldsymbol{p}}{2(2\pi\hbar)^3 p} h_{\mu\nu}(\boldsymbol{p}) \\
&\quad \times \left[e^{-\frac{i}{\hbar}\tilde{p}\cdot(\tilde{x}_1-\tilde{x}_2)} \frac{e^{-is(t_1-t_2)}}{s+i\epsilon} + e^{-\frac{i}{\hbar}\tilde{p}\cdot(\tilde{x}_2-\tilde{x}_1)} \frac{e^{-is(t_2-t_1)}}{s+i\epsilon} \right] \\
&= -\frac{\hbar^2 c}{2\pi i}\int_{-\infty}^{\infty} ds \int \frac{d\boldsymbol{p}}{2(2\pi\hbar)^3 p} \frac{h_{\mu\nu}(\boldsymbol{p})}{s+i\epsilon} \\
&\quad \times [e^{\frac{i}{\hbar}(-cp(t_1-t_2)+\boldsymbol{p}\cdot(\boldsymbol{x}_1-\boldsymbol{x}_2))} e^{-is(t_1-t_2)} + e^{\frac{i}{\hbar}(-cp(t_2-t_1)+\boldsymbol{p}\cdot(\boldsymbol{x}_2-\boldsymbol{x}_1))} e^{-is(t_2-t_1)}] \\
&= -\frac{\hbar^2 c}{2\pi i}\int \frac{d\boldsymbol{p}}{2(2\pi\hbar)^3 p} h_{\mu\nu}(\boldsymbol{p}) e^{-\frac{i}{\hbar}\boldsymbol{p}\cdot(\boldsymbol{x}_1-\boldsymbol{x}_2)} \\
&\quad \times \int_{-\infty}^{\infty} ds \left[\frac{e^{\frac{i}{\hbar}(-cp-\hbar s)(t_1-t_2)}}{s+i\epsilon} + \frac{e^{\frac{i}{\hbar}(cp+\hbar s)(t_1-t_2)}}{s+i\epsilon} \right].
\end{aligned}$$

We change integration variables in the first integral ($s \to p_0 \equiv -cp - \hbar s$) and in the second integral ($s \to p_0 \equiv cp + \hbar s$) to obtain

$$\begin{aligned}
&\langle \mathrm{vac}|T[\mathcal{A}_\mu(\tilde{x}_1)\mathcal{A}_\nu(\tilde{x}_2)]|\mathrm{vac}\rangle \\
&= -\frac{\hbar^2 c}{2\pi i}\int_{-\infty}^{\infty} dp_0 \int \frac{d\boldsymbol{p}}{2(2\pi\hbar)^3 p} h_{\mu\nu}(\boldsymbol{p}) e^{-\frac{i}{\hbar}\boldsymbol{p}\cdot(\boldsymbol{x}_1-\boldsymbol{x}_2)} \left[\frac{e^{\frac{i}{\hbar}p_0(t_1-t_2)}}{-cp-p_0+i\epsilon} + \frac{e^{\frac{i}{\hbar}p_0(t_1-t_2)}}{-cp+p_0+i\epsilon} \right]
\end{aligned}$$

$$= -\frac{\hbar^2 c^2}{2\pi i} \int_{-\infty}^{\infty} dp_0 \int \frac{d\boldsymbol{p}}{(2\pi\hbar)^3} e^{\frac{i}{\hbar}p_0(t_1-t_2)} e^{-\frac{ic}{\hbar}\boldsymbol{p}\cdot(\boldsymbol{x}_1-\boldsymbol{x}_2)} \frac{h_{\mu\nu}(\boldsymbol{p})}{\tilde{p}^2 + i\epsilon}$$

$$= -\frac{\hbar^2 c^2}{2\pi i} \int \frac{d^4 p}{(2\pi\hbar)^3} e^{\frac{i}{\hbar}\tilde{p}\cdot(\tilde{x}_1-\tilde{x}_2)} \frac{h_{\mu\nu}(\boldsymbol{p})}{\tilde{p}^2 + i\epsilon}, \tag{C.23}$$

where we denoted $\tilde{p}^2 \equiv p_0^2 - c^2 p^2$ and $d^4 p \equiv dp_0 d\boldsymbol{p}$.

The factor $h_{\mu\nu}(\boldsymbol{p})$ in (C.23) was calculated in (C.21) in the Coulomb gauge. However, as we explained in Subsection 3.2.3, in applications it is more convenient to use the Feynman gauge in which this matrix is replaced by the metric tensor $h_{\mu\nu}(\boldsymbol{p}) \to -\eta_{\mu\nu}$. This is how we get our final formula for the photon propagator,

$$\langle \mathrm{vac}|T[\mathcal{A}_\mu(\tilde{x}_1)\mathcal{A}_\nu(\tilde{x}_2)]|\mathrm{vac}\rangle = \frac{\hbar^2 c^2}{2\pi i} \int \frac{d^4 p}{(2\pi\hbar)^3} e^{\frac{i}{\hbar}\tilde{p}\cdot(\tilde{x}_1-\tilde{x}_2)} \frac{\eta_{\mu\nu}}{\tilde{p}^2 + i\epsilon}. \tag{C.24}$$

C.6 Poincaré transformations of photon field

Now we approach a very important step, namely, the determination of the photon field's transformations with respect to the noninteracting representation of the Poincaré group in the Fock space [18]. Note that we defined the coefficient functions $e_\mu(\boldsymbol{p},\tau)$ in Appendix C.2 in the hope of getting a covariant transformation law (C.1). Such a goal was achieved in the case of the massive electron–positron field in Appendix B.8. It turns out that for massless photons the situation is more complicated. Although actions of translations and rotations do agree with the requirement (C.1), i. e.,

$$e^{-\frac{i}{\hbar}\boldsymbol{J}_0\cdot\boldsymbol{\varphi}} \mathcal{A}_0(t,\boldsymbol{x}) e^{\frac{i}{\hbar}\boldsymbol{J}_0\cdot\boldsymbol{\varphi}} = \mathcal{A}_0(t,\boldsymbol{\varphi}\boldsymbol{x}),$$

$$e^{-\frac{i}{\hbar}\boldsymbol{J}_0\cdot\boldsymbol{\varphi}} \boldsymbol{\mathcal{A}}(t,\boldsymbol{x}) e^{\frac{i}{\hbar}\boldsymbol{J}_0\cdot\boldsymbol{\varphi}} = \boldsymbol{\varphi}^{-1}\boldsymbol{\mathcal{A}}(t,\boldsymbol{\varphi}\boldsymbol{x}),$$

$$e^{-\frac{i}{\hbar}\boldsymbol{P}_0\cdot\boldsymbol{r}} e^{\frac{i}{\hbar}H_0\tau} \mathcal{A}_\mu(t,\boldsymbol{x}) e^{-\frac{i}{\hbar}H_0\tau} e^{\frac{i}{\hbar}\boldsymbol{P}_0\cdot\boldsymbol{r}} = \mathcal{A}_\mu(t+\tau,\boldsymbol{x}+\boldsymbol{r}), \tag{C.25}$$

boost transformations deviate from our expectation, i. e.,

$$e^{-\frac{ic}{\hbar}\boldsymbol{K}_0\cdot\boldsymbol{\theta}} \mathcal{A}_\mu(\tilde{x}) e^{\frac{ic}{\hbar}\boldsymbol{K}_0\cdot\boldsymbol{\theta}} \neq \sum_\nu (\tilde{\boldsymbol{\theta}}^{-1})_\mu^{\ \nu} \mathcal{A}_\nu(\tilde{\boldsymbol{\theta}}\tilde{x}), \tag{C.26}$$

where $\tilde{\boldsymbol{\theta}}$ is the 4×4 boost matrix (**1**-J.10). In order to see this difference, we first use equations (1.47)–(1.48) and write

$$U_0(\boldsymbol{\theta};\boldsymbol{0};\boldsymbol{0};0)\mathcal{A}_\mu(\tilde{x})U_0^{-1}(\boldsymbol{\theta};\boldsymbol{0};\boldsymbol{0};0)$$

$$= \frac{\hbar c}{(2\pi\hbar)^{3/2}} \int \frac{d\boldsymbol{p}}{\sqrt{2pc}} \sum_\tau \Big(e^{-\frac{i}{\hbar}\tilde{p}\cdot\tilde{x}} e_\mu(\boldsymbol{p},\tau) e^{-\frac{ic}{\hbar}\boldsymbol{K}_0\cdot\boldsymbol{\theta}} c_{\boldsymbol{p}\tau} e^{\frac{ic}{\hbar}\boldsymbol{K}_0\cdot\boldsymbol{\theta}}$$

$$+ e^{\frac{i}{\hbar}\tilde{p}\cdot\tilde{x}} e_\mu^*(\boldsymbol{p},\tau) e^{-\frac{ic}{\hbar}\boldsymbol{K}_0\cdot\boldsymbol{\theta}} c_{\boldsymbol{p}\tau}^\dagger e^{\frac{ic}{\hbar}\boldsymbol{K}_0\cdot\boldsymbol{\theta}}\Big)$$

$$= \frac{\hbar c}{(2\pi\hbar)^{3/2}} \int \frac{d\boldsymbol{p}}{\sqrt{2pc}} \sqrt{\frac{|\boldsymbol{\theta p}|}{p}} \sum_\tau (e^{-\frac{i}{\hbar}\tilde{p}\cdot\tilde{x}} e_\mu(\boldsymbol{p},\tau) e^{-i\tau\varphi_W(\boldsymbol{p},\boldsymbol{\theta})} c_{(\boldsymbol{\theta p})\tau}$$
$$+ e^{\frac{i}{\hbar}\tilde{p}\cdot\tilde{x}} e_\mu^*(\boldsymbol{p},\tau) e^{i\tau\varphi_W(\boldsymbol{p},\boldsymbol{\theta})} c^\dagger_{(\boldsymbol{\theta p})\tau}). \tag{C.27}$$

Next we take equation (C.4) for the momentum $\boldsymbol{\theta p}$, i. e.,

$$\tilde{e}(\boldsymbol{\theta p},\tau) = \Lambda_{\boldsymbol{\theta p}} \tilde{e}(\boldsymbol{\kappa},\tau),$$

and multiply both sides by $\tilde{\boldsymbol{\theta}}^{-1}$ from the left, so

$$\tilde{\boldsymbol{\theta}}^{-1} \tilde{e}(\boldsymbol{\theta p},\tau) = \Lambda_{\boldsymbol{p}} (\Lambda_{\boldsymbol{p}}^{-1} \circ \tilde{\boldsymbol{\theta}}^{-1} \circ \Lambda_{\boldsymbol{\theta p}}) \tilde{e}(\boldsymbol{\kappa},\tau). \tag{C.28}$$

The term in parentheses is the familiar little group element, which keeps the standard vector $\boldsymbol{\kappa}$ unchanged and rotates by the Wigner angle $-\varphi_W(\boldsymbol{p},\boldsymbol{\theta})$ in the κ-space (see Section **1**-5.4). So, we can use representation (**1**-5.62) to get

$$(\Lambda_{\boldsymbol{p}}^{-1} \circ \tilde{\boldsymbol{\theta}}^{-1} \circ \Lambda_{\boldsymbol{\theta p}}) \tilde{e}(\boldsymbol{\kappa},\tau) = \tilde{\Sigma}(X_1, X_2, -\varphi_W) \tilde{e}(\boldsymbol{\kappa},\tau)$$
$$= \begin{bmatrix} 1 + (X_1^2 + X_2^2)/2 & X_1 & X_2 & -(X_1^2 + X_2^2)/2 \\ X_1 c_W - X_2 s_W & c_W & -s_W & -X_1 c_W + X_2 s_W \\ X_1 s_W + X_2 c_W & s_W & c_W & -X_1 s_W - X_2 c_W \\ (X_1^2 + X_2^2)/2 & X_1 & X_2 & 1 - (X_1^2 + X_2^2)/2 \end{bmatrix} \begin{bmatrix} 0 \\ 1 \\ i\tau \\ 0 \end{bmatrix}$$
$$= e^{-i\tau\varphi_W(\boldsymbol{p},\boldsymbol{\theta})} \begin{bmatrix} 0 \\ 1 \\ i\tau \\ 0 \end{bmatrix} + (X_1 + i\tau X_2) \begin{bmatrix} 1 \\ 0 \\ 0 \\ 1 \end{bmatrix}$$
$$= e^{-i\tau\varphi_W(\boldsymbol{p},\boldsymbol{\theta})} \tilde{e}(\boldsymbol{\kappa},\tau) + \frac{X_1 + i\tau X_2}{c} \tilde{\kappa}, \tag{C.29}$$

where $c_W \equiv \cos\varphi_W(\boldsymbol{p},\boldsymbol{\theta})$, $s_W \equiv \sin\varphi_W(\boldsymbol{p},\boldsymbol{\theta})$, $\tilde{\kappa} = (c, 0, 0, c)$ is the standard energy–momentum 4-vector and X_1, X_2 are some functions of $\boldsymbol{\theta}$ and $\boldsymbol{p}$. Our next goal is to remove these unknown functions from the field transformation formula. Denoting

$$X_\tau(\boldsymbol{p},\boldsymbol{\theta}) = \frac{X_1 + i\tau X_2}{c}, \tag{C.30}$$

we obtain from (C.28)–(C.30)[3]

$$(\tilde{\boldsymbol{\theta}}^{-1} \tilde{e}(\boldsymbol{\theta p},\tau))_\mu \equiv (\tilde{\boldsymbol{\theta}}^{-1})_\mu^{\ \nu} e_\nu(\boldsymbol{\theta p},\tau)$$
$$= (\Lambda_{\boldsymbol{p}})_\mu^{\ \nu} (e^{-i\tau\varphi_W(\boldsymbol{p},\boldsymbol{\theta})} e_\nu(\boldsymbol{\kappa},\tau) + X_\tau(\boldsymbol{p},\boldsymbol{\theta}) \kappa_\nu)$$
$$= e^{-i\tau\varphi_W(\boldsymbol{p},\boldsymbol{\theta})} e_\mu(\boldsymbol{p},\tau) + X_\tau(\boldsymbol{p},\boldsymbol{\theta}) p_\mu, \tag{C.31}$$

3 As usual, we assume summation over repeated indices ν.

where $p_\mu = (cp, cp_x, cp_y, cp_z)$ is the energy–momentum 4-vector corresponding to the photon's momentum $\boldsymbol{p}$. By setting $\mu = 0$ and taking into account (C.11), we also get

$$(\tilde{\boldsymbol{\theta}}^{-1})_0^{\ \nu} e_\nu(\boldsymbol{\theta p}, \tau) = e^{-i\tau\varphi_W(\boldsymbol{p},\boldsymbol{\theta})} e_0(\boldsymbol{p}, \tau) + X_\tau(\boldsymbol{p}, \boldsymbol{\theta}) p_0 = X_\tau(\boldsymbol{p}, \boldsymbol{\theta}) cp,$$

$$e^{-i\tau\varphi_W(\boldsymbol{p},\boldsymbol{\theta})} e_\mu(\boldsymbol{p}, \tau) = (\tilde{\boldsymbol{\theta}}^{-1})_\mu^{\ \nu} e_\nu(\boldsymbol{\theta p}, \tau) - X_\tau(\boldsymbol{p}, \boldsymbol{\theta}) p_\mu$$
$$= \left[(\tilde{\boldsymbol{\theta}}^{-1})_\mu^{\ \nu} - \frac{p_\mu}{cp}(\tilde{\boldsymbol{\theta}}^{-1})_0^{\ \nu}\right] e_\nu(\boldsymbol{\theta p}, \tau).$$

Since $\tilde{\boldsymbol{\theta}}$ from (**1**-J.10) are real matrices, the complex conjugate of this equality is

$$e^{i\tau\varphi_W(\boldsymbol{p},\boldsymbol{\theta})} e_\mu^*(\boldsymbol{p}, \tau) = \left[(\tilde{\boldsymbol{\theta}}^{-1})_\mu^{\ \nu} - \frac{p_\mu}{cp}(\tilde{\boldsymbol{\theta}}^{-1})_0^{\ \nu}\right] e_\nu^*(\boldsymbol{\theta p}, \tau)$$

and transformation (C.27) can be written in its final form[4]

$$\begin{aligned}
&e^{-\frac{i}{\hbar}\boldsymbol{K}_0\cdot\boldsymbol{\theta}} \mathcal{A}_\mu(\tilde{x}) e^{\frac{i}{\hbar}\boldsymbol{K}_0\cdot\boldsymbol{\theta}} \\
&= \frac{\hbar c}{(2\pi\hbar)^{3/2}} \int \frac{d\boldsymbol{p}}{\sqrt{2pc}} \sqrt{\frac{|\boldsymbol{\theta p}|}{p}} \sum_{\tau=-1}^{1} \Bigg(e^{-\frac{i}{\hbar}\tilde{p}\cdot\tilde{x}} \left[(\tilde{\boldsymbol{\theta}}^{-1})_\mu^{\ \nu} - \frac{p_\mu}{cp}(\tilde{\boldsymbol{\theta}}^{-1})_0^{\ \nu}\right] e_\nu(\boldsymbol{\theta p}, \tau) c_{\boldsymbol{\theta p}\tau} \\
&\quad + e^{\frac{i}{\hbar}\tilde{p}\cdot\tilde{x}} \left[(\tilde{\boldsymbol{\theta}}^{-1})_\mu^{\ \nu} - \frac{p_\mu}{cp}(\tilde{\boldsymbol{\theta}}^{-1})_0^{\ \nu}\right] e_\nu^*(\boldsymbol{\theta p}, \tau) c_{\boldsymbol{\theta p}\tau}^\dagger \Bigg) \\
&= \frac{\hbar\sqrt{c}}{\sqrt{2}(2\pi\hbar)^{3/2}} \int \frac{d(\boldsymbol{\theta p})}{|\boldsymbol{\theta p}|} \sqrt{|\boldsymbol{\theta p}|} \sum_{\tau=-1}^{1} \left[(\tilde{\boldsymbol{\theta}}^{-1})_\mu^{\ \nu} - \frac{p_\mu}{cp}(\tilde{\boldsymbol{\theta}}^{-1})_0^{\ \nu}\right] \\
&\quad \times (e^{-\frac{i}{\hbar}\tilde{p}\cdot\tilde{x}} e_\nu(\boldsymbol{\theta p}, \tau) c_{\boldsymbol{\theta p}\tau} + e^{\frac{i}{\hbar}\tilde{p}\cdot\tilde{x}} e_\nu^*(\boldsymbol{\theta p}, \tau) c_{\boldsymbol{\theta p}\tau}^\dagger) \\
&= (\tilde{\boldsymbol{\theta}}^{-1})_\mu^{\ \nu} \left[\frac{\hbar c}{(2\pi\hbar)^{3/2}} \int \frac{d\boldsymbol{p}}{\sqrt{2pc}} \sum_{\tau=-1}^{1} (e^{-\frac{i}{\hbar}\tilde{p}\cdot\tilde{\boldsymbol{\theta}}\tilde{x}} e_\nu(\boldsymbol{p}, \tau) c_{\boldsymbol{p}\tau} + e^{\frac{i}{\hbar}\tilde{p}\cdot\tilde{\boldsymbol{\theta}}\tilde{x}} e_\nu^*(\boldsymbol{p}, \tau) c_{\boldsymbol{p}\tau}^\dagger) \right] \\
&\quad - \frac{\hbar}{(2\pi\hbar)^{3/2}} \int \frac{d\boldsymbol{p}}{\sqrt{2pc}} \sum_{\tau=-1}^{1} \frac{(\tilde{\boldsymbol{\theta}}^{-1} p)_\mu}{|\boldsymbol{\theta}^{-1}\boldsymbol{p}|} (\tilde{\boldsymbol{\theta}}^{-1})_0^{\ \nu} \\
&\quad \times [e^{-\frac{i}{\hbar}\tilde{\boldsymbol{\theta}}^{-1}\tilde{p}\cdot\tilde{x}} e_\nu(\boldsymbol{p}, \tau) c_{\boldsymbol{p}\tau} + e^{\frac{i}{\hbar}\tilde{\boldsymbol{\theta}}^{-1}\tilde{p}\cdot\tilde{x}} e_\nu^*(\boldsymbol{p}, \tau) c_{\boldsymbol{p}\tau}^\dagger] \\
&= (\tilde{\boldsymbol{\theta}}^{-1})_\mu^{\ \nu} \mathcal{A}_\nu(\tilde{\boldsymbol{\theta}}\tilde{x}) + \Omega_\mu(\tilde{x}, \boldsymbol{\theta}).
\end{aligned} \tag{C.32}$$

So, we see that property (C.1) is not satisfied for boosts. In addition to the desired covariant term $\tilde{\boldsymbol{\theta}}^{-1}\tilde{\mathcal{A}}(\tilde{\boldsymbol{\theta}}\tilde{x})$, we get the unwanted correction

$$\begin{aligned}
\Omega_\mu(\tilde{x}, \boldsymbol{\theta}) \equiv &- \frac{\hbar}{(2\pi\hbar)^{3/2}} \int \frac{d\boldsymbol{p}}{\sqrt{2pc}} \sum_{\tau=-1}^{1} \frac{(\tilde{\boldsymbol{\theta}}^{-1} p)_\mu}{|\boldsymbol{\theta}^{-1}\boldsymbol{p}|} (\tilde{\boldsymbol{\theta}}^{-1})_0^{\ \nu} \\
&\times [e^{-\frac{i}{\hbar}\tilde{\boldsymbol{\theta}}^{-1}\tilde{p}\cdot\tilde{x}} e_\nu(\boldsymbol{p}, \tau) c_{\boldsymbol{p}\tau} + e^{\frac{i}{\hbar}\tilde{\boldsymbol{\theta}}^{-1}\tilde{p}\cdot\tilde{x}} e_\nu^*(\boldsymbol{p}, \tau) c_{\boldsymbol{p}\tau}^\dagger],
\end{aligned} \tag{C.33}$$

4 Here we used the Lorentz invariance of the integration measure $d\boldsymbol{p}/(cp)$ (**1**-5.31) and equation (**1**-J.6).

which, by the way, can be expressed as a 4-gradient as follows:

$$\Omega_\mu(\tilde{x}, \boldsymbol{\theta}) = \partial_\mu \frac{i\hbar^2 c}{(2\pi\hbar)^{3/2}} \int \frac{d\boldsymbol{p}}{\sqrt{2pc}} \sum_{\tau=-1}^{1} \frac{(\tilde{\boldsymbol{\theta}}^{-1})_0^{\nu}}{|\boldsymbol{\theta}^{-1}\boldsymbol{p}|}$$
$$\times [e^{-\frac{i}{\hbar}\tilde{\boldsymbol{\theta}}^{-1}\tilde{p}\cdot\tilde{x}} e_\nu(\boldsymbol{p},\tau) c_{\boldsymbol{p}\tau} - e^{\frac{i}{\hbar}\tilde{\boldsymbol{\theta}}^{-1}\tilde{p}\cdot\tilde{x}} e_\nu^*(\boldsymbol{p},\tau) c_{\boldsymbol{p}\tau}^\dagger]. \quad \text{(C.34)}$$

The presence of this correction is the reason why interaction operators in the theory with massless photons (QED) cannot be constructed by simple rules from Subsection 3.1.2. For QED we need a more complex construction (explained in Subsection 3.1.3), in which the interacting boost has a more complicated form (3.17).

From

$$\lim_{\theta \to 0} (\tilde{\boldsymbol{\theta}}^{-1})_0^{\nu} e_\nu(\boldsymbol{p},\tau) = \delta_{0\nu} e_\nu(\boldsymbol{p},\tau) = e_0(\boldsymbol{p},\tau) = 0, \quad \text{(C.35)}$$

we obtain the following useful property:

$$\tilde{\Omega}(\tilde{x}, \boldsymbol{0}) = 0. \quad \text{(C.36)}$$

D QED interaction in terms of particle operators

D.1 Current density

In QED, an important role is played by the so-called *current density* operator, which is defined as the following sum of electron–positron $j^{\mu}_{ep}(\tilde{x})$ and proton–antiproton $j^{\mu}_{pa}(\tilde{x})$ components

$$\begin{aligned} j^{\mu}(\tilde{x}) &= j^{\mu}_{ep}(\tilde{x}) + j^{\mu}_{pa}(\tilde{x}) \\ &\equiv -e\overline{\psi}(\tilde{x})\gamma^{\mu}\psi(\tilde{x}) + e\overline{\Psi}(\tilde{x})\gamma^{\mu}\Psi(\tilde{x}), \end{aligned} \tag{D.1}$$

where e is the proton's charge, gamma matrices γ^{μ} are taken from definitions (B.5)–(B.6) and quantum fields $\psi(\tilde{x})$, $\overline{\psi}(\tilde{x})$, $\Psi(\tilde{x})$ and $\overline{\Psi}(\tilde{x})$ are taken from Appendix B.4.[1] Let us consider the electron–positron part $j^{\mu}_{ep}(\tilde{x})$ of the current density and derive three important properties of this operator function.[2] First, we claim that $j^{\mu}_{ep}(\tilde{x})$ transforms as a 4-vector function on the Minkowski space–time. In particular, it is not difficult to show that space–time translations simply shift the arguments, i. e.,

$$e^{-\frac{i}{\hbar}\boldsymbol{P}_0\cdot\boldsymbol{r}}e^{\frac{i}{\hbar}H_0 t}j^{\mu}_{ep}(\tilde{x})e^{-\frac{i}{\hbar}H_0 t}e^{\frac{i}{\hbar}\boldsymbol{P}_0\cdot\boldsymbol{r}} = j^{\mu}_{ep}(\tilde{x}+\tilde{a}), \tag{D.2}$$

where $\tilde{a} = (t, \boldsymbol{r}/c)$.

Verification of the boost transformation is a bit more complicated. We use equations (B.31), (B.33) and (B.70) to obtain

$$\begin{aligned} e^{-\frac{ic}{\hbar}\boldsymbol{K}_0\cdot\boldsymbol{\theta}}j^{\mu}_{ep}(\tilde{x})e^{\frac{ic}{\hbar}\boldsymbol{K}_0\cdot\boldsymbol{\theta}} &= -ee^{-\frac{ic}{\hbar}\boldsymbol{K}_0\cdot\boldsymbol{\theta}}\psi^{\dagger}(\tilde{x})\gamma^0\gamma^{\mu}\psi(\tilde{x})e^{\frac{ic}{\hbar}\boldsymbol{K}_0\cdot\boldsymbol{\theta}} \\ &= -ee^{-\frac{ic}{\hbar}\boldsymbol{K}_0\cdot\boldsymbol{\theta}}\psi^{\dagger}(\tilde{x})e^{\frac{ic}{\hbar}\boldsymbol{K}_0\cdot\boldsymbol{\theta}}\gamma^0\gamma^{\mu}e^{-\frac{ic}{\hbar}\boldsymbol{K}_0\cdot\boldsymbol{\theta}}\psi(\tilde{x})e^{\frac{ic}{\hbar}\boldsymbol{K}_0\cdot\boldsymbol{\theta}} \\ &= -e\psi^{\dagger}(\tilde{\boldsymbol{\theta}}\tilde{x})\mathcal{D}^{\dagger}(\boldsymbol{\theta}^{-1})\gamma^0\gamma^{\mu}\mathcal{D}(\boldsymbol{\theta}^{-1})\psi(\tilde{\boldsymbol{\theta}}\tilde{x}) \\ &= -e\psi^{\dagger}(\tilde{\boldsymbol{\theta}}\tilde{x})\mathcal{D}(\boldsymbol{\theta}^{-1})\gamma^0\mathcal{D}(\boldsymbol{\theta}^{-1})\mathcal{D}(\boldsymbol{\theta})\gamma^{\mu}\mathcal{D}(\boldsymbol{\theta}^{-1})\psi(\tilde{\boldsymbol{\theta}}\tilde{x}) \\ &= -e\psi^{\dagger}(\tilde{\boldsymbol{\theta}}\tilde{x})\gamma^0\mathcal{D}(\boldsymbol{\theta})\gamma^{\mu}\mathcal{D}(\boldsymbol{\theta}^{-1})\psi(\tilde{\boldsymbol{\theta}}\tilde{x}) \\ &= -e\sum_{\nu=0}^{3}\psi^{\dagger}(\tilde{\boldsymbol{\theta}}\tilde{x})\gamma^0(\tilde{\boldsymbol{\theta}}^{-1})^{\mu}_{\ \nu}\gamma^{\nu}\psi(\tilde{\boldsymbol{\theta}}\tilde{x}) \\ &= \sum_{\nu=0}^{3}(\tilde{\boldsymbol{\theta}}^{-1})^{\mu}_{\ \nu}j^{\nu}_{ep}(\tilde{\boldsymbol{\theta}}\tilde{x}). \end{aligned} \tag{D.3}$$

---

1 Note that $\psi(\tilde{x})$ is a four-component column and $\overline{\psi}(\tilde{x})$ is a four-component row, so the product $\overline{\psi}(\tilde{x})\gamma^{\mu}\psi(\tilde{x})$ is a scalar in the 4D space of Dirac indices. We would like to remind the reader that in our interpretation quantum fields are formal mathematical objects and that the "current density" operator has no relationship to electric currents measured in laboratories.

2 The proton–antiproton part $j^{\mu}_{pa}(\tilde{x})$ has similar properties.

https://doi.org/10.1515/9783110493207-008

From this and (**1**-J.13) we obtain a useful commutator,

$$
\begin{aligned}
[K_{0z}, \mathrm{j}^0_{\mathrm{ep}}(\tilde{x})] &= \frac{i\hbar}{c} \lim_{\theta\to 0} \frac{d}{d\theta} e^{-\frac{ic}{\hbar}K_{0z}\theta} \mathrm{j}^0_{\mathrm{ep}}(\tilde{x}) e^{\frac{ic}{\hbar}K_{0z}\theta} \\
&= \frac{i\hbar}{c} \lim_{\theta\to 0} \frac{d}{d\theta} \left[\mathrm{j}^0_{\mathrm{ep}}\left(t\cosh\theta + \frac{z}{c}\sinh\theta, x, y, z\ \cosh\theta + ct\sinh\theta\right)\cosh\theta \right. \\
&\quad \left. - \mathrm{j}^z_{\mathrm{ep}}\left(t\cosh\theta + \frac{z}{c}\sinh\theta, x, y, z\cosh\theta + ct\sinh\theta\right)\sinh\theta \right] \\
&= i\hbar\left(\frac{z}{c^2}\frac{d}{dt} + t\frac{d}{dz}\right)\mathrm{j}^0_{\mathrm{ep}}(\tilde{x}) - \frac{i\hbar}{c}\mathrm{j}^z_{\mathrm{ep}}(\tilde{x}).
\end{aligned}
\tag{D.4}
$$

Second, the current density satisfies the *continuity equation*, which can be proven with the help of Dirac equations (B.90) and (B.92),

$$
\begin{aligned}
\frac{\partial}{\partial t}\mathrm{j}^0_{\mathrm{ep}}(\tilde{x}) &= -e\frac{\partial}{\partial t}(\overline{\psi}(\tilde{x})\gamma^0\psi(\tilde{x})) \\
&= -e\left(\frac{\partial}{\partial t}\overline{\psi}(\tilde{x})\gamma^0\right)\psi(\tilde{x}) - e\overline{\psi}(\tilde{x})\left(\gamma^0\frac{\partial}{\partial t}\psi(\tilde{x})\right) \\
&= e\left(c\frac{\partial\overline{\psi}(\tilde{x})}{\partial \boldsymbol{x}}\cdot\boldsymbol{\gamma} + \frac{im_ec^2}{\hbar}\overline{\psi}(\tilde{x})\right)\psi(\tilde{x}) + e\overline{\psi}(\tilde{x})\left(c\boldsymbol{\gamma}\cdot\frac{\partial}{\partial\boldsymbol{x}}\psi(\tilde{x}) - \frac{im_ec^2}{\hbar}\psi(\tilde{x})\right) \\
&= ec\frac{\partial\overline{\psi}(\tilde{x})}{\partial\boldsymbol{x}}\cdot\boldsymbol{\gamma}\psi(\tilde{x}) + ec\overline{\psi}(\tilde{x})\boldsymbol{\gamma}\cdot\frac{\partial\psi(\tilde{x})}{\partial\boldsymbol{x}} \\
&= ec\frac{\partial}{\partial\boldsymbol{x}}\cdot(\overline{\psi}(\tilde{x})\boldsymbol{\gamma}\psi(\tilde{x})) \\
&= -c\frac{\partial}{\partial\boldsymbol{x}}\cdot\mathbf{j}_{\mathrm{ep}}(\tilde{x}).
\end{aligned}
\tag{D.5}
$$

Third, from equations (B.84)–(B.85) it follows that components of the current density commute at space-like intervals, i. e.,

$$
[\mathrm{j}^\mu_{\mathrm{ep}}(t,\boldsymbol{x}), \mathrm{j}^\nu_{\mathrm{ep}}(t,\boldsymbol{y})] = 0, \quad \text{if } \boldsymbol{x} \neq \boldsymbol{y}. \tag{D.6}
$$

Using expressions for the fields (B.60)–(B.63), we can also write operator (D.1) as a normally ordered polynomial in creation and annihilation operators, i. e.,[3]

$$
\begin{aligned}
\mathrm{j}^\mu(\tilde{x}) &= -e\overline{\psi}(\tilde{x})\gamma^\mu\psi(\tilde{x}) + e\overline{\Psi}(\tilde{x})\gamma^\mu\Psi(\tilde{x}) \\
&= e\int\frac{d\boldsymbol{p}d\boldsymbol{p}'}{(2\pi\hbar)^3} \\
&\quad \times (-[e^{\frac{i}{\hbar}\tilde{p}\cdot\tilde{x}}\overline{A}^\dagger_a(\boldsymbol{p}) + e^{-\frac{i}{\hbar}\tilde{p}\cdot\tilde{x}}\overline{B}_a(\boldsymbol{p})]\gamma^\mu_{ab}[e^{-\frac{i}{\hbar}\tilde{p}'\cdot\tilde{x}}A_b(\boldsymbol{p}') + e^{\frac{i}{\hbar}\tilde{p}'\cdot\tilde{x}}B^\dagger_b(\boldsymbol{p}')] \\
&\quad + [e^{\frac{i}{\hbar}\tilde{P}\cdot\tilde{x}}\overline{D}^\dagger_a(\boldsymbol{p}) + e^{-\frac{i}{\hbar}\tilde{P}\cdot\tilde{x}}\overline{F}_a(\boldsymbol{p})]\gamma^\mu_{ab}[e^{-\frac{i}{\hbar}\tilde{P}'\cdot\tilde{x}}D_b(\boldsymbol{p}') + e^{\frac{i}{\hbar}\tilde{P}'\cdot\tilde{x}}F^\dagger_b(\boldsymbol{p}')])
\end{aligned}
$$

3 We assume summation over Dirac indices a and b.

$$
\begin{aligned}
&= e\int \frac{d\mathbf{p}d\mathbf{p}'}{(2\pi\hbar)^3}\gamma^\mu_{ab} \\
&\times (-\overline{A}^\dagger_a(\mathbf{p})A_b(\mathbf{p}')e^{-\frac{i}{\hbar}(\tilde{p}'-\tilde{p})\cdot\tilde{x}} - \overline{A}^\dagger_a(\mathbf{p})B^\dagger_b(\mathbf{p}')e^{\frac{i}{\hbar}(\tilde{p}'+\tilde{p})\cdot\tilde{x}} - \overline{B}_a(\mathbf{p})A_b(\mathbf{p}')e^{-\frac{i}{\hbar}(\tilde{p}'+\tilde{p})\cdot\tilde{x}} \\
&- \overline{B}_a(\mathbf{p})B^\dagger_b(\mathbf{p}')e^{\frac{i}{\hbar}(\tilde{p}'-\tilde{p})\cdot\tilde{x}} + \overline{D}^\dagger_a(\mathbf{p})D_b(\mathbf{p}')e^{-\frac{i}{\hbar}(\tilde{P}'-\tilde{P})\cdot\tilde{x}} + \overline{D}^\dagger_a(\mathbf{p})F^\dagger_b(\mathbf{p}')e^{+\frac{i}{\hbar}(\tilde{P}'+\tilde{P})\cdot\tilde{x}} \\
&+ \overline{F}_a(\mathbf{p})D_b(\mathbf{p}')e^{-\frac{i}{\hbar}(\tilde{P}'+\tilde{P})\cdot\tilde{x}} + \overline{F}_a(\mathbf{p})F^\dagger_b(\mathbf{p}')e^{\frac{i}{\hbar}(\tilde{P}'-\tilde{P})\cdot\tilde{x}}) \\
&= e\int \frac{d\mathbf{p}d\mathbf{p}'}{(2\pi\hbar)^3}\gamma^\mu_{ab} \\
&\times (-\overline{A}^\dagger_a(\mathbf{p})A_b(\mathbf{p}')e^{-\frac{i}{\hbar}(\tilde{p}'-\tilde{p})\cdot\tilde{x}} - \overline{A}^\dagger_a(\mathbf{p})B^\dagger_b(\mathbf{p}')e^{\frac{i}{\hbar}(\tilde{p}'+\tilde{p})\cdot\tilde{x}} - \overline{B}_a(\mathbf{p})A_b(\mathbf{p}')e^{-\frac{i}{\hbar}(\tilde{p}'+\tilde{p})\cdot\tilde{x}} \\
&+ B^\dagger_b(\mathbf{p}')\overline{B}_a(\mathbf{p})e^{\frac{i}{\hbar}(\tilde{p}'-\tilde{p})\cdot\tilde{x}} + \overline{D}^\dagger_a(\mathbf{p})D_b(\mathbf{p}')e^{-\frac{i}{\hbar}(\tilde{P}'-\tilde{P})\cdot\tilde{x}} + \overline{D}^\dagger_a(\mathbf{p})F^\dagger_b(\mathbf{p}')e^{+\frac{i}{\hbar}(\tilde{P}'+\tilde{P})\cdot\tilde{x}} \\
&+ \overline{F}_a(\mathbf{p})D_b(\mathbf{p}')e^{-\frac{i}{\hbar}(\tilde{P}'+\tilde{P})\cdot\tilde{x}} - F^\dagger_b(\mathbf{p}')\overline{F}_a(\mathbf{p})e^{\frac{i}{\hbar}(\tilde{P}'-\tilde{P})\cdot\tilde{x}} \\
&- \{\overline{B}_a(\mathbf{p}), B^\dagger_b(\mathbf{p}')\}e^{\frac{i}{\hbar}(\tilde{p}'-\tilde{p})\cdot\tilde{x}} + \{\overline{F}_a(\mathbf{p}), F^\dagger_b(\mathbf{p}')\}e^{\frac{i}{\hbar}(\tilde{P}'-\tilde{P})\cdot\tilde{x}}).
\end{aligned}
$$

Let us now show that the last two terms cancel out. Using the anticommutator (B.87) and properties of the gamma matrices, we rewrite these terms as

$$
\begin{aligned}
& e\int \frac{d\mathbf{p}d\mathbf{p}'}{(2\pi\hbar)^3}\gamma^\mu_{ab}\delta(\mathbf{p}-\mathbf{p}') \\
&\times\left(-\frac{(\gamma^0\omega_{\mathbf{p}} + \boldsymbol{\gamma}\cdot\mathbf{p}c - m_ec^2)_{ba}}{2\omega_{\mathbf{p}}}e^{\frac{i}{\hbar}(\tilde{p}'-\tilde{p})\cdot x} + \frac{(\gamma^0\Omega_{\mathbf{p}} + \boldsymbol{\gamma}\cdot\mathbf{p}c - m_pc^2)_{ba}}{2\Omega_{\mathbf{p}}}e^{\frac{i}{\hbar}(\tilde{P}'-\tilde{P})\cdot x}\right) \\
&= e\int \frac{d\mathbf{p}}{(2\pi\hbar)^3}\gamma^\mu_{aa}\left(\frac{m_ec^2}{2\omega_{\mathbf{p}}} - \frac{m_pc^2}{2\Omega_{\mathbf{p}}}\right) + e\int \frac{d\mathbf{p}}{(2\pi\hbar)^3}(\gamma^\mu\boldsymbol{\gamma})_{aa}\cdot\left(-\frac{\mathbf{p}c}{2\omega_{\mathbf{p}}} + \frac{\mathbf{p}c}{2\Omega_{\mathbf{p}}}\right) \\
&+ e\int \frac{d\mathbf{p}}{(2\pi\hbar)^3}(\gamma^\mu\gamma^0)_{aa}\left(-\frac{1}{2} + \frac{1}{2}\right) \qquad \text{(D.7)} \\
&= e\,\mathrm{Tr}(\gamma^\mu)\int \frac{d\mathbf{p}}{(2\pi\hbar)^3}\left(\frac{m_ec^2}{2\omega_{\mathbf{p}}} - \frac{m_pc^2}{2\Omega_{\mathbf{p}}}\right) + ec\,\mathrm{Tr}(\gamma^\mu\boldsymbol{\gamma})\cdot\int \frac{d\mathbf{p}\mathbf{p}}{(2\pi\hbar)^3}\left(-\frac{1}{2\omega_{\mathbf{p}}} + \frac{1}{2\Omega_{\mathbf{p}}}\right).
\end{aligned}
$$

The first term drops out due to (B.11). The second integral vanishes, because the integrand is an odd function of $\mathbf{p}$. Finally, the normally ordered current density operator has the form

$$
\begin{aligned}
\mathrm{j}^\mu(\tilde{x}) &= e\int \frac{d\mathbf{p}d\mathbf{p}'}{(2\pi\hbar)^3}\gamma^\mu_{ab} \\
&\times (-\overline{A}^\dagger_a(\mathbf{p})A_b(\mathbf{p}')e^{-\frac{i}{\hbar}(\tilde{p}'-\tilde{p})\cdot\tilde{x}} - \overline{A}^\dagger_a(\mathbf{p})B^\dagger_b(\mathbf{p}')e^{\frac{i}{\hbar}(\tilde{p}'+\tilde{p})\cdot\tilde{x}} - \overline{B}_a(\mathbf{p})A_b(\mathbf{p}')e^{-\frac{i}{\hbar}(\tilde{p}'+\tilde{p})\cdot\tilde{x}} \\
&+ B^\dagger_b(\mathbf{p}')\overline{B}_a(\mathbf{p})e^{\frac{i}{\hbar}(\tilde{p}'-\tilde{p})\cdot\tilde{x}} + \overline{D}^\dagger_a(\mathbf{p})D_b(\mathbf{p}')e^{-\frac{i}{\hbar}(\tilde{P}'-\tilde{P})\cdot\tilde{x}} + \overline{D}^\dagger_a(\mathbf{p})F^\dagger_b(\mathbf{p}')e^{+\frac{i}{\hbar}(\tilde{P}'+\tilde{P})\cdot\tilde{x}} \\
&+ \overline{F}_a(\mathbf{p})D_b(\mathbf{p}')e^{-\frac{i}{\hbar}(\tilde{P}'+\tilde{P})\cdot\tilde{x}} - F^\dagger_b(\mathbf{p}')\overline{F}_a(\mathbf{p})e^{\frac{i}{\hbar}(\tilde{P}'-\tilde{P})\cdot\tilde{x}}). \qquad \text{(D.8)}
\end{aligned}
$$

D.2 First-order interaction in QED

Substituting (D.8) and (C.17) in (3.14), we obtain the first-order QED interaction expressed through creation and annihilation operators. We have

$$
\begin{aligned}
V_1 &= \int d\boldsymbol{x} j^{\mu}(0,\boldsymbol{x})\mathcal{A}_{\mu}(0,\boldsymbol{x}) \\
&= \frac{e}{(2\pi\hbar)^{9/2}} \int d\boldsymbol{x} d\boldsymbol{p} d\boldsymbol{p}' d\boldsymbol{k} (-\overline{A}_a^{\dagger}(\boldsymbol{p}) A_b(\boldsymbol{p}') e^{-\frac{i}{\hbar}(\boldsymbol{p}'-\boldsymbol{p})\cdot\boldsymbol{x}} + \cdots) \\
&\quad \times (e^{-\frac{i}{\hbar}\boldsymbol{k}\cdot\boldsymbol{x}} \mathcal{C}_{ab}(\boldsymbol{k}) + e^{\frac{i}{\hbar}\boldsymbol{k}\cdot\boldsymbol{x}} \mathcal{C}_{ab}^{\dagger}(\boldsymbol{k})) \\
&= \frac{e}{(2\pi\hbar)^{3/2}} \int d\boldsymbol{k} d\boldsymbol{p} \\
&\quad \times [-\overline{A}_a^{\dagger}(\boldsymbol{p}+\boldsymbol{k}) A_b(\boldsymbol{p}) \mathcal{C}_{ab}(\boldsymbol{k}) - \overline{A}_a^{\dagger}(\boldsymbol{p}-\boldsymbol{k}) A_b(\boldsymbol{p}) \mathcal{C}_{ab}^{\dagger}(\boldsymbol{k}) + \overline{D}_a^{\dagger}(\boldsymbol{p}+\boldsymbol{k}) D_b(\boldsymbol{p}) \mathcal{C}_{ab}(\boldsymbol{k}) \\
&\quad + \overline{D}_a^{\dagger}(\boldsymbol{p}-\boldsymbol{k}) D_b(\boldsymbol{p}) \mathcal{C}_{ab}^{\dagger}(\boldsymbol{k}) + B_b^{\dagger}(\boldsymbol{p}+\boldsymbol{k}) \overline{B}_a(\boldsymbol{p}) \mathcal{C}_{ab}(\boldsymbol{k}) + B_b^{\dagger}(\boldsymbol{p}-\boldsymbol{k}) \overline{B}_a(\boldsymbol{p}) \mathcal{C}_{ab}^{\dagger}(\boldsymbol{k}) \\
&\quad - F_b^{\dagger}(\boldsymbol{p}+\boldsymbol{k}) \overline{F}_a(\boldsymbol{p}) \mathcal{C}_{ab}(\boldsymbol{k}) - F_b^{\dagger}(\boldsymbol{p}-\boldsymbol{k}) \overline{F}_a(\boldsymbol{p}) \mathcal{C}_{ab}^{\dagger}(\boldsymbol{k}) - \overline{A}_a^{\dagger}(\boldsymbol{p}+\boldsymbol{k}) B_b^{\dagger}(\boldsymbol{p}) \mathcal{C}_{ab}(\boldsymbol{k}) \\
&\quad - \overline{A}_a^{\dagger}(\boldsymbol{p}-\boldsymbol{k}) B_b^{\dagger}(\boldsymbol{p}) \mathcal{C}_{ab}^{\dagger}(\boldsymbol{k}) - A_b(\boldsymbol{p}+\boldsymbol{k}) \overline{B}_a(\boldsymbol{p}) \mathcal{C}_{ab}(\boldsymbol{k}) - A_b(\boldsymbol{p}-\boldsymbol{k}) \overline{B}_a(\boldsymbol{p}) \mathcal{C}_{ab}^{\dagger}(\boldsymbol{k}) \\
&\quad + \overline{D}_a^{\dagger}(\boldsymbol{p}+\boldsymbol{k}) F_b^{\dagger}(\boldsymbol{p}) \mathcal{C}_{ab}(\boldsymbol{k}) + \overline{D}_a^{\dagger}(\boldsymbol{p}-\boldsymbol{k}) F_b^{\dagger}(\boldsymbol{p}) \mathcal{C}_{ab}^{\dagger}(\boldsymbol{k}) + D_b(\boldsymbol{p}+\boldsymbol{k}) \overline{F}_a(\boldsymbol{p}) \mathcal{C}_{ab}(\boldsymbol{k}) \\
&\quad + D_b(\boldsymbol{p}-\boldsymbol{k}) \overline{F}_a(\boldsymbol{p}) \mathcal{C}_{ab}^{\dagger}(\boldsymbol{k})].
\end{aligned} \tag{D.9}
$$

This operator is of the purely *unphys* type.

D.3 Second-order interaction in QED

The second-order interaction potential (3.15) is expressed through creation and annihilation operators by a rather cumbersome formula. We write

$$
\begin{aligned}
V_2 &= \int d\boldsymbol{x} d\boldsymbol{y} j^0(0,\boldsymbol{x}) \frac{1}{8\pi|\boldsymbol{x}-\boldsymbol{y}|} j^0(0,\boldsymbol{y}) \\
&= \frac{e^2}{(2\pi\hbar)^6} \sum_{abcd} \gamma_{ab}^0 \gamma_{cd}^0 \int d\boldsymbol{x} d\boldsymbol{y} \int d\boldsymbol{p} d\boldsymbol{p}' d\boldsymbol{q} d\boldsymbol{q}' \frac{1}{8\pi|\boldsymbol{x}-\boldsymbol{y}|} \\
&\quad \times [-\overline{A}_a^{\dagger}(\boldsymbol{p}) A_b(\boldsymbol{p}') e^{-\frac{i}{\hbar}(\boldsymbol{p}'-\boldsymbol{p})\cdot\boldsymbol{x}} - \overline{A}_a^{\dagger}(\boldsymbol{p}) B_b^{\dagger}(\boldsymbol{p}') e^{\frac{i}{\hbar}(\boldsymbol{p}'+\boldsymbol{p})\cdot\boldsymbol{x}} - \overline{B}_a(\boldsymbol{p}) A_b(\boldsymbol{p}') e^{-\frac{i}{\hbar}(\boldsymbol{p}'+\boldsymbol{p})\cdot\boldsymbol{x}} \\
&\quad - \overline{B}_a(\boldsymbol{p}) B_b^{\dagger}(\boldsymbol{p}') e^{\frac{i}{\hbar}(\boldsymbol{p}'-\boldsymbol{p})\cdot\boldsymbol{x}} + \overline{D}_a^{\dagger}(\boldsymbol{p}) D_b(\boldsymbol{p}') e^{-\frac{i}{\hbar}(\boldsymbol{p}'-\boldsymbol{p})\cdot\boldsymbol{x}} + \overline{D}_a^{\dagger}(\boldsymbol{p}) F_b^{\dagger}(\boldsymbol{p}') e^{\frac{i}{\hbar}(\boldsymbol{p}'+\boldsymbol{p})\cdot\boldsymbol{x}} \\
&\quad + \overline{F}_a(\boldsymbol{p}) D_b(\boldsymbol{p}') e^{-\frac{i}{\hbar}(\boldsymbol{p}'+\boldsymbol{p})\cdot\boldsymbol{x}} + \overline{F}_a(\boldsymbol{p}) F_b^{\dagger}(\boldsymbol{p}') e^{\frac{i}{\hbar}(\boldsymbol{p}'-\boldsymbol{p})\cdot\boldsymbol{x}}] \\
&\quad \times [-\overline{A}_c^{\dagger}(\boldsymbol{q}) A_d(\boldsymbol{q}') e^{-\frac{i}{\hbar}(\boldsymbol{q}'-\boldsymbol{q})\cdot\boldsymbol{y}} - \overline{A}_c^{\dagger}(\boldsymbol{q}) B_d^{\dagger}(\boldsymbol{q}') e^{\frac{i}{\hbar}(\boldsymbol{q}'+\boldsymbol{q})\cdot\boldsymbol{y}} - \overline{B}_c(\boldsymbol{q}) A_d(\boldsymbol{q}') e^{-\frac{i}{\hbar}(\boldsymbol{q}'+\boldsymbol{q})\cdot\boldsymbol{y}} \\
&\quad - \overline{B}_c(\boldsymbol{q}) B_d^{\dagger}(\boldsymbol{q}') e^{\frac{i}{\hbar}(\boldsymbol{q}'-\boldsymbol{q})\cdot\boldsymbol{y}} + \overline{D}_c^{\dagger}(\boldsymbol{q}) D_d(\boldsymbol{q}') e^{-\frac{i}{\hbar}(\boldsymbol{q}'-\boldsymbol{q})\cdot\boldsymbol{y}} + \overline{D}_c^{\dagger}(\boldsymbol{q}) F_d^{\dagger}(\boldsymbol{q}') e^{\frac{i}{\hbar}(\boldsymbol{q}'+\boldsymbol{q})\cdot\boldsymbol{y}} \\
&\quad + \overline{F}_c(\boldsymbol{q}) D_d(\boldsymbol{q}') e^{-\frac{i}{\hbar}(\boldsymbol{q}'+\boldsymbol{q})\cdot\boldsymbol{y}} + \overline{F}_c(\boldsymbol{q}) F_d^{\dagger}(\boldsymbol{q}') e^{\frac{i}{\hbar}(\boldsymbol{q}'-\boldsymbol{q})\cdot\boldsymbol{y}}]
\end{aligned}
$$

$$
\begin{aligned}
&= \frac{e^2}{(2\pi\hbar)^6} \sum_{abcd} \int d\boldsymbol{x} d\boldsymbol{y} \int d\boldsymbol{p} d\boldsymbol{p}' d\boldsymbol{q} d\boldsymbol{q}' \frac{\gamma^0_{ab}\gamma^0_{cd}}{8\pi|\boldsymbol{x}-\boldsymbol{y}|} \\
&\times [\overline{A}^\dagger_a(\boldsymbol{p}) A_b(\boldsymbol{p}') \overline{A}^\dagger_c(\boldsymbol{q}) A_d(\boldsymbol{q}') e^{-\frac{i}{\hbar}(\boldsymbol{q}'-\boldsymbol{q})\cdot\boldsymbol{y}} e^{-\frac{i}{\hbar}(\boldsymbol{p}'-\boldsymbol{p})\cdot\boldsymbol{x}} \\
&+ \overline{A}^\dagger_a(\boldsymbol{p}) A_b(\boldsymbol{p}') \overline{A}^\dagger_c(\boldsymbol{q}) B^\dagger_d(\boldsymbol{q}') e^{\frac{i}{\hbar}(\boldsymbol{q}'+\boldsymbol{q})\cdot\boldsymbol{y}} e^{-\frac{i}{\hbar}(\boldsymbol{p}'-\boldsymbol{p})\cdot\boldsymbol{x}} \\
&+ \overline{A}^\dagger_a(\boldsymbol{p}) A_b(\boldsymbol{p}') \overline{B}_c(\boldsymbol{q}) A_d(\boldsymbol{q}') e^{-\frac{i}{\hbar}(\boldsymbol{q}'+\boldsymbol{q})\cdot\boldsymbol{y}} e^{-\frac{i}{\hbar}(\boldsymbol{p}'-\boldsymbol{p})\cdot\boldsymbol{x}} \\
&+ \overline{A}^\dagger_a(\boldsymbol{p}) A_b(\boldsymbol{p}') \overline{B}_c(\boldsymbol{q}) B^\dagger_d(\boldsymbol{q}') e^{\frac{i}{\hbar}(\boldsymbol{q}'-\boldsymbol{q})\cdot\boldsymbol{y}} e^{-\frac{i}{\hbar}(\boldsymbol{p}'-\boldsymbol{p})\cdot\boldsymbol{x}} \\
&- \overline{A}^\dagger_a(\boldsymbol{p}) A_b(\boldsymbol{p}') \overline{D}^\dagger_c(\boldsymbol{q}) D_d(\boldsymbol{q}') e^{-\frac{i}{\hbar}(\boldsymbol{q}'-\boldsymbol{q})\cdot\boldsymbol{y}} e^{-\frac{i}{\hbar}(\boldsymbol{p}'-\boldsymbol{p})\cdot\boldsymbol{x}} \\
&- \overline{A}^\dagger_a(\boldsymbol{p}) A_b(\boldsymbol{p}') \overline{D}^\dagger_c(\boldsymbol{q}) F^\dagger_d(\boldsymbol{q}') e^{\frac{i}{\hbar}(\boldsymbol{q}'+\boldsymbol{q})\cdot\boldsymbol{y}} e^{-\frac{i}{\hbar}(\boldsymbol{p}'-\boldsymbol{p})\cdot\boldsymbol{x}} \\
&- \overline{A}^\dagger_a(\boldsymbol{p}) A_b(\boldsymbol{p}') \overline{F}_c(\boldsymbol{q}) D_d(\boldsymbol{q}') e^{-\frac{i}{\hbar}(\boldsymbol{q}'+\boldsymbol{q})\cdot\boldsymbol{y}} e^{-\frac{i}{\hbar}(\boldsymbol{p}'-\boldsymbol{p})\cdot\boldsymbol{x}} \\
&- \overline{A}^\dagger_a(\boldsymbol{p}) A_b(\boldsymbol{p}') \overline{F}_c(\boldsymbol{q}) F^\dagger_d(\boldsymbol{q}') e^{\frac{i}{\hbar}(\boldsymbol{q}'-\boldsymbol{q})\cdot\boldsymbol{y}} e^{-\frac{i}{\hbar}(\boldsymbol{p}'-\boldsymbol{p})\cdot\boldsymbol{x}} \\
&+ \overline{A}^\dagger_a(\boldsymbol{p}) B^\dagger_b(\boldsymbol{p}') \overline{A}^\dagger_c(\boldsymbol{q}) A_d(\boldsymbol{q}') e^{-\frac{i}{\hbar}(\boldsymbol{q}'-\boldsymbol{q})\cdot\boldsymbol{y}} e^{\frac{i}{\hbar}(\boldsymbol{p}'+\boldsymbol{p})\cdot\boldsymbol{x}} \\
&+ \overline{A}^\dagger_a(\boldsymbol{p}) B^\dagger_b(\boldsymbol{p}') \overline{A}^\dagger_c(\boldsymbol{q}) B^\dagger_d(\boldsymbol{q}') e^{\frac{i}{\hbar}(\boldsymbol{q}'+\boldsymbol{q})\cdot\boldsymbol{y}} e^{\frac{i}{\hbar}(\boldsymbol{p}'+\boldsymbol{p})\cdot\boldsymbol{x}} \\
&+ \overline{A}^\dagger_a(\boldsymbol{p}) B^\dagger_b(\boldsymbol{p}') \overline{B}_c(\boldsymbol{q}) A_d(\boldsymbol{q}') e^{-\frac{i}{\hbar}(\boldsymbol{q}'+\boldsymbol{q})\cdot\boldsymbol{y}} e^{\frac{i}{\hbar}(\boldsymbol{p}'+\boldsymbol{p})\cdot\boldsymbol{x}} \\
&+ \overline{A}^\dagger_a(\boldsymbol{p}) B^\dagger_b(\boldsymbol{p}') \overline{B}_c(\boldsymbol{q}) B^\dagger_d(\boldsymbol{q}') e^{\frac{i}{\hbar}(\boldsymbol{q}'-\boldsymbol{q})\cdot\boldsymbol{y}} e^{\frac{i}{\hbar}(\boldsymbol{p}'+\boldsymbol{p})\cdot\boldsymbol{x}} \\
&- \overline{A}^\dagger_a(\boldsymbol{p}) B^\dagger_b(\boldsymbol{p}') \overline{D}^\dagger_c(\boldsymbol{q}) D_d(\boldsymbol{q}') e^{-\frac{i}{\hbar}(\boldsymbol{q}'-\boldsymbol{q})\cdot\boldsymbol{y}} e^{\frac{i}{\hbar}(\boldsymbol{p}'+\boldsymbol{p})\cdot\boldsymbol{x}} \\
&- \overline{A}^\dagger_a(\boldsymbol{p}) B^\dagger_b(\boldsymbol{p}') \overline{D}^\dagger_c(\boldsymbol{q}) F^\dagger_d(\boldsymbol{q}') e^{\frac{i}{\hbar}(\boldsymbol{q}'+\boldsymbol{q})\cdot\boldsymbol{y}} e^{\frac{i}{\hbar}(\boldsymbol{p}'+\boldsymbol{p})\cdot\boldsymbol{x}} \\
&- \overline{A}^\dagger_a(\boldsymbol{p}) B^\dagger_b(\boldsymbol{p}') \overline{F}_c(\boldsymbol{q}) D_d(\boldsymbol{q}') e^{-\frac{i}{\hbar}(\boldsymbol{q}'+\boldsymbol{q})\cdot\boldsymbol{y}} e^{\frac{i}{\hbar}(\boldsymbol{p}'+\boldsymbol{p})\cdot\boldsymbol{x}} \\
&- \overline{A}^\dagger_a(\boldsymbol{p}) B^\dagger_b(\boldsymbol{p}') \overline{F}_c(\boldsymbol{q}) F^\dagger_d(\boldsymbol{q}') e^{\frac{i}{\hbar}(\boldsymbol{q}'-\boldsymbol{q})\cdot\boldsymbol{y}} e^{\frac{i}{\hbar}(\boldsymbol{p}'+\boldsymbol{p})\cdot\boldsymbol{x}} \\
&+ \overline{B}_a(\boldsymbol{p}) A_b(\boldsymbol{p}') \overline{A}^\dagger_c(\boldsymbol{q}) A_d(\boldsymbol{q}') e^{-\frac{i}{\hbar}(\boldsymbol{q}'-\boldsymbol{q})\cdot\boldsymbol{y}} e^{-\frac{i}{\hbar}(\boldsymbol{p}'+\boldsymbol{p})\cdot\boldsymbol{x}} \\
&+ \overline{B}_a(\boldsymbol{p}) A_b(\boldsymbol{p}') \overline{A}^\dagger_c(\boldsymbol{q}) B^\dagger_d(\boldsymbol{q}') e^{\frac{i}{\hbar}(\boldsymbol{q}'+\boldsymbol{q})\cdot\boldsymbol{y}} e^{-\frac{i}{\hbar}(\boldsymbol{p}'+\boldsymbol{p})\cdot\boldsymbol{x}} \\
&+ \overline{B}_a(\boldsymbol{p}) A_b(\boldsymbol{p}') \overline{B}_c(\boldsymbol{q}) A_d(\boldsymbol{q}') e^{-\frac{i}{\hbar}(\boldsymbol{q}'+\boldsymbol{q})\cdot\boldsymbol{y}} e^{-\frac{i}{\hbar}(\boldsymbol{p}'+\boldsymbol{p})\cdot\boldsymbol{x}} \\
&+ \overline{B}_a(\boldsymbol{p}) A_b(\boldsymbol{p}') \overline{B}_c(\boldsymbol{q}) B^\dagger_d(\boldsymbol{q}') e^{\frac{i}{\hbar}(\boldsymbol{q}'-\boldsymbol{q})\cdot\boldsymbol{y}} e^{-\frac{i}{\hbar}(\boldsymbol{p}'+\boldsymbol{p})\cdot\boldsymbol{x}} \\
&- \overline{B}_a(\boldsymbol{p}) A_b(\boldsymbol{p}') \overline{D}^\dagger_c(\boldsymbol{q}) D_d(\boldsymbol{q}') e^{-\frac{i}{\hbar}(\boldsymbol{q}'-\boldsymbol{q})\cdot\boldsymbol{y}} e^{-\frac{i}{\hbar}(\boldsymbol{p}'+\boldsymbol{p})\cdot\boldsymbol{x}} \\
&- \overline{B}_a(\boldsymbol{p}) A_b(\boldsymbol{p}') \overline{D}^\dagger_c(\boldsymbol{q}) F^\dagger_d(\boldsymbol{q}') e^{\frac{i}{\hbar}(\boldsymbol{q}'+\boldsymbol{q})\cdot\boldsymbol{y}} e^{-\frac{i}{\hbar}(\boldsymbol{p}'+\boldsymbol{p})\cdot\boldsymbol{x}} \\
&- \overline{B}_a(\boldsymbol{p}) A_b(\boldsymbol{p}') \overline{F}_c(\boldsymbol{q}) D_d(\boldsymbol{q}') e^{-\frac{i}{\hbar}(\boldsymbol{q}'+\boldsymbol{q})\cdot\boldsymbol{y}} e^{-\frac{i}{\hbar}(\boldsymbol{p}'+\boldsymbol{p})\cdot\boldsymbol{x}} \\
&- \overline{B}_a(\boldsymbol{p}) A_b(\boldsymbol{p}') \overline{F}_c(\boldsymbol{q}) F^\dagger_d(\boldsymbol{q}') e^{\frac{i}{\hbar}(\boldsymbol{q}'-\boldsymbol{q})\cdot\boldsymbol{y}} e^{-\frac{i}{\hbar}(\boldsymbol{p}'+\boldsymbol{p})\cdot\boldsymbol{x}} \\
&+ \overline{B}_a(\boldsymbol{p}) B^\dagger_b(\boldsymbol{p}') \overline{A}^\dagger_c(\boldsymbol{q}) A_d(\boldsymbol{q}') e^{-\frac{i}{\hbar}(\boldsymbol{q}'-\boldsymbol{q})\cdot\boldsymbol{y}} e^{\frac{i}{\hbar}(\boldsymbol{p}'-\boldsymbol{p})\cdot\boldsymbol{x}} \\
&+ \overline{B}_a(\boldsymbol{p}) B^\dagger_b(\boldsymbol{p}') \overline{A}^\dagger_c(\boldsymbol{q}) B^\dagger_d(\boldsymbol{q}') e^{\frac{i}{\hbar}(\boldsymbol{q}'+\boldsymbol{q})\cdot\boldsymbol{y}} e^{\frac{i}{\hbar}(\boldsymbol{p}'-\boldsymbol{p})\cdot\boldsymbol{x}} \\
&+ \overline{B}_a(\boldsymbol{p}) B^\dagger_b(\boldsymbol{p}') \overline{B}_c(\boldsymbol{q}) A_d(\boldsymbol{q}') e^{-\frac{i}{\hbar}(\boldsymbol{q}'+\boldsymbol{q})\cdot\boldsymbol{y}} e^{\frac{i}{\hbar}(\boldsymbol{p}'-\boldsymbol{p})\cdot\boldsymbol{x}}
\end{aligned}
$$

$$
\begin{aligned}
&+\overline{B}_a(\boldsymbol{p})B_b^\dagger(\boldsymbol{p}')\overline{B}_c(\boldsymbol{q})B_d^\dagger(\boldsymbol{q}')e^{\frac{i}{\hbar}(\boldsymbol{q}'-\boldsymbol{q})\cdot\boldsymbol{y}}e^{\frac{i}{\hbar}(\boldsymbol{p}'-\boldsymbol{p})\cdot\boldsymbol{x}}\\
&-\overline{B}_a(\boldsymbol{p})B_b^\dagger(\boldsymbol{p}')\overline{D}_c^\dagger(\boldsymbol{q})D_d(\boldsymbol{q}')e^{-\frac{i}{\hbar}(\boldsymbol{q}'-\boldsymbol{q})\cdot\boldsymbol{y}}e^{\frac{i}{\hbar}(\boldsymbol{p}'-\boldsymbol{p})\cdot\boldsymbol{x}}\\
&-\overline{B}_a(\boldsymbol{p})B_b^\dagger(\boldsymbol{p}')\overline{D}_c^\dagger(\boldsymbol{q})F_d^\dagger(\boldsymbol{q}')e^{\frac{i}{\hbar}(\boldsymbol{q}'+\boldsymbol{q})\cdot\boldsymbol{y}}e^{\frac{i}{\hbar}(\boldsymbol{p}'-\boldsymbol{p})\cdot\boldsymbol{x}}\\
&-\overline{B}_a(\boldsymbol{p})B_b^\dagger(\boldsymbol{p}')\overline{F}_c(\boldsymbol{q})D_d(\boldsymbol{q}')e^{-\frac{i}{\hbar}(\boldsymbol{q}'+\boldsymbol{q})\cdot\boldsymbol{y}}e^{\frac{i}{\hbar}(\boldsymbol{p}'-\boldsymbol{p})\cdot\boldsymbol{x}}\\
&-\overline{B}_a(\boldsymbol{p})B_b^\dagger(\boldsymbol{p}')\overline{F}_c(\boldsymbol{q})F_d^\dagger(\boldsymbol{q}')e^{\frac{i}{\hbar}(\boldsymbol{q}'-\boldsymbol{q})\cdot\boldsymbol{y}}e^{\frac{i}{\hbar}(\boldsymbol{p}'-\boldsymbol{p})\cdot\boldsymbol{x}}\\
&-\overline{D}_a^\dagger(\boldsymbol{p})D_b(\boldsymbol{p}')\overline{A}_c^\dagger(\boldsymbol{q})A_d(\boldsymbol{q}')e^{-\frac{i}{\hbar}(\boldsymbol{q}'-\boldsymbol{q})\cdot\boldsymbol{y}}e^{-\frac{i}{\hbar}(\boldsymbol{p}'-\boldsymbol{p})\cdot\boldsymbol{x}}\\
&-\overline{D}_a^\dagger(\boldsymbol{p})D_b(\boldsymbol{p}')\overline{A}_c^\dagger(\boldsymbol{q})B_d^\dagger(\boldsymbol{q}')e^{\frac{i}{\hbar}(\boldsymbol{q}'+\boldsymbol{q})\cdot\boldsymbol{y}}e^{-\frac{i}{\hbar}(\boldsymbol{p}'-\boldsymbol{p})\cdot\boldsymbol{x}}\\
&-\overline{D}_a^\dagger(\boldsymbol{p})D_b(\boldsymbol{p}')\overline{B}_c(\boldsymbol{q})A_d(\boldsymbol{q}')e^{-\frac{i}{\hbar}(\boldsymbol{q}'+\boldsymbol{q})\cdot\boldsymbol{y}}e^{-\frac{i}{\hbar}(\boldsymbol{p}'-\boldsymbol{p})\cdot\boldsymbol{x}}\\
&-\overline{D}_a^\dagger(\boldsymbol{p})D_b(\boldsymbol{p}')\overline{B}_c(\boldsymbol{q})B_d^\dagger(\boldsymbol{q}')e^{\frac{i}{\hbar}(\boldsymbol{q}'-\boldsymbol{q})\cdot\boldsymbol{y}}e^{-\frac{i}{\hbar}(\boldsymbol{p}'-\boldsymbol{p})\cdot\boldsymbol{x}}\\
&+\overline{D}_a^\dagger(\boldsymbol{p})D_b(\boldsymbol{p}')\overline{D}_c^\dagger(\boldsymbol{q})D_d(\boldsymbol{q}')e^{-\frac{i}{\hbar}(\boldsymbol{q}'-\boldsymbol{q})\cdot\boldsymbol{y}}e^{-\frac{i}{\hbar}(\boldsymbol{p}'-\boldsymbol{p})\cdot\boldsymbol{x}}\\
&+\overline{D}_a^\dagger(\boldsymbol{p})D_b(\boldsymbol{p}')\overline{D}_c^\dagger(\boldsymbol{q})F_d^\dagger(\boldsymbol{q}')e^{\frac{i}{\hbar}(\boldsymbol{q}'+\boldsymbol{q})\cdot\boldsymbol{y}}e^{-\frac{i}{\hbar}(\boldsymbol{p}'-\boldsymbol{p})\cdot\boldsymbol{x}}\\
&+\overline{D}_a^\dagger(\boldsymbol{p})D_b(\boldsymbol{p}')\overline{F}_c(\boldsymbol{q})D_d(\boldsymbol{q}')e^{-\frac{i}{\hbar}(\boldsymbol{q}'+\boldsymbol{q})\cdot\boldsymbol{y}}e^{-\frac{i}{\hbar}(\boldsymbol{p}'-\boldsymbol{p})\cdot\boldsymbol{x}}\\
&+\overline{D}_a^\dagger(\boldsymbol{p})D_b(\boldsymbol{p}')\overline{F}_c(\boldsymbol{q})F_d^\dagger(\boldsymbol{q}')e^{\frac{i}{\hbar}(\boldsymbol{q}'-\boldsymbol{q})\cdot\boldsymbol{y}}e^{-\frac{i}{\hbar}(\boldsymbol{p}'-\boldsymbol{p})\cdot\boldsymbol{x}}\\
&-\overline{D}_a^\dagger(\boldsymbol{p})F_b^\dagger(\boldsymbol{p}')\overline{A}_c^\dagger(\boldsymbol{q})A_d(\boldsymbol{q}')e^{-\frac{i}{\hbar}(\boldsymbol{q}'-\boldsymbol{q})\cdot\boldsymbol{y}}e^{\frac{i}{\hbar}(\boldsymbol{p}'+\boldsymbol{p})\cdot\boldsymbol{x}}\\
&-\overline{D}_a^\dagger(\boldsymbol{p})F_b^\dagger(\boldsymbol{p}')\overline{A}_c^\dagger(\boldsymbol{q})B_d^\dagger(\boldsymbol{q}')e^{\frac{i}{\hbar}(\boldsymbol{q}'+\boldsymbol{q})\cdot\boldsymbol{y}}e^{\frac{i}{\hbar}(\boldsymbol{p}'+\boldsymbol{p})\cdot\boldsymbol{x}}\\
&-\overline{D}_a^\dagger(\boldsymbol{p})F_b^\dagger(\boldsymbol{p}')\overline{B}_c(\boldsymbol{q})A_d(\boldsymbol{q}')e^{-\frac{i}{\hbar}(\boldsymbol{q}'+\boldsymbol{q})\cdot\boldsymbol{y}}e^{\frac{i}{\hbar}(\boldsymbol{p}'+\boldsymbol{p})\cdot\boldsymbol{x}}\\
&-\overline{D}_a^\dagger(\boldsymbol{p})F_b^\dagger(\boldsymbol{p}')\overline{B}_c(\boldsymbol{q})B_d^\dagger(\boldsymbol{q}')e^{\frac{i}{\hbar}(\boldsymbol{q}'-\boldsymbol{q})\cdot\boldsymbol{y}}e^{\frac{i}{\hbar}(\boldsymbol{p}'+\boldsymbol{p})\cdot\boldsymbol{x}}\\
&+\overline{D}_a^\dagger(\boldsymbol{p})F_b^\dagger(\boldsymbol{p}')\overline{D}_c^\dagger(\boldsymbol{q})D_d(\boldsymbol{q}')e^{-\frac{i}{\hbar}(\boldsymbol{q}'-\boldsymbol{q})\cdot\boldsymbol{y}}e^{\frac{i}{\hbar}(\boldsymbol{p}'+\boldsymbol{p})\cdot\boldsymbol{x}}\\
&+\overline{D}_a^\dagger(\boldsymbol{p})F_b^\dagger(\boldsymbol{p}')\overline{D}_c^\dagger(\boldsymbol{q})F_d^\dagger(\boldsymbol{q}')e^{\frac{i}{\hbar}(\boldsymbol{q}'+\boldsymbol{q})\cdot\boldsymbol{y}}e^{\frac{i}{\hbar}(\boldsymbol{p}'+\boldsymbol{p})\cdot\boldsymbol{x}}\\
&+\overline{D}_a^\dagger(\boldsymbol{p})F_b^\dagger(\boldsymbol{p}')\overline{F}_c(\boldsymbol{q})D_d(\boldsymbol{q}')e^{-\frac{i}{\hbar}(\boldsymbol{q}'+\boldsymbol{q})\cdot\boldsymbol{y}}e^{\frac{i}{\hbar}(\boldsymbol{p}'+\boldsymbol{p})\cdot\boldsymbol{x}}\\
&+\overline{D}_a^\dagger(\boldsymbol{p})F_b^\dagger(\boldsymbol{p}')\overline{F}_c(\boldsymbol{q})F_d^\dagger(\boldsymbol{q}')e^{\frac{i}{\hbar}(\boldsymbol{q}'-\boldsymbol{q})\cdot\boldsymbol{y}}e^{\frac{i}{\hbar}(\boldsymbol{p}'+\boldsymbol{p})\cdot\boldsymbol{x}}\\
&-\overline{F}_a(\boldsymbol{p})D_b(\boldsymbol{p}')\overline{A}_c^\dagger(\boldsymbol{q})A_d(\boldsymbol{q}')e^{-\frac{i}{\hbar}(\boldsymbol{q}'-\boldsymbol{q})\cdot\boldsymbol{y}}e^{-\frac{i}{\hbar}(\boldsymbol{p}'+\boldsymbol{p})\cdot\boldsymbol{x}}\\
&-\overline{F}_a(\boldsymbol{p})D_b(\boldsymbol{p}')\overline{A}_c^\dagger(\boldsymbol{q})B_d^\dagger(\boldsymbol{q}')e^{\frac{i}{\hbar}(\boldsymbol{q}'+\boldsymbol{q})\cdot\boldsymbol{y}}e^{-\frac{i}{\hbar}(\boldsymbol{p}'+\boldsymbol{p})\cdot\boldsymbol{x}}\\
&-\overline{F}_a(\boldsymbol{p})D_b(\boldsymbol{p}')\overline{B}_c(\boldsymbol{q})A_d(\boldsymbol{q}')e^{-\frac{i}{\hbar}(\boldsymbol{q}'+\boldsymbol{q})\cdot\boldsymbol{y}}e^{-\frac{i}{\hbar}(\boldsymbol{p}'+\boldsymbol{p})\cdot\boldsymbol{x}}\\
&-\overline{F}_a(\boldsymbol{p})D_b(\boldsymbol{p}')\overline{B}_c(\boldsymbol{q})B_d^\dagger(\boldsymbol{q}')e^{\frac{i}{\hbar}(\boldsymbol{q}'-\boldsymbol{q})\cdot\boldsymbol{y}}e^{-\frac{i}{\hbar}(\boldsymbol{p}'+\boldsymbol{p})\cdot\boldsymbol{x}}\\
&+\overline{F}_a(\boldsymbol{p})D_b(\boldsymbol{p}')\overline{D}_c^\dagger(\boldsymbol{q})D_d(\boldsymbol{q}')e^{-\frac{i}{\hbar}(\boldsymbol{q}'-\boldsymbol{q})\cdot\boldsymbol{y}}e^{-\frac{i}{\hbar}(\boldsymbol{p}'+\boldsymbol{p})\cdot\boldsymbol{x}}\\
&+\overline{F}_a(\boldsymbol{p})D_b(\boldsymbol{p}')\overline{D}_c^\dagger(\boldsymbol{q})F_d^\dagger(\boldsymbol{q}')e^{\frac{i}{\hbar}(\boldsymbol{q}'+\boldsymbol{q})\cdot\boldsymbol{y}}e^{-\frac{i}{\hbar}(\boldsymbol{p}'+\boldsymbol{p})\cdot\boldsymbol{x}}\\
&+\overline{F}_a(\boldsymbol{p})D_b(\boldsymbol{p}')\overline{F}_c(\boldsymbol{q})D_d(\boldsymbol{q}')e^{-\frac{i}{\hbar}(\boldsymbol{q}'+\boldsymbol{q})\cdot\boldsymbol{y}}e^{-\frac{i}{\hbar}(\boldsymbol{p}'+\boldsymbol{p})\cdot\boldsymbol{x}}
\end{aligned}
$$

$$
\begin{aligned}
&+\overline{F}_a(\boldsymbol{p})D_b(\boldsymbol{p}')\overline{F}_c(\boldsymbol{q})F_d^\dagger(\boldsymbol{q}')e^{\frac{i}{\hbar}(\boldsymbol{q}'-\boldsymbol{q})\cdot\boldsymbol{y}}e^{-\frac{i}{\hbar}(\boldsymbol{p}'+\boldsymbol{p})\cdot\boldsymbol{x}}\\
&-\overline{F}_a(\boldsymbol{p})F_b^\dagger(\boldsymbol{p}')\overline{A}_c^\dagger(\boldsymbol{q})A_d(\boldsymbol{q}')e^{-\frac{i}{\hbar}(\boldsymbol{q}'-\boldsymbol{q})\cdot\boldsymbol{y}}e^{\frac{i}{\hbar}(\boldsymbol{p}'-\boldsymbol{p})\cdot\boldsymbol{x}}\\
&-\overline{F}_a(\boldsymbol{p})F_b^\dagger(\boldsymbol{p}')\overline{A}_c^\dagger(\boldsymbol{q})B_d^\dagger(\boldsymbol{q}')e^{\frac{i}{\hbar}(\boldsymbol{q}'+\boldsymbol{q})\cdot\boldsymbol{y}}e^{\frac{i}{\hbar}(\boldsymbol{p}'-\boldsymbol{p})\cdot\boldsymbol{x}}\\
&-\overline{F}_a(\boldsymbol{p})F_b^\dagger(\boldsymbol{p}')\overline{B}_c(\boldsymbol{q})A_d(\boldsymbol{q}')e^{-\frac{i}{\hbar}(\boldsymbol{q}'+\boldsymbol{q})\cdot\boldsymbol{y}}e^{\frac{i}{\hbar}(\boldsymbol{p}'-\boldsymbol{p})\cdot\boldsymbol{x}}\\
&-\overline{F}_a(\boldsymbol{p})F_b^\dagger(\boldsymbol{p}')\overline{B}_c(\boldsymbol{q})B_d^\dagger(\boldsymbol{q}')e^{\frac{i}{\hbar}(\boldsymbol{q}'-\boldsymbol{q})\cdot\boldsymbol{y}}e^{\frac{i}{\hbar}(\boldsymbol{p}'-\boldsymbol{p})\cdot\boldsymbol{x}}\\
&+\overline{F}_a(\boldsymbol{p})F_b^\dagger(\boldsymbol{p}')\overline{D}_c^\dagger(\boldsymbol{q})D_d(\boldsymbol{q}')e^{-\frac{i}{\hbar}(\boldsymbol{q}'-\boldsymbol{q})\cdot\boldsymbol{y}}e^{\frac{i}{\hbar}(\boldsymbol{p}'-\boldsymbol{p})\cdot\boldsymbol{x}}\\
&+\overline{F}_a(\boldsymbol{p})F_b^\dagger(\boldsymbol{p}')\overline{D}_c^\dagger(\boldsymbol{q})F_d^\dagger(\boldsymbol{q}')e^{\frac{i}{\hbar}(\boldsymbol{q}'+\boldsymbol{q})\cdot\boldsymbol{y}}e^{\frac{i}{\hbar}(\boldsymbol{p}'-\boldsymbol{p})\cdot\boldsymbol{x}}\\
&+\overline{F}_a(\boldsymbol{p})F_b^\dagger(\boldsymbol{p}')\overline{F}_c(\boldsymbol{q})D_d(\boldsymbol{q}')e^{-\frac{i}{\hbar}(\boldsymbol{q}'+\boldsymbol{q})\cdot\boldsymbol{y}}e^{\frac{i}{\hbar}(\boldsymbol{p}'-\boldsymbol{p})\cdot\boldsymbol{x}}\\
&+\overline{F}_a(\boldsymbol{p})F_b^\dagger(\boldsymbol{p}')\overline{F}_c(\boldsymbol{q})F_d^\dagger(\boldsymbol{q}')e^{\frac{i}{\hbar}(\boldsymbol{q}'-\boldsymbol{q})\cdot\boldsymbol{y}}e^{\frac{i}{\hbar}(\boldsymbol{p}'-\boldsymbol{p})\cdot\boldsymbol{x}}\Big].
\end{aligned}
\tag{D.10}
$$

We should bring this polynomial to a normal order, i. e., move all creation operators to the left. The resulting expression will contain *phys*, *unphys* and *renorm* terms.[4]

Among the *unphys* terms we can notice some cancellations. For example, let us bring to the normal order the 12th term in (D.10), so we have

$$
\begin{aligned}
&\frac{e^2}{(2\pi\hbar)^6}\sum_{abcd}\int d\boldsymbol{x}d\boldsymbol{y}\int d\boldsymbol{p}d\boldsymbol{p}'d\boldsymbol{q}d\boldsymbol{q}'\frac{\gamma^0_{ab}\gamma^0_{cd}}{8\pi|\boldsymbol{x}-\boldsymbol{y}|}\\
&\quad\times\overline{A}_a^\dagger(\boldsymbol{p})B_b^\dagger(\boldsymbol{p}')\overline{B}_c(\boldsymbol{q})B_d^\dagger(\boldsymbol{q}')e^{\frac{i}{\hbar}(\boldsymbol{q}'-\boldsymbol{q})\cdot\boldsymbol{y}}e^{\frac{i}{\hbar}(\boldsymbol{p}'+\boldsymbol{p})\cdot\boldsymbol{x}}\\
&=-\frac{e^2}{(2\pi\hbar)^6}\sum_{abcd}\int d\boldsymbol{x}d\boldsymbol{y}\int d\boldsymbol{p}d\boldsymbol{p}'d\boldsymbol{q}d\boldsymbol{q}'\frac{\gamma^0_{ab}\gamma^0_{cd}}{8\pi|\boldsymbol{x}-\boldsymbol{y}|}\\
&\quad\times\overline{A}_a^\dagger(\boldsymbol{p})B_b^\dagger(\boldsymbol{p}')B_d^\dagger(\boldsymbol{q}')\overline{B}_c(\boldsymbol{q})e^{\frac{i}{\hbar}(\boldsymbol{q}'-\boldsymbol{q})\cdot\boldsymbol{y}}e^{\frac{i}{\hbar}(\boldsymbol{p}'+\boldsymbol{p})\cdot\boldsymbol{x}}\\
&\quad+\frac{e^2}{(2\pi\hbar)^6}\sum_{abcd}\int d\boldsymbol{x}d\boldsymbol{y}\int d\boldsymbol{p}d\boldsymbol{p}'d\boldsymbol{q}d\boldsymbol{q}'\frac{\gamma^0_{ab}\gamma^0_{cd}}{8\pi|\boldsymbol{x}-\boldsymbol{y}|}\\
&\quad\times\overline{A}_a^\dagger(\boldsymbol{p})B_b^\dagger(\boldsymbol{p}')\{\overline{B}_c(\boldsymbol{q}),B_d^\dagger(\boldsymbol{q}')\}e^{\frac{i}{\hbar}(\boldsymbol{q}'-\boldsymbol{q})\cdot\boldsymbol{y}}e^{\frac{i}{\hbar}(\boldsymbol{p}'+\boldsymbol{p})\cdot\boldsymbol{x}}.
\end{aligned}
$$

For the second term on the right-hand side we use (B.87), (B.11)–(B.12), (A.1) and obtain

$$
\begin{aligned}
&\frac{e^2}{(2\pi\hbar)^6}\sum_{abcd}\int d\boldsymbol{x}d\boldsymbol{y}\int d\boldsymbol{p}d\boldsymbol{p}'d\boldsymbol{q}d\boldsymbol{q}'\frac{\gamma^0_{ab}\gamma^0_{cd}}{8\pi|\boldsymbol{x}-\boldsymbol{y}|}\\
&\quad\times\overline{A}_a^\dagger(\boldsymbol{p})B_b^\dagger(\boldsymbol{p}')\frac{1}{2\omega_{\boldsymbol{q}}}(\gamma^0\omega_{\boldsymbol{q}}-\boldsymbol{\gamma}\cdot\boldsymbol{q}c-m_ec^2)_{cd}\delta(\boldsymbol{q}'-\boldsymbol{q})e^{\frac{i}{\hbar}(\boldsymbol{q}'-\boldsymbol{q})\cdot\boldsymbol{y}}e^{\frac{i}{\hbar}(\boldsymbol{p}'+\boldsymbol{p})\cdot\boldsymbol{x}}\\
&=\frac{e^2}{(2\pi\hbar)^6}\sum_{ab}\int d\boldsymbol{x}d\boldsymbol{y}\int d\boldsymbol{p}d\boldsymbol{p}'d\boldsymbol{q}\frac{\gamma^0_{ab}}{8\pi|\boldsymbol{x}-\boldsymbol{y}|}
\end{aligned}
$$

4 Integrals in the *renorm* terms diverge. This is another evidence of the renormalization problems discussed in Chapter 4. In the remainder of this appendix, we omit the *renorm* part of V_2.

$$\times \overline{A}_a^\dagger(\boldsymbol{p})B_b^\dagger(\boldsymbol{p}')\frac{1}{2\omega_{\boldsymbol{q}}}(\omega_{\boldsymbol{q}}\,\mathrm{Tr}(\gamma^0\gamma^0) - c\boldsymbol{q}\cdot\mathrm{Tr}(\gamma^0\boldsymbol{\gamma}) - m_ec^2\,\mathrm{Tr}(\gamma^0))e^{\frac{i}{\hbar}(\boldsymbol{p}'+\boldsymbol{p})\cdot\boldsymbol{x}}$$
$$= \frac{2e^2}{(2\pi\hbar)^6}\sum_{ab}\int d\boldsymbol{x}d\boldsymbol{y}\int d\boldsymbol{p}d\boldsymbol{p}'d\boldsymbol{q}\frac{\gamma^0_{ab}}{8\pi|\boldsymbol{x}-\boldsymbol{y}|}\overline{A}_a^\dagger(\boldsymbol{p})B_b^\dagger(\boldsymbol{p}')e^{\frac{i}{\hbar}(\boldsymbol{p}'+\boldsymbol{p})\cdot\boldsymbol{x}}$$
$$= \frac{e^2\hbar^2}{(2\pi\hbar)^3}\sum_{ab}\int d\boldsymbol{p}d\boldsymbol{p}'d\boldsymbol{q}\gamma^0_{ab}\overline{A}_a^\dagger(\boldsymbol{p})B_b^\dagger(\boldsymbol{p}')\frac{\delta(\boldsymbol{p}'+\boldsymbol{p})}{(\boldsymbol{p}'+\boldsymbol{p})^2}. \tag{D.11}$$

This integral diverges. However, there are three other divergent terms in (D.10) that arise in a similar manner from $-A^\dagger B^\dagger FF^\dagger + BB^\dagger A^\dagger B^\dagger - FF^\dagger A^\dagger B^\dagger$. Added together these four infinities yield zero.

Taking into account the above results and using anticommutators like (B.86) and (B.87), we obtain interaction (D.10) in a normally ordered form, i. e.,

$$\begin{aligned}
V_2 = \frac{e^2}{(2\pi\hbar)^6}\sum_{abcd}\int d\boldsymbol{x}d\boldsymbol{y}\int d\boldsymbol{p}d\boldsymbol{p}'d\boldsymbol{q}d\boldsymbol{q}'\frac{\gamma^0_{ab}\gamma^0_{cd}}{8\pi|\boldsymbol{x}-\boldsymbol{y}|}\\
\times [-\overline{A}_a^\dagger(\boldsymbol{p})\overline{A}_c^\dagger(\boldsymbol{q})A_b(\boldsymbol{p}')A_d(\boldsymbol{q}')e^{-\frac{i}{\hbar}(\boldsymbol{q}'-\boldsymbol{q})\cdot\boldsymbol{y}}e^{-\frac{i}{\hbar}(\boldsymbol{p}'-\boldsymbol{p})\cdot\boldsymbol{x}}\\
-\overline{A}_a^\dagger(\boldsymbol{p})\overline{A}_c^\dagger(\boldsymbol{q})A_b(\boldsymbol{p}')B_d^\dagger(\boldsymbol{q}')e^{\frac{i}{\hbar}(\boldsymbol{q}'+\boldsymbol{q})\cdot\boldsymbol{y}}e^{-\frac{i}{\hbar}(\boldsymbol{p}'-\boldsymbol{p})\cdot\boldsymbol{x}}\\
+\overline{A}_a^\dagger(\boldsymbol{p})A_b(\boldsymbol{p}')A_d(\boldsymbol{q}')\overline{B}_c(\boldsymbol{q})e^{-\frac{i}{\hbar}(\boldsymbol{q}'+\boldsymbol{q})\cdot\boldsymbol{y}}e^{-\frac{i}{\hbar}(\boldsymbol{p}'-\boldsymbol{p})\cdot\boldsymbol{x}}\\
-\overline{A}_a^\dagger(\boldsymbol{p})A_b(\boldsymbol{p}')B_d^\dagger(\boldsymbol{q}')\overline{B}_c(\boldsymbol{q})e^{\frac{i}{\hbar}(\boldsymbol{q}'-\boldsymbol{q})\cdot\boldsymbol{y}}e^{-\frac{i}{\hbar}(\boldsymbol{p}'-\boldsymbol{p})\cdot\boldsymbol{x}}\\
-\overline{A}_a^\dagger(\boldsymbol{p})A_b(\boldsymbol{p}')\overline{D}_c^\dagger(\boldsymbol{q})D_d(\boldsymbol{q}')e^{-\frac{i}{\hbar}(\boldsymbol{q}'-\boldsymbol{q})\cdot\boldsymbol{y}}e^{-\frac{i}{\hbar}(\boldsymbol{p}'-\boldsymbol{p})\cdot\boldsymbol{x}}\\
-\overline{A}_a^\dagger(\boldsymbol{p})A_b(\boldsymbol{p}')\overline{D}_c^\dagger(\boldsymbol{q})F_d^\dagger(\boldsymbol{q}')e^{\frac{i}{\hbar}(\boldsymbol{q}'+\boldsymbol{q})\cdot\boldsymbol{y}}e^{-\frac{i}{\hbar}(\boldsymbol{p}'-\boldsymbol{p})\cdot\boldsymbol{x}}\\
-\overline{A}_a^\dagger(\boldsymbol{p})A_b(\boldsymbol{p}')D_d(\boldsymbol{q}')\overline{F}_c(\boldsymbol{q})e^{-\frac{i}{\hbar}(\boldsymbol{q}'+\boldsymbol{q})\cdot\boldsymbol{y}}e^{-\frac{i}{\hbar}(\boldsymbol{p}'-\boldsymbol{p})\cdot\boldsymbol{x}}\\
+\overline{A}_a^\dagger(\boldsymbol{p})A_b(\boldsymbol{p}')F_d^\dagger(\boldsymbol{q}')\overline{F}_c(\boldsymbol{q})e^{\frac{i}{\hbar}(\boldsymbol{q}'-\boldsymbol{q})\cdot\boldsymbol{y}}e^{-\frac{i}{\hbar}(\boldsymbol{p}'-\boldsymbol{p})\cdot\boldsymbol{x}}\\
+\overline{A}_a^\dagger(\boldsymbol{p})\overline{A}_c^\dagger(\boldsymbol{q})A_d(\boldsymbol{q}')B_b^\dagger(\boldsymbol{p}')e^{-\frac{i}{\hbar}(\boldsymbol{q}'-\boldsymbol{q})\cdot\boldsymbol{y}}e^{\frac{i}{\hbar}(\boldsymbol{p}'+\boldsymbol{p})\cdot\boldsymbol{x}}\\
+\overline{A}_a^\dagger(\boldsymbol{p})\overline{A}_c^\dagger(\boldsymbol{q})B_b^\dagger(\boldsymbol{p}')B_d^\dagger(\boldsymbol{q}')e^{\frac{i}{\hbar}(\boldsymbol{q}'+\boldsymbol{q})\cdot\boldsymbol{y}}e^{\frac{i}{\hbar}(\boldsymbol{p}'+\boldsymbol{p})\cdot\boldsymbol{x}}\\
+\overline{A}_a^\dagger(\boldsymbol{p})A_d(\boldsymbol{q}')B_b^\dagger(\boldsymbol{p}')\overline{B}_c(\boldsymbol{q})e^{-\frac{i}{\hbar}(\boldsymbol{q}'+\boldsymbol{q})\cdot\boldsymbol{y}}e^{\frac{i}{\hbar}(\boldsymbol{p}'+\boldsymbol{p})\cdot\boldsymbol{x}}\\
-\overline{A}_a^\dagger(\boldsymbol{p})B_b^\dagger(\boldsymbol{p}')B_d^\dagger(\boldsymbol{q}')\overline{B}_c(\boldsymbol{q})e^{\frac{i}{\hbar}(\boldsymbol{q}'-\boldsymbol{q})\cdot\boldsymbol{y}}e^{\frac{i}{\hbar}(\boldsymbol{p}'+\boldsymbol{p})\cdot\boldsymbol{x}}\\
-\overline{A}_a^\dagger(\boldsymbol{p})B_b^\dagger(\boldsymbol{p}')\overline{D}_c^\dagger(\boldsymbol{q})D_d(\boldsymbol{q}')e^{-\frac{i}{\hbar}(\boldsymbol{q}'-\boldsymbol{q})\cdot\boldsymbol{y}}e^{\frac{i}{\hbar}(\boldsymbol{p}'+\boldsymbol{p})\cdot\boldsymbol{x}}\\
-\overline{A}_a^\dagger(\boldsymbol{p})B_b^\dagger(\boldsymbol{p}')\overline{D}_c^\dagger(\boldsymbol{q})F_d^\dagger(\boldsymbol{q}')e^{\frac{i}{\hbar}(\boldsymbol{q}'+\boldsymbol{q})\cdot\boldsymbol{y}}e^{\frac{i}{\hbar}(\boldsymbol{p}'+\boldsymbol{p})\cdot\boldsymbol{x}}\\
-\overline{A}_a^\dagger(\boldsymbol{p})B_b^\dagger(\boldsymbol{p}')D_d(\boldsymbol{q}')\overline{F}_c(\boldsymbol{q})e^{-\frac{i}{\hbar}(\boldsymbol{q}'+\boldsymbol{q})\cdot\boldsymbol{y}}e^{\frac{i}{\hbar}(\boldsymbol{p}'+\boldsymbol{p})\cdot\boldsymbol{x}}\\
+\overline{A}_a^\dagger(\boldsymbol{p})B_b^\dagger(\boldsymbol{p}')F_d^\dagger(\boldsymbol{q}')\overline{F}_c(\boldsymbol{q})e^{\frac{i}{\hbar}(\boldsymbol{q}'-\boldsymbol{q})\cdot\boldsymbol{y}}e^{\frac{i}{\hbar}(\boldsymbol{p}'+\boldsymbol{p})\cdot\boldsymbol{x}}\\
-\overline{A}_c^\dagger(\boldsymbol{q})A_b(\boldsymbol{p}')A_d(\boldsymbol{q}')\overline{B}_a(\boldsymbol{p})e^{-\frac{i}{\hbar}(\boldsymbol{q}'-\boldsymbol{q})\cdot\boldsymbol{y}}e^{-\frac{i}{\hbar}(\boldsymbol{p}'+\boldsymbol{p})\cdot\boldsymbol{x}}\\
+\overline{A}_c^\dagger(\boldsymbol{q})A_b(\boldsymbol{p}')B_d^\dagger(\boldsymbol{q}')\overline{B}_a(\boldsymbol{p})e^{\frac{i}{\hbar}(\boldsymbol{q}'+\boldsymbol{q})\cdot\boldsymbol{y}}e^{-\frac{i}{\hbar}(\boldsymbol{p}'+\boldsymbol{p})\cdot\boldsymbol{x}}
\end{aligned}$$

$$+A_b(\boldsymbol{p}')A_d(\boldsymbol{q}')\overline{B}_a(\boldsymbol{p})\overline{B}_c(\boldsymbol{q})e^{-\frac{i}{\hbar}(\boldsymbol{q}'+\boldsymbol{q})\cdot\boldsymbol{y}}e^{-\frac{i}{\hbar}(\boldsymbol{p}'+\boldsymbol{p})\cdot\boldsymbol{x}}$$

$$+A_b(\boldsymbol{p}')B_d^\dagger(\boldsymbol{q}')\overline{B}_a(\boldsymbol{p})\overline{B}_c(\boldsymbol{q})e^{\frac{i}{\hbar}(\boldsymbol{q}'-\boldsymbol{q})\cdot\boldsymbol{y}}e^{-\frac{i}{\hbar}(\boldsymbol{p}'+\boldsymbol{p})\cdot\boldsymbol{x}}$$

$$-A_b(\boldsymbol{p}')\overline{B}_a(\boldsymbol{p})\overline{D}_c^\dagger(\boldsymbol{q})D_d(\boldsymbol{q}')e^{-\frac{i}{\hbar}(\boldsymbol{q}'-\boldsymbol{q})\cdot\boldsymbol{y}}e^{-\frac{i}{\hbar}(\boldsymbol{p}'+\boldsymbol{p})\cdot\boldsymbol{x}}$$

$$-A_b(\boldsymbol{p}')\overline{B}_a(\boldsymbol{p})\overline{D}_c^\dagger(\boldsymbol{q})F_d^\dagger(\boldsymbol{q}')e^{\frac{i}{\hbar}(\boldsymbol{q}'+\boldsymbol{q})\cdot\boldsymbol{y}}e^{-\frac{i}{\hbar}(\boldsymbol{p}'+\boldsymbol{p})\cdot\boldsymbol{x}}$$

$$-A_b(\boldsymbol{p}')\overline{B}_a(\boldsymbol{p})D_d(\boldsymbol{q}')\overline{F}_c(\boldsymbol{q})e^{-\frac{i}{\hbar}(\boldsymbol{q}'+\boldsymbol{q})\cdot\boldsymbol{y}}e^{-\frac{i}{\hbar}(\boldsymbol{p}'+\boldsymbol{p})\cdot\boldsymbol{x}}$$

$$+A_b(\boldsymbol{p}')\overline{B}_a(\boldsymbol{p})F_d^\dagger(\boldsymbol{q}')\overline{F}_c(\boldsymbol{q})e^{\frac{i}{\hbar}(\boldsymbol{q}'-\boldsymbol{q})\cdot\boldsymbol{y}}e^{-\frac{i}{\hbar}(\boldsymbol{p}'+\boldsymbol{p})\cdot\boldsymbol{x}}$$

$$-\overline{A}_c^\dagger(\boldsymbol{q})A_d(\boldsymbol{q}')B_b^\dagger(\boldsymbol{p}')\overline{B}_a(\boldsymbol{p})e^{-\frac{i}{\hbar}(\boldsymbol{q}'-\boldsymbol{q})\cdot\boldsymbol{y}}e^{\frac{i}{\hbar}(\boldsymbol{p}'-\boldsymbol{p})\cdot\boldsymbol{x}}$$

$$+\overline{A}_c^\dagger(\boldsymbol{q})B_b^\dagger(\boldsymbol{p}')B_d^\dagger(\boldsymbol{q}')\overline{B}_a(\boldsymbol{p})e^{\frac{i}{\hbar}(\boldsymbol{q}'+\boldsymbol{q})\cdot\boldsymbol{y}}e^{\frac{i}{\hbar}(\boldsymbol{p}'-\boldsymbol{p})\cdot\boldsymbol{x}}$$

$$-A_d(\boldsymbol{q}')B_b^\dagger(\boldsymbol{p}')\overline{B}_a(\boldsymbol{p})\overline{B}_c(\boldsymbol{q})e^{-\frac{i}{\hbar}(\boldsymbol{q}'+\boldsymbol{q})\cdot\boldsymbol{y}}e^{\frac{i}{\hbar}(\boldsymbol{p}'-\boldsymbol{p})\cdot\boldsymbol{x}}$$

$$-B_b^\dagger(\boldsymbol{p}')B_d^\dagger(\boldsymbol{q}')\overline{B}_a(\boldsymbol{p})\overline{B}_c(\boldsymbol{q})e^{\frac{i}{\hbar}(\boldsymbol{q}'-\boldsymbol{q})\cdot\boldsymbol{y}}e^{\frac{i}{\hbar}(\boldsymbol{p}'-\boldsymbol{p})\cdot\boldsymbol{x}}$$

$$+B_b^\dagger(\boldsymbol{p}')\overline{B}_a(\boldsymbol{p})\overline{D}_c^\dagger(\boldsymbol{q})D_d(\boldsymbol{q}')e^{-\frac{i}{\hbar}(\boldsymbol{q}'-\boldsymbol{q})\cdot\boldsymbol{y}}e^{\frac{i}{\hbar}(\boldsymbol{p}'-\boldsymbol{p})\cdot\boldsymbol{x}}$$

$$+B_b^\dagger(\boldsymbol{p}')\overline{B}_a(\boldsymbol{p})\overline{D}_c^\dagger(\boldsymbol{q})F_d^\dagger(\boldsymbol{q}')e^{\frac{i}{\hbar}\boldsymbol{q}'+\boldsymbol{q})\cdot\boldsymbol{y}}e^{\frac{i}{\hbar}(\boldsymbol{p}'-\boldsymbol{p})\cdot\boldsymbol{x}}$$

$$+B_b^\dagger(\boldsymbol{p}')\overline{B}_a(\boldsymbol{p})D_d(\boldsymbol{q}')\overline{F}_c(\boldsymbol{q})e^{-\frac{i}{\hbar}(\boldsymbol{q}'+\boldsymbol{q})\cdot\boldsymbol{y}}e^{\frac{i}{\hbar}(\boldsymbol{p}'-\boldsymbol{p})\cdot\boldsymbol{x}}$$

$$-B_b^\dagger(\boldsymbol{p}')\overline{B}_a(\boldsymbol{p})F_d^\dagger(\boldsymbol{q}'\overline{F}_c(\boldsymbol{q})e^{\frac{i}{\hbar}(\boldsymbol{q}'-\boldsymbol{q})\cdot\boldsymbol{y}}e^{\frac{i}{\hbar}(\boldsymbol{p}'-\boldsymbol{p})\cdot\boldsymbol{x}}$$

$$-\overline{A}_c^\dagger(\boldsymbol{q})A_d(\boldsymbol{q}')\overline{D}_a^\dagger(\boldsymbol{p})D_b(\boldsymbol{p}')e^{-\frac{i}{\hbar}(\boldsymbol{q}'-\boldsymbol{q})\cdot\boldsymbol{y}}e^{-\frac{i}{\hbar}(\boldsymbol{p}'-\boldsymbol{p})\cdot\boldsymbol{x}}$$

$$-\overline{A}_c^\dagger(\boldsymbol{q})B_d^\dagger(\boldsymbol{q}')\overline{D}_a^\dagger(\boldsymbol{p})D_b(\boldsymbol{p}')e^{\frac{i}{\hbar}(\boldsymbol{q}'+\boldsymbol{q})\cdot\boldsymbol{y}}e^{-\frac{i}{\hbar}(\boldsymbol{p}'-\boldsymbol{p})\cdot\boldsymbol{x}}$$

$$-A_d(\boldsymbol{q}')\overline{B}_c(\boldsymbol{q})\overline{D}_a^\dagger(\boldsymbol{p})D_b(\boldsymbol{p}')e^{-\frac{i}{\hbar}(\boldsymbol{q}'+\boldsymbol{q})\cdot\boldsymbol{y}}e^{-\frac{i}{\hbar}(\boldsymbol{p}'-\boldsymbol{p})\cdot\boldsymbol{x}}$$

$$+B_d^\dagger(\boldsymbol{q}')\overline{B}_c(\boldsymbol{q})\overline{D}_a^\dagger(\boldsymbol{p})D_b(\boldsymbol{p}')e^{\frac{i}{\hbar}(\boldsymbol{q}'-\boldsymbol{q})\cdot\boldsymbol{y}}e^{-\frac{i}{\hbar}(\boldsymbol{p}'-\boldsymbol{p})\cdot\boldsymbol{x}}$$

$$-\overline{D}_a^\dagger(\boldsymbol{q})\overline{D}_c^\dagger(\boldsymbol{p})D_b(\boldsymbol{q}')D_d(\boldsymbol{p}')e^{-\frac{i}{\hbar}(\boldsymbol{q}'-\boldsymbol{q})\cdot\boldsymbol{y}}e^{-\frac{i}{\hbar}(\boldsymbol{p}'-\boldsymbol{p})\cdot\boldsymbol{x}}$$

$$-\overline{D}_a^\dagger(\boldsymbol{p})\overline{D}_c^\dagger(\boldsymbol{q})D_b(\boldsymbol{p}')F_d^\dagger(\boldsymbol{q}')e^{\frac{i}{\hbar}(\boldsymbol{q}'+\boldsymbol{q})\cdot\boldsymbol{y}}e^{-\frac{i}{\hbar}(\boldsymbol{p}'-\boldsymbol{p})\cdot\boldsymbol{x}}$$

$$+\overline{D}_a^\dagger(\boldsymbol{p})D_b(\boldsymbol{p}')D_d(\boldsymbol{q}')\overline{F}_c(\boldsymbol{q})e^{-\frac{i}{\hbar}(\boldsymbol{q}'+\boldsymbol{q})\cdot\boldsymbol{y}}e^{-\frac{i}{\hbar}(\boldsymbol{p}'-\boldsymbol{p})\cdot\boldsymbol{x}}$$

$$-\overline{D}_a^\dagger(\boldsymbol{p})D_b(\boldsymbol{p}')F_d^\dagger(\boldsymbol{q}')\overline{F}_c(\boldsymbol{q})e^{\frac{i}{\hbar}(\boldsymbol{q}'-\boldsymbol{q})\cdot\boldsymbol{y}}e^{-\frac{i}{\hbar}(\boldsymbol{p}'-\boldsymbol{p})\cdot\boldsymbol{x}}$$

$$-\overline{A}_c^\dagger(\boldsymbol{q})A_d(\boldsymbol{q}')\overline{D}_a^\dagger(\boldsymbol{p})F_b^\dagger(\boldsymbol{p}')e^{-\frac{i}{\hbar}(\boldsymbol{q}'-\boldsymbol{q})\cdot\boldsymbol{y}}e^{\frac{i}{\hbar}(\boldsymbol{p}'+\boldsymbol{p})\cdot\boldsymbol{x}}$$

$$-\overline{A}_c^\dagger(\boldsymbol{q})B_d^\dagger(\boldsymbol{q}')\overline{D}_a^\dagger(\boldsymbol{p})F_b^\dagger(\boldsymbol{p}')e^{\frac{i}{\hbar}(\boldsymbol{q}'+\boldsymbol{q})\cdot\boldsymbol{y}}e^{\frac{i}{\hbar}(\boldsymbol{p}'+\boldsymbol{p})\cdot\boldsymbol{x}}$$

$$-A_d(\boldsymbol{q}')\overline{B}_c(\boldsymbol{q})\overline{D}_a^\dagger(\boldsymbol{p})F_b^\dagger(\boldsymbol{p}')e^{-\frac{i}{\hbar}(\boldsymbol{q}'+\boldsymbol{q})\cdot\boldsymbol{y}}e^{\frac{i}{\hbar}(\boldsymbol{p}'+\boldsymbol{p})\cdot\boldsymbol{x}}$$

$$+B_d^\dagger(\boldsymbol{q}')\overline{B}_c(\boldsymbol{q})\overline{D}_a^\dagger(\boldsymbol{p})F_b^\dagger(\boldsymbol{p}')e^{\frac{i}{\hbar}(\boldsymbol{q}'-\boldsymbol{q})\cdot\boldsymbol{y}}e^{\frac{i}{\hbar}(\boldsymbol{p}'+\boldsymbol{p})\cdot\boldsymbol{x}}$$

$$+\overline{D}_a^\dagger(\boldsymbol{p})\overline{D}_c^\dagger(\boldsymbol{q})D_d(\boldsymbol{q}')F_b^\dagger(\boldsymbol{p}')e^{-\frac{i}{\hbar}(\boldsymbol{q}'-\boldsymbol{q})\cdot\boldsymbol{y}}e^{\frac{i}{\hbar}(\boldsymbol{p}'+\boldsymbol{p})\cdot\boldsymbol{x}}$$

$$+\overline{D}_a^\dagger(\boldsymbol{p})\overline{D}_c^\dagger(\boldsymbol{q})F_b^\dagger(\boldsymbol{p}')F_d^\dagger(\boldsymbol{q}')e^{\frac{i}{\hbar}(\boldsymbol{q}'+\boldsymbol{q})\cdot\boldsymbol{y}}e^{\frac{i}{\hbar}(\boldsymbol{p}'+\boldsymbol{p})\cdot\boldsymbol{x}}$$

$$
\begin{aligned}
&+\overline{D}_a^\dagger(\mathbf{p})D_d(\mathbf{q}')F_b^\dagger(\mathbf{p}')\overline{F}_c(\mathbf{q})e^{-\frac{i}{\hbar}(\mathbf{q}'+\mathbf{q})\cdot\mathbf{y}}e^{\frac{i}{\hbar}(\mathbf{p}'+\mathbf{p})\cdot\mathbf{x}}\\
&-\overline{D}_a^\dagger(\mathbf{p})F_b^\dagger(\mathbf{p}')F_d^\dagger(\mathbf{q}')\overline{F}_c(\mathbf{q})e^{\frac{i}{\hbar}(\mathbf{q}'-\mathbf{q})\cdot\mathbf{y}}e^{\frac{i}{\hbar}(\mathbf{p}'+\mathbf{p})\cdot\mathbf{x}}\\
&-\overline{A}_c^\dagger(\mathbf{q})A_d(\mathbf{q}')D_b(\mathbf{p}')\overline{F}_a(\mathbf{p})e^{-\frac{i}{\hbar}(\mathbf{q}'-\mathbf{q})\cdot\mathbf{y}}e^{-\frac{i}{\hbar}(\mathbf{p}'+\mathbf{p})\cdot\mathbf{x}}\\
&-\overline{A}_c^\dagger(\mathbf{q})B_d^\dagger(\mathbf{q}')D_b(\mathbf{p}')\overline{F}_a(\mathbf{p})e^{\frac{i}{\hbar}(\mathbf{q}'+\mathbf{q})\cdot\mathbf{y}}e^{-\frac{i}{\hbar}(\mathbf{p}'+\mathbf{p})\cdot\mathbf{x}}\\
&-A_d(\mathbf{q}')\overline{B}_c(\mathbf{q})D_b(\mathbf{p}')\overline{F}_a(\mathbf{p})e^{-\frac{i}{\hbar}(\mathbf{q}'+\mathbf{q})\cdot\mathbf{y}}e^{-\frac{i}{\hbar}(\mathbf{p}'+\mathbf{p})\cdot\mathbf{x}}\\
&+B_b^\dagger(\mathbf{q}')\overline{B}_c(\mathbf{q})D_b(\mathbf{p}')\overline{F}_a(\mathbf{p})e^{\frac{i}{\hbar}(\mathbf{q}'-\mathbf{q})\cdot\mathbf{y}}e^{-\frac{i}{\hbar}(\mathbf{p}'+\mathbf{p})\cdot\mathbf{x}}\\
&-\overline{D}_c^\dagger(\mathbf{q})D_b(\mathbf{p}')D_d(\mathbf{q}')\overline{F}_a(\mathbf{p})e^{-\frac{i}{\hbar}(\mathbf{q}'-\mathbf{q})\cdot\mathbf{y}}e^{-\frac{i}{\hbar}(\mathbf{p}'+\mathbf{p})\cdot\mathbf{x}}\\
&+\overline{D}_c^\dagger(\mathbf{q})D_b(\mathbf{p}')F_d^\dagger(\mathbf{q}')\overline{F}_a(\mathbf{p})e^{\frac{i}{\hbar}(\mathbf{q}'+\mathbf{q})\cdot\mathbf{y}}e^{-\frac{i}{\hbar}(\mathbf{p}'+\mathbf{p})\cdot\mathbf{x}}\\
&+D_b(\mathbf{p}')D_d(\mathbf{q}')\overline{F}_a(\mathbf{p})\overline{F}_c(\mathbf{q})e^{-\frac{i}{\hbar}(\mathbf{q}'+\mathbf{q})\cdot\mathbf{y}}e^{-\frac{i}{\hbar}(\mathbf{p}'+\mathbf{p})\cdot\mathbf{x}}\\
&+D_b(\mathbf{p}')F_d^\dagger(\mathbf{q}')\overline{F}_a(\mathbf{p})\overline{F}_c(\mathbf{q})e^{\frac{i}{\hbar}(\mathbf{q}'-\mathbf{q})\cdot\mathbf{y}}e^{-\frac{i}{\hbar}(\mathbf{p}'+\mathbf{p})\cdot\mathbf{x}}\\
&+\overline{A}_c^\dagger(\mathbf{q})A_d(\mathbf{q}')F_b^\dagger(\mathbf{p}')\overline{F}_a(\mathbf{p})e^{-\frac{i}{\hbar}(\mathbf{q}'-\mathbf{q})\cdot\mathbf{y}}e^{\frac{i}{\hbar}(\mathbf{p}'-\mathbf{p})\cdot\mathbf{x}}\\
&+\overline{A}_c^\dagger(\mathbf{q})B_d^\dagger(\mathbf{q}')F_b^\dagger(\mathbf{p}')\overline{F}_a(\mathbf{p})e^{\frac{i}{\hbar}(\mathbf{q}'+\mathbf{q})\cdot\mathbf{y}}e^{\frac{i}{\hbar}(\mathbf{p}'-\mathbf{p})\cdot\mathbf{x}}\\
&+A_d(\mathbf{q}')\overline{B}_c(\mathbf{q})F_b^\dagger(\mathbf{p}')\overline{F}_a(\mathbf{p})e^{-\frac{i}{\hbar}(\mathbf{q}'+\mathbf{q})\cdot\mathbf{y}}e^{\frac{i}{\hbar}(\mathbf{p}'-\mathbf{p})\cdot\mathbf{x}}\\
&-B_d^\dagger(\mathbf{q}')\overline{B}_c(\mathbf{q})F_b^\dagger(\mathbf{p}')\overline{F}_a(\mathbf{p})e^{\frac{i}{\hbar}(\mathbf{q}'-\mathbf{q})\cdot\mathbf{y}}e^{\frac{i}{\hbar}(\mathbf{p}'-\mathbf{p})\cdot\mathbf{x}}\\
&-\overline{D}_c^\dagger(\mathbf{q})D_d(\mathbf{q}')F_b^\dagger(\mathbf{p}')\overline{F}_a(\mathbf{p})e^{-\frac{i}{\hbar}(\mathbf{q}'-\mathbf{q})\cdot\mathbf{y}}e^{\frac{i}{\hbar}(\mathbf{p}'-\mathbf{p})\cdot\mathbf{x}}\\
&+\overline{D}_c^\dagger(\mathbf{q})F_b^\dagger(\mathbf{p}')F_d^\dagger(\mathbf{q}')\overline{F}_a(\mathbf{p})e^{\frac{i}{\hbar}(\mathbf{q}'+\mathbf{q})\cdot\mathbf{y}}e^{\frac{i}{\hbar}(\mathbf{p}'-\mathbf{p})\cdot\mathbf{x}}\\
&-D_d(\mathbf{q}')F_b^\dagger(\mathbf{p}')\overline{F}_a(\mathbf{p})\overline{F}_c(\mathbf{q})e^{-\frac{i}{\hbar}(\mathbf{q}'+\mathbf{q})\cdot\mathbf{y}}e^{\frac{i}{\hbar}(\mathbf{p}'-\mathbf{p})\cdot\mathbf{x}}\\
&-F_b^\dagger(\mathbf{p}')F_d^\dagger(\mathbf{q}')\overline{F}_a(\mathbf{p})\overline{F}_c(\mathbf{q})e^{\frac{i}{\hbar}(\mathbf{q}'-\mathbf{q})\cdot\mathbf{y}}e^{\frac{i}{\hbar}(\mathbf{p}'-\mathbf{p})\cdot\mathbf{x}}\Big].
\end{aligned}
$$

Next we change summation indices $a \leftrightarrow c$ and integration variables $\mathbf{x} \leftrightarrow \mathbf{y}$ and $\mathbf{p} \leftrightarrow \mathbf{q}$ to get the following simplified expression:

$$
\begin{aligned}
V_2 = \frac{e^2}{(2\pi\hbar)^6}\sum_{abcd}\int d\mathbf{x}d\mathbf{y}\int d\mathbf{p}d\mathbf{p}'d\mathbf{q}d\mathbf{q}'\,\frac{\gamma^0_{ab}\gamma^0_{cd}}{8\pi|\mathbf{x}-\mathbf{y}|}\\
\times\Big[&-\overline{A}_a^\dagger(\mathbf{p})\overline{A}_c^\dagger(\mathbf{q})A_b(\mathbf{p}')A_d(\mathbf{q}')e^{-\frac{i}{\hbar}(\mathbf{q}'-\mathbf{q})\cdot\mathbf{y}}e^{-\frac{i}{\hbar}(\mathbf{p}'-\mathbf{p})\cdot\mathbf{x}}\\
&+2\overline{A}_a^\dagger(\mathbf{p})A_b(\mathbf{p}')A_d(\mathbf{q}')\overline{B}_c(\mathbf{q})e^{-\frac{i}{\hbar}(\mathbf{q}'+\mathbf{q})\cdot\mathbf{y}}e^{-\frac{i}{\hbar}(\mathbf{p}'-\mathbf{p})\cdot\mathbf{x}}\\
&-2\overline{A}_a^\dagger(\mathbf{p})A_b(\mathbf{p}')B_d^\dagger(\mathbf{q}')\overline{B}_c(\mathbf{q})e^{\frac{i}{\hbar}(\mathbf{q}'-\mathbf{q})\cdot\mathbf{y}}e^{-\frac{i}{\hbar}(\mathbf{p}'-\mathbf{p})\cdot\mathbf{x}}\\
&-2\overline{A}_a^\dagger(\mathbf{p})A_b(\mathbf{p}')\overline{D}_c^\dagger(\mathbf{q})D_d(\mathbf{q}')e^{-\frac{i}{\hbar}(\mathbf{q}'-\mathbf{q})\cdot\mathbf{y}}e^{-\frac{i}{\hbar}(\mathbf{p}'-\mathbf{p})\cdot\mathbf{x}}\\
&-2\overline{A}_a^\dagger(\mathbf{p})A_b(\mathbf{p}')\overline{D}_c^\dagger(\mathbf{q})F_d^\dagger(\mathbf{q}')e^{\frac{i}{\hbar}(\mathbf{q}'+\mathbf{q})\cdot\mathbf{y}}e^{-\frac{i}{\hbar}(\mathbf{p}'-\mathbf{p})\cdot\mathbf{x}}\\
&-2\overline{A}_a^\dagger(\mathbf{p})A_b(\mathbf{p}')D_d(\mathbf{q}')\overline{F}_c(\mathbf{q})e^{-\frac{i}{\hbar}(\mathbf{q}'+\mathbf{q})\cdot\mathbf{y}}e^{-\frac{i}{\hbar}(\mathbf{p}'-\mathbf{p})\cdot\mathbf{x}}\\
&+2\overline{A}_a^\dagger(\mathbf{p})A_b(\mathbf{p}')F_d^\dagger(\mathbf{q}')\overline{F}_c(\mathbf{q})e^{\frac{i}{\hbar}(\mathbf{q}'-\mathbf{q})\cdot\mathbf{y}}e^{-\frac{i}{\hbar}(\mathbf{p}'-\mathbf{p})\cdot\mathbf{x}}
\end{aligned}
$$

$$
\begin{aligned}
&+2\overline{A}_a^\dagger(\boldsymbol{p})\overline{A}_c^\dagger(\boldsymbol{q})A_d(\boldsymbol{q}')B_b^\dagger(\boldsymbol{p}')e^{-\frac{i}{\hbar}(\boldsymbol{q}'-\boldsymbol{q})\cdot\boldsymbol{y}}e^{\frac{i}{\hbar}(\boldsymbol{p}'+\boldsymbol{p})\cdot\boldsymbol{x}}\\
&+\overline{A}_a^\dagger(\boldsymbol{p})\overline{A}_c^\dagger(\boldsymbol{q})B_b^\dagger(\boldsymbol{p}')B_d^\dagger(\boldsymbol{q}')e^{\frac{i}{\hbar}(\boldsymbol{q}'+\boldsymbol{q})\cdot\boldsymbol{y}}e^{\frac{i}{\hbar}(\boldsymbol{p}'+\boldsymbol{p})\cdot\boldsymbol{x}}\\
&+2\overline{A}_a^\dagger(\boldsymbol{p})A_d(\boldsymbol{q}')B_b^\dagger(\boldsymbol{p}')\overline{B}_c(\boldsymbol{q})e^{-\frac{i}{\hbar}(\boldsymbol{q}'+\boldsymbol{q})\cdot\boldsymbol{y}}e^{\frac{i}{\hbar}(\boldsymbol{p}'+\boldsymbol{p})\cdot\boldsymbol{x}}\\
&-2\overline{A}_a^\dagger(\boldsymbol{p})B_b^\dagger(\boldsymbol{p}')B_d^\dagger(\boldsymbol{q}')\overline{B}_c(\boldsymbol{q})e^{\frac{i}{\hbar}(\boldsymbol{q}'-\boldsymbol{q})\cdot\boldsymbol{y}}e^{\frac{i}{\hbar}(\boldsymbol{p}'+\boldsymbol{p})\cdot\boldsymbol{x}}\\
&-2\overline{A}_a^\dagger(\boldsymbol{p})B_b^\dagger(\boldsymbol{p}')\overline{D}_c^\dagger(\boldsymbol{q})D_d(\boldsymbol{q}')e^{-\frac{i}{\hbar}(\boldsymbol{q}'-\boldsymbol{q})\cdot\boldsymbol{y}}e^{\frac{i}{\hbar}(\boldsymbol{p}'+\boldsymbol{p})\cdot\boldsymbol{x}}\\
&-2\overline{A}_a^\dagger(\boldsymbol{p})B_b^\dagger(\boldsymbol{p}')\overline{D}_c^\dagger(\boldsymbol{q})F_d^\dagger(\boldsymbol{q}')e^{\frac{i}{\hbar}(\boldsymbol{q}'+\boldsymbol{q})\cdot\boldsymbol{y}}e^{\frac{i}{\hbar}(\boldsymbol{p}'+\boldsymbol{p})\cdot\boldsymbol{x}}\\
&-2\overline{A}_a^\dagger(\boldsymbol{p})B_b^\dagger(\boldsymbol{p}')D_d(\boldsymbol{q}')\overline{F}_c(\boldsymbol{q})e^{-\frac{i}{\hbar}(\boldsymbol{q}'+\boldsymbol{q})\cdot\boldsymbol{y}}e^{\frac{i}{\hbar}(\boldsymbol{p}'+\boldsymbol{p})\cdot\boldsymbol{x}}\\
&+2\overline{A}_a^\dagger(\boldsymbol{p})B_b^\dagger(\boldsymbol{p}')F_d^\dagger(\boldsymbol{q}')\overline{F}_c(\boldsymbol{q})e^{\frac{i}{\hbar}(\boldsymbol{q}'-\boldsymbol{q})\cdot\boldsymbol{y}}e^{\frac{i}{\hbar}(\boldsymbol{p}'+\boldsymbol{p})\cdot\boldsymbol{x}}\\
&+A_b(\boldsymbol{p}')A_d(\boldsymbol{q}')\overline{B}_a(\boldsymbol{p})\overline{B}_c(\boldsymbol{q})e^{-\frac{i}{\hbar}(\boldsymbol{q}'+\boldsymbol{q})\cdot\boldsymbol{y}}e^{-\frac{i}{\hbar}(\boldsymbol{p}'+\boldsymbol{p})\cdot\boldsymbol{x}}\\
&+2A_b(\boldsymbol{p}')B_d^\dagger(\boldsymbol{q}')\overline{B}_a(\boldsymbol{p})\overline{B}_c(\boldsymbol{q})e^{\frac{i}{\hbar}(\boldsymbol{q}'-\boldsymbol{q})\cdot\boldsymbol{y}}e^{-\frac{i}{\hbar}(\boldsymbol{p}'+\boldsymbol{p})\cdot\boldsymbol{x}}\\
&-2A_b(\boldsymbol{p}')\overline{B}_a(\boldsymbol{p})\overline{D}_c^\dagger(\boldsymbol{q})D_d(\boldsymbol{q}')e^{-\frac{i}{\hbar}(\boldsymbol{q}'-\boldsymbol{q})\cdot\boldsymbol{y}}e^{-\frac{i}{\hbar}(\boldsymbol{p}'+\boldsymbol{p})\cdot\boldsymbol{x}}\\
&-2A_b(\boldsymbol{p}')\overline{B}_a(\boldsymbol{p})\overline{D}_c^\dagger(\boldsymbol{q})F_d^\dagger(\boldsymbol{q}')e^{\frac{i}{\hbar}(\boldsymbol{q}'+\boldsymbol{q})\cdot\boldsymbol{y}}e^{-\frac{i}{\hbar}(\boldsymbol{p}'+\boldsymbol{p})\cdot\boldsymbol{x}}\\
&-2\overline{A}_b(\boldsymbol{p}')B_a(\boldsymbol{p})D_d(\boldsymbol{q}')\overline{F}_c(\boldsymbol{q})e^{-\frac{i}{\hbar}(\boldsymbol{q}'+\boldsymbol{q})\cdot\boldsymbol{y}}e^{-\frac{i}{\hbar}(\boldsymbol{p}'+\boldsymbol{p})\cdot\boldsymbol{x}}\\
&+2\overline{A}_b(\boldsymbol{p}')B_a(\boldsymbol{p})F_d^\dagger(\boldsymbol{q}')\overline{F}_c(\boldsymbol{q})e^{\frac{i}{\hbar}(\boldsymbol{q}'-\boldsymbol{q})\cdot\boldsymbol{y}}e^{-\frac{i}{\hbar}(\boldsymbol{p}'+\boldsymbol{p})\cdot\boldsymbol{x}}\\
&-B_b^\dagger(\boldsymbol{p}')B_d^\dagger(\boldsymbol{q}')\overline{B}_a(\boldsymbol{p})\overline{B}_c(\boldsymbol{q})e^{\frac{i}{\hbar}(\boldsymbol{q}'-\boldsymbol{q})\cdot\boldsymbol{y}}e^{\frac{i}{\hbar}(\boldsymbol{p}'-\boldsymbol{p})\cdot\boldsymbol{x}}\\
&+2B_b^\dagger(\boldsymbol{p}')\overline{B}_a(\boldsymbol{p})\overline{D}_c^\dagger(\boldsymbol{q})D_d(\boldsymbol{q}')e^{-\frac{i}{\hbar}(\boldsymbol{q}'-\boldsymbol{q})\cdot\boldsymbol{y}}e^{\frac{i}{\hbar}(\boldsymbol{p}'-\boldsymbol{p})\cdot\boldsymbol{x}}\\
&+2B_b^\dagger(\boldsymbol{p}')\overline{B}_a(\boldsymbol{p})\overline{D}_c^\dagger(\boldsymbol{q})F_d^\dagger(\boldsymbol{q}')e^{\frac{i}{\hbar}(\boldsymbol{q}'+\boldsymbol{q})\cdot\boldsymbol{y}}e^{\frac{i}{\hbar}(\boldsymbol{p}'-\boldsymbol{p})\cdot\boldsymbol{x}}\\
&+2B_b^\dagger(\boldsymbol{p}')\overline{B}_a(\boldsymbol{p})D_d(\boldsymbol{q}')\overline{F}_c(\boldsymbol{q})e^{-\frac{i}{\hbar}(\boldsymbol{q}'+\boldsymbol{q})\cdot\boldsymbol{y}}e^{\frac{i}{\hbar}(\boldsymbol{p}'-\boldsymbol{p})\cdot\boldsymbol{x}}\\
&-2B_b^\dagger(\boldsymbol{p}')\overline{B}_a(\boldsymbol{p})F_d^\dagger(\boldsymbol{q}')\overline{F}_c(\boldsymbol{q})e^{\frac{i}{\hbar}(\boldsymbol{q}'-\boldsymbol{q})\cdot\boldsymbol{y}}e^{\frac{i}{\hbar}(\boldsymbol{p}'-\boldsymbol{p})\cdot\boldsymbol{x}}\\
&-\overline{D}_a^\dagger(\boldsymbol{p})\overline{D}_c^\dagger(\boldsymbol{q})D_b(\boldsymbol{p}')D_d(\boldsymbol{q}')e^{-\frac{i}{\hbar}(\boldsymbol{q}'-\boldsymbol{q})\cdot\boldsymbol{y}}e^{-\frac{i}{\hbar}(\boldsymbol{p}'-\boldsymbol{p})\cdot\boldsymbol{x}}\\
&-2\overline{D}_a^\dagger(\boldsymbol{p})\overline{D}_c^\dagger(\boldsymbol{q})D_b(\boldsymbol{p}')F_d^\dagger(\boldsymbol{q}')e^{\frac{i}{\hbar}(\boldsymbol{q}'+\boldsymbol{q})\cdot\boldsymbol{y}}e^{-\frac{i}{\hbar}(\boldsymbol{p}'-\boldsymbol{p})\cdot\boldsymbol{x}}\\
&+2\overline{D}_a^\dagger(\boldsymbol{p})D_b(\boldsymbol{p}')D_d(\boldsymbol{q}')\overline{F}_c(\boldsymbol{q})e^{-\frac{i}{\hbar}(\boldsymbol{q}'+\boldsymbol{q})\cdot\boldsymbol{y}}e^{-\frac{i}{\hbar}(\boldsymbol{p}'-\boldsymbol{p})\cdot\boldsymbol{x}}\\
&-2\overline{D}_a^\dagger(\boldsymbol{p})D_b(\boldsymbol{p}')F_c^\dagger(\boldsymbol{q}')\overline{F}_d(\boldsymbol{q})e^{\frac{i}{\hbar}(\boldsymbol{q}'-\boldsymbol{q})\cdot\boldsymbol{y}}e^{-\frac{i}{\hbar}(\boldsymbol{p}'-\boldsymbol{p})\cdot\boldsymbol{x}}\\
&+\overline{D}_a^\dagger(\boldsymbol{p})\overline{D}_c^\dagger(\boldsymbol{q})F_b^\dagger(\boldsymbol{p}')F_d^\dagger(\boldsymbol{q}')e^{\frac{i}{\hbar}(\boldsymbol{q}'+\boldsymbol{q})\cdot\boldsymbol{y}}e^{\frac{i}{\hbar}(\boldsymbol{p}'+\boldsymbol{p})\cdot\boldsymbol{x}}\\
&+2\overline{D}_a^\dagger(\boldsymbol{p})D_d(\boldsymbol{q}')F_b^\dagger(\boldsymbol{p}')\overline{F}_c(\boldsymbol{q})e^{-\frac{i}{\hbar}(\boldsymbol{q}'+\boldsymbol{q})\cdot\boldsymbol{y}}e^{\frac{i}{\hbar}(\boldsymbol{p}'+\boldsymbol{p})\cdot\boldsymbol{x}}\\
&-2\overline{D}_a^\dagger(\boldsymbol{p})F_b^\dagger(\boldsymbol{p}')F_d^\dagger(\boldsymbol{q}')\overline{F}_c(\boldsymbol{q})e^{\frac{i}{\hbar}(\boldsymbol{q}'-\boldsymbol{q})\cdot\boldsymbol{y}}e^{\frac{i}{\hbar}(\boldsymbol{p}'+\boldsymbol{p})\cdot\boldsymbol{x}}\\
&+D_b(\boldsymbol{p}')D_d(\boldsymbol{q}')\overline{F}_a(\boldsymbol{p})\overline{F}_c(\boldsymbol{q})e^{-\frac{i}{\hbar}(\boldsymbol{q}'+\boldsymbol{q})\cdot\boldsymbol{y}}e^{-\frac{i}{\hbar}(\boldsymbol{p}'+\boldsymbol{p})\cdot\boldsymbol{x}}
\end{aligned}
$$

$$+ 2D_b(\boldsymbol{p}')F_d^\dagger(\boldsymbol{q}')\overline{F}_a(\boldsymbol{p})\overline{F}_c(\boldsymbol{q})e^{\frac{i}{\hbar}(\boldsymbol{q}'-\boldsymbol{q})\cdot\boldsymbol{y}}e^{-\frac{i}{\hbar}(\boldsymbol{p}'+\boldsymbol{p})\cdot\boldsymbol{x}}$$
$$- F_b^\dagger(\boldsymbol{p}')F_d^\dagger(\boldsymbol{q}')\overline{F}_a(\boldsymbol{p})\overline{F}_c(\boldsymbol{q})e^{\frac{i}{\hbar}(\boldsymbol{q}'-\boldsymbol{q})\cdot\boldsymbol{y}}e^{\frac{i}{\hbar}(\boldsymbol{p}'-\boldsymbol{p})\cdot\boldsymbol{x}}].$$

Integrals with respect to $\boldsymbol{x}$ and $\boldsymbol{y}$ are calculated by formula (A.2), i. e.,

$$
\begin{aligned}
V_2 = {}& \frac{e^2\hbar^2}{2(2\pi\hbar)^3}\sum_{abcd}\int d\boldsymbol{p}d\boldsymbol{p}'d\boldsymbol{q}d\boldsymbol{q}'\gamma_{ab}^0\gamma_{cd}^0 \\
&\times\Big(-\overline{A}_a^\dagger(\boldsymbol{p})\overline{A}_c^\dagger(\boldsymbol{q})A_b(\boldsymbol{p}')A_d(\boldsymbol{q}')\delta(\boldsymbol{q}'-\boldsymbol{q}+\boldsymbol{p}'-\boldsymbol{p})\frac{1}{|\boldsymbol{q}'-\boldsymbol{q}|^2} \\
&+ 2\overline{A}_a^\dagger(\boldsymbol{p})A_b(\boldsymbol{p}')A_d(\boldsymbol{q}')\overline{B}_c(\boldsymbol{q})\delta(\boldsymbol{q}'+\boldsymbol{q}+\boldsymbol{p}'-\boldsymbol{p})\frac{1}{|\boldsymbol{q}'+\boldsymbol{q}|^2} \\
&- 2\overline{A}_a^\dagger(\boldsymbol{p})A_b(\boldsymbol{p}')B_d^\dagger(\boldsymbol{q}')\overline{B}_c(\boldsymbol{q})\delta(\boldsymbol{q}'-\boldsymbol{q}-\boldsymbol{p}'+\boldsymbol{p})\frac{1}{|\boldsymbol{q}'-\boldsymbol{q}|^2} \\
&- 2\overline{A}_a^\dagger(\boldsymbol{p})A_b(\boldsymbol{p}')\overline{D}_c^\dagger(\boldsymbol{q})D_d(\boldsymbol{q}')\delta(\boldsymbol{q}'-\boldsymbol{q}+\boldsymbol{p}'-\boldsymbol{p})\frac{1}{|\boldsymbol{q}'-\boldsymbol{q}|^2} \\
&- 2\overline{A}_a^\dagger(\boldsymbol{p})A_b(\boldsymbol{p}')\overline{D}_c^\dagger(\boldsymbol{q})F_d^\dagger(\boldsymbol{q}')\delta(\boldsymbol{q}'+\boldsymbol{q}-\boldsymbol{p}'+\boldsymbol{p})\frac{1}{|\boldsymbol{q}'+\boldsymbol{q}|^2} \\
&- 2\overline{A}_a^\dagger(\boldsymbol{p})A_b(\boldsymbol{p}')D_d(\boldsymbol{q}')\overline{F}_c(\boldsymbol{q})\delta(\boldsymbol{q}'+\boldsymbol{q}+\boldsymbol{p}'-\boldsymbol{p})\frac{1}{|\boldsymbol{q}'+\boldsymbol{q}|^2} \\
&+ 2\overline{A}_a^\dagger(\boldsymbol{p})A_b(\boldsymbol{p}')F_d^\dagger(\boldsymbol{q}')\overline{F}_c(\boldsymbol{q})\delta(\boldsymbol{q}'-\boldsymbol{q}-\boldsymbol{p}'+\boldsymbol{p})\frac{1}{|\boldsymbol{q}'-\boldsymbol{q}|^2} \\
&+ 2\overline{A}_a^\dagger(\boldsymbol{p})\overline{A}_c^\dagger(\boldsymbol{q})A_d(\boldsymbol{q}')B_b^\dagger(\boldsymbol{p}')\delta(\boldsymbol{q}'-\boldsymbol{q}-\boldsymbol{p}'-\boldsymbol{p})\frac{1}{|\boldsymbol{q}'-\boldsymbol{q}|^2} \\
&+ \overline{A}_a^\dagger(\boldsymbol{p})\overline{A}_c^\dagger(\boldsymbol{q})B_b^\dagger(\boldsymbol{p}')B_d^\dagger(\boldsymbol{q}')\delta(\boldsymbol{q}'+\boldsymbol{q}+\boldsymbol{p}'+\boldsymbol{p})\frac{1}{|\boldsymbol{q}'+\boldsymbol{q}|^2} \\
&+ 2\overline{A}_a^\dagger(\boldsymbol{p})A_d(\boldsymbol{q}')B_b^\dagger(\boldsymbol{p}')\overline{B}_c(\boldsymbol{q})\delta(\boldsymbol{q}'+\boldsymbol{q}-\boldsymbol{p}'-\boldsymbol{p})\frac{1}{|\boldsymbol{q}'+\boldsymbol{q}|^2} \\
&- 2\overline{A}_a^\dagger(\boldsymbol{p})B_b^\dagger(\boldsymbol{p}')B_d^\dagger(\boldsymbol{q}')\overline{B}_c(\boldsymbol{q})\delta(\boldsymbol{q}'-\boldsymbol{q}+\boldsymbol{p}'+\boldsymbol{p})\frac{1}{|\boldsymbol{q}'-\boldsymbol{q}|^2} \\
&- 2\overline{A}_a^\dagger(\boldsymbol{p})B_b^\dagger(\boldsymbol{p}')\overline{D}_c^\dagger(\boldsymbol{q})D_d(\boldsymbol{q}')\delta(\boldsymbol{q}'-\boldsymbol{q}-\boldsymbol{p}'-\boldsymbol{p})\frac{1}{|\boldsymbol{q}'-\boldsymbol{q}|^2} \\
&- 2\overline{A}_a^\dagger(\boldsymbol{p})B_b^\dagger(\boldsymbol{p}')\overline{D}_c^\dagger(\boldsymbol{q})F_d^\dagger(\boldsymbol{q}')\delta(\boldsymbol{q}'+\boldsymbol{q}+\boldsymbol{p}'+\boldsymbol{p})\frac{1}{|\boldsymbol{q}'+\boldsymbol{q}|^2} \\
&- 2\overline{A}_a^\dagger(\boldsymbol{p})B_b^\dagger(\boldsymbol{p}')D_d(\boldsymbol{q}')\overline{F}_c(\boldsymbol{q})\delta(\boldsymbol{q}'+\boldsymbol{q}-\boldsymbol{p}'-\boldsymbol{p})\frac{1}{|\boldsymbol{q}'+\boldsymbol{q}|^2} \\
&+ 2\overline{A}_a^\dagger(\boldsymbol{p})B_b^\dagger(\boldsymbol{p}')F_d^\dagger(\boldsymbol{q}')\overline{F}_c(\boldsymbol{q})\delta(\boldsymbol{q}'-\boldsymbol{q}+\boldsymbol{p}'+\boldsymbol{p})\frac{1}{|\boldsymbol{q}'-\boldsymbol{q}|^2} \\
&+ A_b(\boldsymbol{p}')A_d(\boldsymbol{q}')\overline{B}_a(\boldsymbol{p})\overline{B}_c(\boldsymbol{q})\delta(\boldsymbol{q}'+\boldsymbol{q}+\boldsymbol{p}'+\boldsymbol{p})\frac{1}{|\boldsymbol{q}'+\boldsymbol{q}|^2} \\
&+ 2A_b(\boldsymbol{p}')B_d^\dagger(\boldsymbol{q}')\overline{B}_a(\boldsymbol{p})\overline{B}_c(\boldsymbol{q})\delta(\boldsymbol{q}'-\boldsymbol{q}-\boldsymbol{p}'-\boldsymbol{p})\frac{1}{|\boldsymbol{q}'-\boldsymbol{q}|^2} \\
&- 2A_b(\boldsymbol{p}')\overline{B}_a(\boldsymbol{p})\overline{D}_c^\dagger(\boldsymbol{q})D_d(\boldsymbol{q}')\delta(\boldsymbol{q}'-\boldsymbol{q}+\boldsymbol{p}'+\boldsymbol{p})\frac{1}{|\boldsymbol{q}'-\boldsymbol{q}|^2}
\end{aligned}
$$

$$
\begin{aligned}
&- 2A_b(\boldsymbol{p}')\overline{B}_a(\boldsymbol{p})\overline{D}_c^\dagger(\boldsymbol{q})F_d^\dagger(\boldsymbol{q}')\delta(\boldsymbol{q}' + \boldsymbol{q} - \boldsymbol{p}' - \boldsymbol{p})\frac{1}{|\boldsymbol{q}' + \boldsymbol{q}|^2} \\
&- 2\overline{A}_b(\boldsymbol{p}')B_a(\boldsymbol{p})D_d(\boldsymbol{q}')\overline{F}_c(\boldsymbol{q})\delta(\boldsymbol{q}' + \boldsymbol{q} + \boldsymbol{p}' + \boldsymbol{p})\frac{1}{|\boldsymbol{q}' + \boldsymbol{q}|^2} \\
&+ 2\overline{A}_b(\boldsymbol{p}')B_a(\boldsymbol{p})F_d^\dagger(\boldsymbol{q}')\overline{F}_c(\boldsymbol{q})\delta(\boldsymbol{q}' - \boldsymbol{q} - \boldsymbol{p}' - \boldsymbol{p})\frac{1}{|\boldsymbol{q}' - \boldsymbol{q}|^2} \\
&- B_b^\dagger(\boldsymbol{p}')B_d^\dagger(\boldsymbol{q}')\overline{B}_a(\boldsymbol{p})\overline{B}_c(\boldsymbol{q})\delta(\boldsymbol{q}' - \boldsymbol{q} + \boldsymbol{p}' - \boldsymbol{p})\frac{1}{|\boldsymbol{q}' - \boldsymbol{q}|^2} \\
&+ 2B_b^\dagger(\boldsymbol{p}')\overline{B}_a(\boldsymbol{p})\overline{D}_c^\dagger(\boldsymbol{q})D_d(\boldsymbol{q}')\delta(\boldsymbol{q}' - \boldsymbol{q} - \boldsymbol{p}' + \boldsymbol{p})\frac{1}{|\boldsymbol{q}' - \boldsymbol{q}|^2} \\
&+ 2B_b^\dagger(\boldsymbol{p}')\overline{B}_a(\boldsymbol{p})\overline{D}_c^\dagger(\boldsymbol{q})F_d^\dagger(\boldsymbol{q}')\delta(\boldsymbol{q}' + \boldsymbol{q} + \boldsymbol{p}' - \boldsymbol{p})\frac{1}{|\boldsymbol{q}' + \boldsymbol{q}|^2} \\
&+ 2B_b^\dagger(\boldsymbol{p}')\overline{B}_a(\boldsymbol{p})D_d(\boldsymbol{q}')\overline{F}_c(\boldsymbol{q})\delta(\boldsymbol{q}' + \boldsymbol{q} - \boldsymbol{p}' + \boldsymbol{p})\frac{1}{|\boldsymbol{q}' + \boldsymbol{q}|^2} \\
&- 2B_b^\dagger(\boldsymbol{p}')\overline{B}_a(\boldsymbol{p})F_d^\dagger(\boldsymbol{q}')\overline{F}_c(\boldsymbol{q})\delta(\boldsymbol{q}' - \boldsymbol{q} + \boldsymbol{p}' - \boldsymbol{p})\frac{1}{|\boldsymbol{q}' - \boldsymbol{q}|^2} \\
&- \overline{D}_a^\dagger(\boldsymbol{p})\overline{D}_c^\dagger(\boldsymbol{q})D_\beta(\boldsymbol{p}')D_d(\boldsymbol{q}')\delta(\boldsymbol{q}' - \boldsymbol{q} + \boldsymbol{p}' - \boldsymbol{p})\frac{1}{|\boldsymbol{q}' - \boldsymbol{q}|^2} \\
&- 2\overline{D}_a^\dagger(\boldsymbol{p})\overline{D}_c^\dagger(\boldsymbol{q})D_b(\boldsymbol{p}')F_d^\dagger(\boldsymbol{q}')\delta(\boldsymbol{q}' + \boldsymbol{q} - \boldsymbol{p}' + \boldsymbol{p})\frac{1}{|\boldsymbol{q}' + \boldsymbol{q}|^2} \\
&+ 2\overline{D}_a^\dagger(\boldsymbol{p})D_b(\boldsymbol{p}')D_d(\boldsymbol{q}')\overline{F}_c(\boldsymbol{q})\delta(\boldsymbol{q}' + \boldsymbol{q} + \boldsymbol{p}' - \boldsymbol{p})\frac{1}{|\boldsymbol{q}' + \boldsymbol{q}|^2} \\
&- 2\overline{D}_a^\dagger(\boldsymbol{p})D_\beta(\boldsymbol{p}')F_c^\dagger(\boldsymbol{q}')\overline{F}_d(\boldsymbol{q})\delta(\boldsymbol{q}' - \boldsymbol{q} - \boldsymbol{p}' + \boldsymbol{p})\frac{1}{|\boldsymbol{q}' - \boldsymbol{q}|^2} \\
&+ \overline{D}_a^\dagger(\boldsymbol{p})\overline{D}_c^\dagger(\boldsymbol{q})F_b^\dagger(\boldsymbol{p}')F_d^\dagger(\boldsymbol{q}')\delta(\boldsymbol{q}' + \boldsymbol{q} + \boldsymbol{p}' + \boldsymbol{p})\frac{1}{|\boldsymbol{q}' + \boldsymbol{q}|^2} \\
&+ 2\overline{D}_a^\dagger(\boldsymbol{p})D_d(\boldsymbol{q}')F_b^\dagger(\boldsymbol{p}')\overline{F}_c(\boldsymbol{q})\delta(\boldsymbol{q}' + \boldsymbol{q} - \boldsymbol{p}' - \boldsymbol{p})\frac{1}{|\boldsymbol{q}' + \boldsymbol{q}|^2} \\
&- 2\overline{D}_a^\dagger(\boldsymbol{p})F_b^\dagger(\boldsymbol{p}')F_d^\dagger(\boldsymbol{q}')\overline{F}_c(\boldsymbol{q})\delta(\boldsymbol{q}' - \boldsymbol{q} + \boldsymbol{p}' + \boldsymbol{p})\frac{1}{|\boldsymbol{q}' - \boldsymbol{q}|^2} \\
&+ D_b(\boldsymbol{p}')D_d(\boldsymbol{q}')\overline{F}_a(\boldsymbol{p})\overline{F}_c(\boldsymbol{q})\delta(\boldsymbol{q}' + \boldsymbol{q} + \boldsymbol{p}' + \boldsymbol{p})\frac{1}{|\boldsymbol{q}' + \boldsymbol{q}|^2} \\
&+ 2D_b(\boldsymbol{p}')F_d^\dagger(\boldsymbol{q}')\overline{F}_a(\boldsymbol{p})\overline{F}_c(\boldsymbol{q})\delta(\boldsymbol{q}' - \boldsymbol{q} - \boldsymbol{p}' - \boldsymbol{p})\frac{1}{|\boldsymbol{q}' - \boldsymbol{q}|^2} \\
&- F_b^\dagger(\boldsymbol{p}')F_d^\dagger(\boldsymbol{q}')\overline{F}_a(\boldsymbol{p})\overline{F}_c(\boldsymbol{q})\delta(\boldsymbol{q}' - \boldsymbol{q} + \boldsymbol{p}' - \boldsymbol{p})\frac{1}{|\boldsymbol{q}' - \boldsymbol{q}|^2}\Bigg).
\end{aligned}
$$

Finally, we integrate with respect to $\boldsymbol{q}'$ and divide V_2 into *phys* and *unphys* parts, as follows

$$V_2 = V_2^{\text{phys}} + V_2^{\text{unp}},$$

$$V_2^{\text{phys}} = \frac{e^2\hbar^2}{2(2\pi\hbar)^3}\sum_{abcd}\int d\boldsymbol{p}d\boldsymbol{p}'d\boldsymbol{q}\gamma_{ab}^0\gamma_{cd}^0$$

$$
\begin{aligned}
&\times \Big(-\overline{A}_a^\dagger(\boldsymbol{p})\overline{A}_c^\dagger(\boldsymbol{q})A_b(\boldsymbol{p}')A_d(\boldsymbol{q}-\boldsymbol{p}'+\boldsymbol{p})\frac{1}{|\boldsymbol{p}'-\boldsymbol{p}|^2} \\
&- 2\overline{A}_a^\dagger(\boldsymbol{p})A_b(\boldsymbol{p}')B_d^\dagger(\boldsymbol{q}+\boldsymbol{p}'-\boldsymbol{p})\overline{B}_c(\boldsymbol{q})\frac{1}{|\boldsymbol{p}'-\boldsymbol{p}|^2} \\
&- 2\overline{A}_a^\dagger(\boldsymbol{p})A_b(\boldsymbol{p}')\overline{D}_c^\dagger(\boldsymbol{q})D_d(\boldsymbol{q}-\boldsymbol{p}'+\boldsymbol{p})\frac{1}{|\boldsymbol{p}'-\boldsymbol{p}|^2} \\
&+ 2\overline{A}_a^\dagger(\boldsymbol{p})A_b(\boldsymbol{p}')F_d^\dagger(+\boldsymbol{q}+\boldsymbol{p}'-\boldsymbol{p})\overline{F}_c(\boldsymbol{q})\frac{1}{|\boldsymbol{p}'-\boldsymbol{p}|^2} \\
&+ 2\overline{A}_a^\dagger(\boldsymbol{p})A_d(-\boldsymbol{q}+\boldsymbol{p}'+\boldsymbol{p})B_b^\dagger(\boldsymbol{p}')\overline{B}_c(\boldsymbol{q})\frac{1}{|\boldsymbol{p}'+\boldsymbol{p}|^2} \\
&- 2\overline{A}_a^\dagger(\boldsymbol{p})B_b^\dagger(\boldsymbol{p}')D_d(-\boldsymbol{q}+\boldsymbol{p}'+\boldsymbol{p})\overline{F}_c(\boldsymbol{q})\frac{1}{|\boldsymbol{p}'+\boldsymbol{p}|^2} \\
&- 2A_b(\boldsymbol{p}')\overline{B}_a(\boldsymbol{p})\overline{D}_c^\dagger(\boldsymbol{q})F_d^\dagger(-\boldsymbol{q}+\boldsymbol{p}'+\boldsymbol{p})\frac{1}{|\boldsymbol{p}'+\boldsymbol{p}|^2} \\
&- B_b^\dagger(\boldsymbol{p}')B_d^\dagger(\boldsymbol{q}-\boldsymbol{p}'+\boldsymbol{p})\overline{B}_a(\boldsymbol{p})\overline{B}_c(\boldsymbol{q})\frac{1}{|\boldsymbol{p}'-\boldsymbol{p}|^2} \\
&+ 2B_b^\dagger(\boldsymbol{p}')\overline{B}_a(\boldsymbol{p})\overline{D}_c^\dagger(\boldsymbol{q})D_d(\boldsymbol{q}+\boldsymbol{p}'-\boldsymbol{p})\frac{1}{|\boldsymbol{p}'-\boldsymbol{p}|^2} \\
&- 2B_b^\dagger(\boldsymbol{p}')\overline{B}_a(\boldsymbol{p})F_d^\dagger(\boldsymbol{q}-\boldsymbol{p}'+\boldsymbol{p})\overline{F}_c(\boldsymbol{q})\frac{1}{|\boldsymbol{p}'-\boldsymbol{p}|^2} \\
&- \overline{D}_a^\dagger(\boldsymbol{p})\overline{D}_c^\dagger(\boldsymbol{q})D_b(\boldsymbol{p}')D_d(\boldsymbol{q}-\boldsymbol{p}'+\boldsymbol{p})\frac{1}{|\boldsymbol{p}'-\boldsymbol{p}|^2} \\
&- 2\overline{D}_a^\dagger(\boldsymbol{p})D_b(\boldsymbol{p}')F_c^\dagger(\boldsymbol{q}+\boldsymbol{p}'-\boldsymbol{p})\overline{F}_d(\boldsymbol{q})\frac{1}{|\boldsymbol{p}'-\boldsymbol{p}|^2} \\
&+ 2\overline{D}_a^\dagger(\boldsymbol{p})D_d(-\boldsymbol{q}+\boldsymbol{p}'+\boldsymbol{p})F_b^\dagger(\boldsymbol{p}')\overline{F}_c(\boldsymbol{q})\frac{1}{|\boldsymbol{p}'+\boldsymbol{p}|^2} \\
&- F_b^\dagger(\boldsymbol{p}')F_d^\dagger(\boldsymbol{q}-\boldsymbol{p}'+\boldsymbol{p})\overline{F}_a(\boldsymbol{p})\overline{F}_c(\boldsymbol{q})\frac{1}{|\boldsymbol{p}'-\boldsymbol{p}|^2} \Big),
\end{aligned}
\tag{D.12}
$$

$$
\begin{aligned}
V_2^{\text{unp}} = &\frac{e^2\hbar^2}{2(2\pi\hbar)^3} \sum_{abcd} \int d\boldsymbol{p}d\boldsymbol{p}'d\boldsymbol{q}\gamma_{ab}^0\gamma_{cd}^0 \\
&\times \Big(2\overline{A}_a^\dagger(\boldsymbol{p})A_b(\boldsymbol{p}')A_d(-\boldsymbol{q}-\boldsymbol{p}'+\boldsymbol{p})\overline{B}_c(\boldsymbol{q})\frac{1}{|\boldsymbol{p}'-\boldsymbol{p}|^2} \\
&- 2\overline{A}_a^\dagger(\boldsymbol{p})A_b(\boldsymbol{p}')\overline{D}_c^\dagger(\boldsymbol{q})F_d^\dagger(-\boldsymbol{q}+\boldsymbol{p}'-\boldsymbol{p})\frac{1}{|\boldsymbol{p}'-\boldsymbol{p}|^2} \\
&- 2\overline{A}_a^\dagger(\boldsymbol{p})A_b(\boldsymbol{p}')D_d(-\boldsymbol{q}-\boldsymbol{p}'+\boldsymbol{p})\overline{F}_c(\boldsymbol{q})\frac{1}{|\boldsymbol{p}'-\boldsymbol{p}|^2} \\
&+ 2\overline{A}_a^\dagger(\boldsymbol{p})\overline{A}_c^\dagger(\boldsymbol{q})A_d(\boldsymbol{q}+\boldsymbol{p}'+\boldsymbol{p})B_b^\dagger(\boldsymbol{p}')\frac{1}{|\boldsymbol{p}'+\boldsymbol{p}|^2} \\
&+ \overline{A}_a^\dagger(\boldsymbol{p})\overline{A}_c^\dagger(\boldsymbol{q})B_b^\dagger(\boldsymbol{p}')B_d^\dagger(-\boldsymbol{q}-\boldsymbol{p}'-\boldsymbol{p})\frac{1}{|\boldsymbol{p}'+\boldsymbol{p}|^2} \\
&- 2\overline{A}_a^\dagger(\boldsymbol{p})B_b^\dagger(\boldsymbol{p}')B_d^\dagger(\boldsymbol{q}-\boldsymbol{p}'-\boldsymbol{p})\overline{B}_c(\boldsymbol{q})\frac{1}{|\boldsymbol{p}'+\boldsymbol{p}|^2}
\end{aligned}
$$

$$
\begin{aligned}
&- 2\overline{A}^\dagger_a(\boldsymbol{p})B^\dagger_b(\boldsymbol{p}')\overline{D}^\dagger_c(\boldsymbol{q})D_d(\boldsymbol{q}+\boldsymbol{p}'+\boldsymbol{p})\frac{1}{|\boldsymbol{p}'+\boldsymbol{p}|^2}\\
&- 2\overline{A}^\dagger_a(\boldsymbol{p})B^\dagger_b(\boldsymbol{p}')\overline{D}^\dagger_c(\boldsymbol{q})F^\dagger_d(-\boldsymbol{q}-\boldsymbol{p}'-\boldsymbol{p})\frac{1}{|\boldsymbol{p}'+\boldsymbol{p}|^2}\\
&+ 2\overline{A}^\dagger_a(\boldsymbol{p})B^\dagger_b(\boldsymbol{p}')F^\dagger_d(\boldsymbol{q}-\boldsymbol{p}'-\boldsymbol{p})\overline{F}_c(\boldsymbol{q})\frac{1}{|\boldsymbol{p}'+\boldsymbol{p}|^2}\\
&+ A_b(\boldsymbol{p}')A_d(-\boldsymbol{q}-\boldsymbol{p}'-\boldsymbol{p})\overline{B}_a(\boldsymbol{p})\overline{B}_c(\boldsymbol{q})\frac{1}{|\boldsymbol{p}'+\boldsymbol{p}|^2}\\
&+ 2A_b(\boldsymbol{p}')B^\dagger_d(\boldsymbol{q}+\boldsymbol{p}'+\boldsymbol{p})\overline{B}_a(\boldsymbol{p})\overline{B}_c(\boldsymbol{q})\frac{1}{|\boldsymbol{p}'+\boldsymbol{p}|^2}\\
&- 2A_b(\boldsymbol{p}')\overline{B}_a(\boldsymbol{p})\overline{D}^\dagger_c(\boldsymbol{q})D_d(\boldsymbol{q}-\boldsymbol{p}'-\boldsymbol{p})\frac{1}{|\boldsymbol{p}'+\boldsymbol{p}|^2}\\
&- 2\overline{A}_b(\boldsymbol{p}')B_a(\boldsymbol{p})D_d(-\boldsymbol{q}-\boldsymbol{p}'-\boldsymbol{p})\overline{F}_c(\boldsymbol{q})\frac{1}{|\boldsymbol{p}'+\boldsymbol{p}|^2}\\
&+ 2\overline{A}_b(\boldsymbol{p}')B_a(\boldsymbol{p})F^\dagger_d(\boldsymbol{q}+\boldsymbol{p}'+\boldsymbol{p})\overline{F}_c(\boldsymbol{q})\frac{1}{|\boldsymbol{p}'+\boldsymbol{p}|^2}\\
&+ 2B^\dagger_b(\boldsymbol{p}')\overline{B}_a(\boldsymbol{p})\overline{D}^\dagger_c(\boldsymbol{q})F^\dagger_d(-\boldsymbol{q}-\boldsymbol{p}'+\boldsymbol{p})\frac{1}{|\boldsymbol{p}'-\boldsymbol{p}|^2}\\
&+ 2B^\dagger_b(\boldsymbol{p}')\overline{B}_a(\boldsymbol{p})D_d(-\boldsymbol{q}+\boldsymbol{p}'-\boldsymbol{p})\overline{F}_c(\boldsymbol{q})\frac{1}{|\boldsymbol{p}'-\boldsymbol{p}|^2}\\
&- 2\overline{D}^\dagger_a(\boldsymbol{p})\overline{D}^\dagger_c(\boldsymbol{q})D_b(\boldsymbol{p}')F^\dagger_d(-\boldsymbol{q}+\boldsymbol{p}'-\boldsymbol{p})\frac{1}{|\boldsymbol{p}'-\boldsymbol{p}|^2}\\
&+ 2\overline{D}^\dagger_a(\boldsymbol{p})D_b(\boldsymbol{p}')D_d(-\boldsymbol{q}-\boldsymbol{p}'+\boldsymbol{p})\overline{F}_c(\boldsymbol{q})\frac{1}{|\boldsymbol{p}'-\boldsymbol{p}|^2}\\
&+ \overline{D}^\dagger_a(\boldsymbol{p})\overline{D}^\dagger_c(\boldsymbol{q})F^\dagger_b(\boldsymbol{p}')F^\dagger_d(-\boldsymbol{q}-\boldsymbol{p}'-\boldsymbol{p})\frac{1}{|\boldsymbol{p}'+\boldsymbol{p}|^2}\\
&- 2\overline{D}^\dagger_a(\boldsymbol{p})F^\dagger_b(\boldsymbol{p}')F^\dagger_d(\boldsymbol{q}-\boldsymbol{p}'-\boldsymbol{p})\overline{F}_c(\boldsymbol{q})\frac{1}{|\boldsymbol{p}'+\boldsymbol{p}|^2}\\
&+ D_b(\boldsymbol{p}')D_d(-\boldsymbol{q}-\boldsymbol{p}'-\boldsymbol{p})\overline{F}_a(\boldsymbol{p})\overline{F}_c(\boldsymbol{q})\frac{1}{|\boldsymbol{p}'+\boldsymbol{p}|^2}\\
&+ 2D_b(\boldsymbol{p}')F^\dagger_d(\boldsymbol{q}+\boldsymbol{p}'+\boldsymbol{p})\overline{F}_a(\boldsymbol{p})\overline{F}_c(\boldsymbol{q})\frac{1}{|\boldsymbol{p}'+\boldsymbol{p}|^2}\Bigg).
\end{aligned}
\tag{D.13}
$$

After all these efforts, we can appreciate the enormous simplifications achieved by introducing the Feynman gauge in Subsection 3.2.3. In that approach, the cumbersome V_2 interaction operator was not present at all, so in calculating the S-matrix it was not necessary to evaluate the commutators or products of multiple terms in (D.12)–(D.13). Moreover, Feynman gauge calculations can be done even without the explicit form of V_1 in (D.9) and without explicit evaluation of related products and/or commutators. All these complex computations turn out to be hidden in convenient Feynman diagrams. Feynman rules immediately convert diagrams into relativistically invariant (loop) integrals. Their calculations will be discussed in Appendix F.

E Relativistic invariance of QFT

E.1 Relativistic invariance of simple QFT

Here we would like to verify that interacting quantum field theory presented in Subsection 3.1.2 is indeed relativistically invariant [21, 19]. In other words, we are going to check Poincaré commutators (**1**-6.26)–(**1**-6.30) for the potential energy and boost operators

$$
\begin{aligned}
V &= \int d\boldsymbol{x} V(0,\boldsymbol{x}),\\
\boldsymbol{Z} &= -\frac{1}{c^2}\int d\boldsymbol{x}\boldsymbol{x} V(0,\boldsymbol{x})
\end{aligned}
\tag{E.1}
$$

postulated in (3.10)–(3.11).

Equation (**1**-6.26) follows directly from the property (3.8) in the case of spatial translations and rotations. The potential boost $\boldsymbol{Z}$ in (E.1) is a 3-vector by construction, so equation (**1**-6.28) is also satisfied. Let us now check the commutator (**1**-6.27). Consider the case $i = j = z$. Using equation (3.8) with $\Lambda = 1$, we obtain

$$
\begin{aligned}
[P_{0z}, Z_z] &= -\frac{i\hbar}{c^2}\lim_{a\to 0}\frac{d}{da}\int d\boldsymbol{x} e^{-\frac{i}{\hbar}P_{0z}a} zV(0,\boldsymbol{x}) e^{\frac{i}{\hbar}P_{0z}a}\\
&= -\frac{i\hbar}{c^2}\lim_{a\to 0}\frac{d}{da}\int d\boldsymbol{x} zV(0,x,y,z+a)\\
&= -\frac{i\hbar}{c^2}\lim_{a\to 0}\frac{d}{da}\int d\boldsymbol{x}(z-a)V(0,x,y,z)\\
&= \frac{i\hbar}{c^2}\int d\boldsymbol{x} V(0,x,y,z) = \frac{i\hbar}{c^2}V,
\end{aligned}
\tag{E.2}
$$

i. e., exactly (**1**-6.27).

Checking equation (**1**-6.30) is a bit more difficult. Again, we restrict ourselves to the case $i = z$ and try to prove[1]

$$
[K_{0z}, V(t)] + [Z_z(t), H_0] - [V(t), Z_z(t)] = 0. \tag{E.3}
$$

To derive the first term on the left-hand side we use (3.8) and

$$
\begin{aligned}
\lim_{\theta\to 0}\frac{d}{d\theta}V(\tilde{\boldsymbol{\theta}}\tilde{x}) &= \lim_{\theta\to 0}\frac{d}{d\theta}V\left(t\cosh\theta + \frac{z}{c}\sinh\theta, x, y, z\cosh\theta + ct\sinh\theta\right)\\
&= \lim_{\theta\to 0}\left(\frac{\partial V}{\partial t}\left(t\sinh\theta + \frac{z}{c}\cosh\theta\right) + \frac{\partial V}{\partial z}(z\sinh\theta + ct\cosh\theta)\right)\\
&= ct\frac{\partial V}{\partial z} + \frac{z}{c}\frac{\partial V}{\partial t},
\end{aligned}
$$

1 In this calculation, it will be convenient to represent condition (**1**-6.30) in a t-dependent form, i. e., multiplied from the left by $\exp(\frac{i}{\hbar}H_0 t)$ and from the right by $\exp(-\frac{i}{\hbar}H_0 t)$, as in (1.60). At the end of calculations, we will set $t = 0$.

https://doi.org/10.1515/9783110493207-009

where $\tilde{\boldsymbol{\theta}}$ is the boost matrix (**1**-J.13). Then

$$\begin{aligned}[K_{0z}, V(t)] &= \frac{i\hbar}{c} \lim_{\theta\to 0} \frac{d}{d\theta} e^{-\frac{ic}{\hbar}K_{0z}\theta} \int d\boldsymbol{x} V(\tilde{x}) e^{\frac{ic}{\hbar}K_{0z}\theta} \\ &= \frac{i\hbar}{c} \lim_{\theta\to 0} \frac{d}{d\theta} \int d\boldsymbol{x} V(\tilde{\boldsymbol{\theta}}\tilde{x}) \\ &= \frac{i\hbar}{c} \int d\boldsymbol{x}\left(ct \frac{\partial V(\boldsymbol{x},t)}{\partial z} + \frac{z}{c}\frac{\partial V(\boldsymbol{x},t)}{\partial t}\right). \end{aligned} \tag{E.4}$$

For the second term in (E.3) we obtain

$$[Z_z(t), H_0] = i\hbar \frac{\partial}{\partial t} Z_z(t) = -\frac{i\hbar}{c^2}\frac{\partial}{\partial t} \int d\boldsymbol{x} z V(t,\boldsymbol{x}). \tag{E.5}$$

Setting $t = 0$, we see that (E.4) and (E.5) cancel out. The last term on the left-hand side of (E.3) vanishes due to (3.9).

Calculation of the remaining nontrivial commutator (**1**-6.29)

$$[K_{0i}, Z_j] + [Z_i, K_{0j}] + [Z_i, Z_j] = 0,$$

is left as an exercise for the reader.

E.2 Relativistic invariance of QED

In this appendix we are going to verify the relativistic invariance of QED. In other words, we will check the commutators (**1**-6.26)–(**1**-6.30) of the Poincaré Lie algebra for the interaction operators (3.13) and (3.17).[2]

The potential energy $V^n(t)$ in (3.13) obviously commutes with operators of total momentum and total angular momentum, as in (**1**-6.26). The potential boost $\boldsymbol{Z}$ in (3.17) is a 3-vector by construction, so equation (**1**-6.28) is also valid.

Next we check the commutator (**1**-6.27). Denote

$$\begin{aligned} V(t,\boldsymbol{x}) &\equiv -\mathbf{j}(t,\boldsymbol{x})\cdot \boldsymbol{\mathcal{A}}(t,\boldsymbol{x}) + \frac{1}{2}\int d\boldsymbol{y} j^0(t,\boldsymbol{x})\mathcal{G}(\boldsymbol{x}-\boldsymbol{y}) j^0(t,\boldsymbol{y}), \\ \mathcal{G}(\boldsymbol{x}) &\equiv \frac{1}{4\pi|\boldsymbol{x}|}, \end{aligned}$$

so that

$$\begin{aligned} V^n(t) &= \int d\boldsymbol{x} V(t,\boldsymbol{x}), \\ \boldsymbol{Z}^n(t) &= -\frac{1}{c^2}\int d\boldsymbol{x}\boldsymbol{x} V(t,\boldsymbol{x}) - \frac{1}{c^2}\int d\boldsymbol{x} j^0(t,\boldsymbol{x})\boldsymbol{C}(t,\boldsymbol{x}), \end{aligned} \tag{E.6}$$

2 It will be more convenient to work with operators in the t-dependent form; see the footnote on page 155. The proof presented here was borrowed from Weinberg's works [21, 19]; see, especially, Appendix B in [20].

where the operator function $\boldsymbol{C}(\tilde{x})$ is given by equation (3.18). Consider, for example, the case $i = j = x$. Then from equations (C.25), (D.2) and (1.47)–(1.48) we obtain

$$
\begin{aligned}
[P_{0x}, Z_x^n(t)] &= i\hbar \lim_{a\to 0} \frac{d}{da} e^{-\frac{i}{\hbar}P_{0x}a} Z_x^n(t) e^{\frac{i}{\hbar}P_{0x}a} \\
&= -\frac{i\hbar}{c^2} \lim_{a\to 0} \frac{d}{da} \int d\boldsymbol{x} e^{-\frac{i}{\hbar}P_{0x}a} [xV(t,\boldsymbol{x}) + \mathrm{j}^0(t,\boldsymbol{x}) C_x(t,\boldsymbol{x})] e^{\frac{i}{\hbar}P_{0x}a} \\
&= -\frac{i\hbar}{c^2} \lim_{a\to 0} \frac{d}{da} \int d\boldsymbol{x} [xV(t,x+a,y,z) + \mathrm{j}^0(t,x+a,y,z) C_x(t,x+a,y,z)] \\
&= -\frac{i\hbar}{c^2} \lim_{a\to 0} \frac{d}{da} \int d\boldsymbol{x} [(x-a)V(t,x,y,z) + \mathrm{j}^0(t,x,y,z) C_x(t,x,y,z)] \\
&= \frac{i\hbar}{c^2} \int d\boldsymbol{x} V(t,x,y,z) = \frac{i\hbar}{c^2} V^n(t),
\end{aligned} \tag{E.7}
$$

i. e., exactly equation (**1**-6.27).

Let us prove equation (**1**-6.30) in the particular case where $i = z$. We rewrite this formula by taking into account that $[Z_z^n(t), H_0] = i\hbar \frac{d}{dt} Z_z^n(t)$, so we have

$$
[K_{0z}, V_1(t)] + [K_{0z}, V_2(t)] + i\hbar \frac{d}{dt} Z_z^n(t) - [V^n(t), Z_z^n(t)] = 0. \tag{E.8}
$$

We shall calculate all four terms on the left-hand side of (E.8) one by one. For the first term we use equations (D.4), (C.32), (C.36), (**1**-J.4) and (**1**-J.6) and obtain

$$
\begin{aligned}
[K_{0z}, V_1(t)] &= \frac{i\hbar}{c} \lim_{\theta\to 0} \frac{d}{d\theta} e^{-\frac{ic}{\hbar}K_{0z}\theta} \int d\boldsymbol{x} \tilde{\mathrm{j}}(\tilde{x}) \cdot \tilde{A}(\tilde{x}) e^{\frac{ic}{\hbar}K_{0z}\theta} \\
&= \frac{i\hbar}{c} \lim_{\theta\to 0} \frac{d}{d\theta} \int d\boldsymbol{x} [\tilde{\boldsymbol{\theta}}^{-1} \tilde{\mathrm{j}}(\tilde{\boldsymbol{\theta}}\tilde{x}) \cdot \tilde{\boldsymbol{\theta}}^{-1} \tilde{A}(\tilde{\boldsymbol{\theta}}\tilde{x}) + \tilde{\boldsymbol{\theta}}^{-1} \tilde{\mathrm{j}}(\tilde{\boldsymbol{\theta}}\tilde{x}) \cdot \tilde{\Omega}(\tilde{x}, \boldsymbol{\theta})] \\
&= \frac{i\hbar}{c} \lim_{\theta\to 0} \frac{d}{d\theta} \int d\boldsymbol{x} [\tilde{\mathrm{j}}(\tilde{\boldsymbol{\theta}}\tilde{x}) \cdot \tilde{A}(\tilde{\boldsymbol{\theta}}\tilde{x}) + \tilde{\boldsymbol{\theta}}^{-1} \tilde{\mathrm{j}}(\tilde{\boldsymbol{\theta}}\tilde{x}) \cdot \tilde{\Omega}(\tilde{x}, \boldsymbol{\theta})] \\
&= \frac{i\hbar}{c} \lim_{\theta\to 0} \int d\boldsymbol{x} \bigg(\frac{d}{d\theta} \tilde{\mathrm{j}}(\tilde{\boldsymbol{\theta}}\tilde{x}) \cdot \tilde{A}(\tilde{x}) + \tilde{\mathrm{j}}(\tilde{x}) \cdot \frac{d}{d\theta} \tilde{A}(\tilde{\boldsymbol{\theta}}\tilde{x}) + \left(\frac{d}{d\theta} \tilde{\boldsymbol{\theta}}^{-1} \right) \tilde{\mathrm{j}}(\tilde{x}) \cdot \tilde{\Omega}(\tilde{x}, \boldsymbol{0}) \\
&\quad + \frac{d}{d\theta} \tilde{\mathrm{j}}(\tilde{\boldsymbol{\theta}}\tilde{x}) \cdot \tilde{\Omega}(\tilde{x}, \boldsymbol{0}) + \tilde{\mathrm{j}}(\tilde{x}) \cdot \frac{d}{d\theta} \tilde{\Omega}(\tilde{x}, \boldsymbol{\theta}) \bigg) \\
&= \frac{i\hbar}{c} \lim_{\theta\to 0} \int d\boldsymbol{x} \left(\frac{d}{d\theta} \tilde{\mathrm{j}}(\tilde{\boldsymbol{\theta}}\tilde{x}) \cdot \tilde{A}(\tilde{x}) + \tilde{\mathrm{j}}(\tilde{x}) \cdot \frac{d}{d\theta} \tilde{A}(\tilde{\boldsymbol{\theta}}\tilde{x}) + \tilde{\mathrm{j}}(\tilde{x}) \cdot \frac{d}{d\theta} \tilde{\Omega}(\tilde{x}, \boldsymbol{\theta}) \right),
\end{aligned} \tag{E.9}
$$

where $\tilde{\Omega}(\tilde{x}, \boldsymbol{\theta})$ is given by formula (C.33) and $\tilde{\boldsymbol{\theta}}$ is the 4×4 boost matrix (**1**-J.13). To simplify this expression we will need the following results:

$$
\begin{aligned}
\lim_{\theta\to 0} \frac{d}{d\theta} \tilde{\mathrm{j}}(\tilde{\boldsymbol{\theta}}\tilde{x}) &= \lim_{\theta\to 0} \frac{d}{d\theta} \tilde{\mathrm{j}} \left(t\cosh\theta + \frac{z}{c}\sinh\theta, x, y, z\cosh\theta + ct\sinh\theta \right) \\
&= \lim_{\theta\to 0} \left[\frac{\partial \tilde{\mathrm{j}}}{\partial t} \left(t\sinh\theta + \frac{z}{c}\cosh\theta \right) + \frac{\partial \tilde{\mathrm{j}}}{\partial z} (z\sinh\theta + ct\cosh\theta) \right] \\
&= \frac{z}{c} \frac{\partial \tilde{\mathrm{j}}}{\partial t} + ct \frac{\partial \tilde{\mathrm{j}}}{\partial z},
\end{aligned} \tag{E.10}
$$

$$\lim_{\theta\to 0}\frac{d}{d\theta}\tilde{A}(\tilde{\boldsymbol{\theta}}\tilde{x}) = \frac{z}{c}\frac{\partial\tilde{A}}{\partial t} + ct\frac{\partial\tilde{A}}{\partial z}. \tag{E.11}$$

Calculation of the term with $d\tilde{\Omega}/d\theta$ is a bit more complicated. We have

$$\begin{aligned}\lim_{\theta\to 0}\frac{d}{d\theta}\Omega_\mu(\tilde{x},\boldsymbol{\theta}) = &-\frac{\hbar}{(2\pi\hbar)^{3/2}}\lim_{\theta\to 0}\frac{d}{d\theta}\int\frac{d\boldsymbol{p}}{\sqrt{2pc}}\sum_{\tau=-1}^{1}\frac{(\tilde{\boldsymbol{\theta}}^{-1}p)_\mu}{|\boldsymbol{\theta}^{-1}\boldsymbol{p}|}\\ &\times(\tilde{\boldsymbol{\theta}}^{-1})_0^{\ \nu}[e^{-\frac{i}{\hbar}\tilde{\boldsymbol{\theta}}^{-1}\tilde{p}\cdot\tilde{x}}e_\nu(\boldsymbol{p},\tau)c_{\boldsymbol{p}\tau} + e^{\frac{i}{\hbar}\tilde{\boldsymbol{\theta}}^{-1}\tilde{p}\cdot\tilde{x}}e^*_\nu(\boldsymbol{p},\tau)c^\dagger_{\boldsymbol{p}\tau}].\end{aligned} \tag{E.12}$$

The only quantities depending on θ are matrices $\tilde{\boldsymbol{\theta}}$. Hence, the derivative on the right-hand side of (E.12) has four terms, i. e., $\frac{d}{d\theta}(\tilde{\boldsymbol{\theta}}^{-1}p)^\mu$, $\frac{d}{d\theta}|\boldsymbol{\theta}^{-1}\boldsymbol{p}|^{-1}$, $\frac{d}{d\theta}(\tilde{\boldsymbol{\theta}}^{-1})_0^{\ \nu}$ and $\frac{d}{d\theta}\exp(\pm i\tilde{\boldsymbol{\theta}}^{-1}\tilde{p}\cdot\tilde{x})$. After the differentiation we have to set $\theta\to 0$. From (C.35) and (**1**-J.13) it then follows that the only nonvanishing contribution is from

$$\begin{aligned}\lim_{\theta\to 0}\frac{d}{d\theta}(\tilde{\boldsymbol{\theta}}^{-1})_0^{\ \nu} &= \lim_{\theta\to 0}\frac{d}{d\theta}(\cosh\theta,0,0,-\sinh\theta) = \lim_{\theta\to 0}(\sinh\theta,0,0,-\cosh\theta)\\ &= (0,0,0,-1).\end{aligned}$$

Therefore

$$\begin{aligned}\lim_{\theta\to 0}\frac{d}{d\theta}\Omega_\mu(\tilde{x},\boldsymbol{\theta}) &= \frac{\hbar}{(2\pi\hbar)^{3/2}}\int\frac{d\boldsymbol{p}p_\mu}{\sqrt{2p^3c}}\times\sum_{\tau=-1}^{1}[e^{-\frac{i}{\hbar}\tilde{p}\cdot\tilde{x}}e_z(\boldsymbol{p},\tau)c_{\boldsymbol{p}\tau} + e^{\frac{i}{\hbar}\tilde{p}\cdot\tilde{x}}e^*_z(\boldsymbol{p},\tau)c^\dagger_{\boldsymbol{p}\tau}]\\ &= -\frac{i\hbar^2\sqrt{c}}{\sqrt{2(2\pi\hbar)^3}}\partial_\mu\int\frac{d\boldsymbol{p}}{p^{3/2}}\sum_{\tau=-1}^{1}[e^{-\frac{i}{\hbar}\tilde{p}\cdot\tilde{x}}e_z(\boldsymbol{p},\tau)c_{\boldsymbol{p}\tau} - e^{\frac{i}{\hbar}\tilde{p}\cdot\tilde{x}}e^*_z(\boldsymbol{p},\tau)c^\dagger_{\boldsymbol{p}\tau}]\\ &= -\partial_\mu C_z(\tilde{x}).\end{aligned} \tag{E.13}$$

Here we used definitions (3.18), (4.21) and took into account that

$$p_\mu e^{\pm\frac{i}{\hbar}\tilde{p}\cdot\tilde{x}} = \pm i\hbar c\partial_\mu e^{\pm\frac{i}{\hbar}\tilde{p}\cdot\tilde{x}}.$$

From (E.13) and the continuity equation (D.5), we conclude that the last term on the right-hand side of (E.9) is equal to[3]

$$\frac{i\hbar}{c}\lim_{\theta\to 0}\int d\boldsymbol{x}\tilde{\jmath}(\tilde{x})\cdot\frac{d}{d\theta}\tilde{\Omega}(\tilde{x},\boldsymbol{\theta}) = -\frac{i\hbar}{c}\int d\boldsymbol{x}\jmath^\mu(t,\boldsymbol{x})\partial_\mu C_z(t,\boldsymbol{x})$$

---

3 Here we have assumed that the fields are damped in remote regions of the space–time $\mathcal{M}$, so that all relevant functions (f,g) of quantum fields also turn to zero at infinity, so we can take the integrals by parts, as in the following example:

$$\begin{aligned}\int_{-\infty}^{\infty}dx\left(\frac{d}{dx}f(\tilde{x})\right)g(\tilde{x}) &= \int_{-\infty}^{\infty}dx\frac{d}{dx}(f(\tilde{x})g(\tilde{x})) - \int_{-\infty}^{\infty}dxf(\tilde{x})\frac{d}{dx}g(\tilde{x})\\ &= f(\tilde{x})g(\tilde{x})|_{x=-\infty}^{x=\infty} - \int_{-\infty}^{\infty}dxf(\tilde{x})\left(\frac{d}{dx}g(\tilde{x})\right) = -\int_{-\infty}^{\infty}dxf(\tilde{x})\left(\frac{d}{dx}g(\tilde{x})\right).\end{aligned}$$

$$
\begin{aligned}
&= \frac{i\hbar}{c^2}\int d\boldsymbol{x}\mathrm{j}^0(t,\boldsymbol{x})\frac{\partial C_z(t,\boldsymbol{x})}{\partial t} + \frac{i\hbar}{c}\int d\boldsymbol{x}\mathbf{j}(t,\boldsymbol{x})\frac{\partial C_z(t,\boldsymbol{x})}{\partial \boldsymbol{x}} \\
&= \frac{i\hbar}{c^2}\int d\boldsymbol{x}\mathrm{j}^0(t,\boldsymbol{x})\frac{\partial C_z(t,\boldsymbol{x})}{\partial t} - \frac{i\hbar}{c}\int d\boldsymbol{x}\frac{\partial \mathbf{j}(t,\boldsymbol{x})}{\partial \boldsymbol{x}}C_z(t,\boldsymbol{x}) \\
&= \frac{i\hbar}{c^2}\int d\boldsymbol{x}\mathrm{j}^0(t,\boldsymbol{x})\frac{\partial C_z(t,\boldsymbol{x})}{\partial t} + \frac{i\hbar}{c^2}\int d\boldsymbol{x}\frac{\partial \mathrm{j}^0(t,\boldsymbol{x})}{\partial t}C_z(t,\boldsymbol{x}) \\
&= \frac{i\hbar}{c^2}\frac{\partial}{\partial t}\int d\boldsymbol{x}\mathrm{j}^0(t,\boldsymbol{x})C_z(t,\boldsymbol{x}).
\end{aligned} \tag{E.14}
$$

Inserting results (E.10), (E.11) and (E.14) into equation (E.9) and setting $t = 0$, we get

$$
\begin{aligned}
[K_{0z}, V_1(t)] &= \frac{i\hbar}{c}\int d\boldsymbol{x}\left(\frac{z}{c}\frac{\partial\tilde{\mathbf{j}}}{\partial t}\cdot\tilde{\mathcal{A}}(\tilde{x}) + \tilde{\mathbf{j}}(\tilde{x})\cdot\frac{z}{c}\frac{\partial\tilde{\mathcal{A}}}{\partial t} + \frac{1}{c}\frac{\partial}{\partial t}(\mathrm{j}^0(\tilde{x})C_z(\tilde{x}))\right) \\
&= \frac{i\hbar}{c^2}\frac{\partial}{\partial t}\int d\boldsymbol{x}[z(\tilde{\mathbf{j}}(\tilde{x})\cdot\tilde{\mathcal{A}}(\tilde{x})) + \mathrm{j}^0(\tilde{x})C_z(\tilde{x})].
\end{aligned} \tag{E.15}
$$

For the second term on the left-hand side of (E.8) we use equations (D.4) and (D.5). Also, we take into account that at the end we will set $t = 0$, so all terms proportional to t can be dropped. Then

$$
\begin{aligned}
[K_{0z}, V_2(t)] &= \frac{1}{2}\int d\boldsymbol{x}d\boldsymbol{x}'[K_{0z},\mathrm{j}^0(t,\boldsymbol{x})]\mathcal{G}(\boldsymbol{x}-\boldsymbol{x}')\mathrm{j}^0(t,\boldsymbol{x}') \\
&\quad + \frac{1}{2}\int d\boldsymbol{x}d\boldsymbol{x}'\mathrm{j}^0(t,\boldsymbol{x})\mathcal{G}(\boldsymbol{x}-\boldsymbol{x}')[K_{0z},\mathrm{j}^0(t,\boldsymbol{x}')] \\
&= \int d\boldsymbol{x}d\boldsymbol{x}'[K_{0z},\mathrm{j}^0(t,\boldsymbol{x})]\mathcal{G}(\boldsymbol{x}-\boldsymbol{x}')\mathrm{j}^0(t,\boldsymbol{x}') \\
&= i\hbar\int d\boldsymbol{x}d\boldsymbol{x}'\left(\frac{z}{c^2}\frac{\partial\mathrm{j}^0(t,\boldsymbol{x})}{\partial t} - \frac{1}{c}\mathrm{j}^z(t,\boldsymbol{x})\right)\mathcal{G}(\boldsymbol{x}-\boldsymbol{x}')\mathrm{j}^0(t,\boldsymbol{x}') \\
&= \frac{i\hbar}{2c^2}\int d\boldsymbol{x}d\boldsymbol{x}'\frac{\partial\mathrm{j}^0(t,\boldsymbol{x})}{\partial t}(z-z')\mathcal{G}(\boldsymbol{x}-\boldsymbol{x}')\mathrm{j}^0(t,\boldsymbol{x}') \\
&\quad + \frac{i\hbar}{2c^2}\int d\boldsymbol{x}d\boldsymbol{x}'\frac{\partial\mathrm{j}^0(t,\boldsymbol{x})}{\partial t}z\mathcal{G}(\boldsymbol{x}-\boldsymbol{x}')\mathrm{j}^0(t,\boldsymbol{x}') \\
&\quad + \frac{i\hbar}{2c^2}\int d\boldsymbol{x}d\boldsymbol{x}'\frac{\partial\mathrm{j}^0(t,\boldsymbol{x})}{\partial t}z'\mathcal{G}(\boldsymbol{x}-\boldsymbol{x}')\mathrm{j}^0(t,\boldsymbol{x}') \\
&\quad - \frac{i\hbar}{c}\int d\boldsymbol{x}d\boldsymbol{x}'\mathrm{j}^z(t,\boldsymbol{x})\mathcal{G}(\boldsymbol{x}-\boldsymbol{x}')\mathrm{j}^0(t,\boldsymbol{x}') \\
&= -\frac{i\hbar}{2c}\int d\boldsymbol{x}d\boldsymbol{x}'\frac{\partial\mathbf{j}(t,\boldsymbol{x})}{\partial\boldsymbol{x}}(z-z')\mathcal{G}(\boldsymbol{x}-\boldsymbol{x}')\mathrm{j}^0(t,\boldsymbol{x}') \\
&\quad + \frac{i\hbar}{2c^2}\int d\boldsymbol{x}d\boldsymbol{x}'\frac{\partial\mathrm{j}^0(t,\boldsymbol{x})}{\partial t}z\mathcal{G}(\boldsymbol{x}-\boldsymbol{x}')\mathrm{j}^0(t,\boldsymbol{x}') \\
&\quad + \frac{i\hbar}{2c^2}\int d\boldsymbol{x}d\boldsymbol{x}'\mathrm{j}^0(t,\boldsymbol{x})z\mathcal{G}(\boldsymbol{x}-\boldsymbol{x}')\frac{\partial\mathrm{j}^0(t,\boldsymbol{x}')}{\partial t} \\
&\quad - \frac{i\hbar}{c}\int d\boldsymbol{x}d\boldsymbol{x}'\mathrm{j}^z(t,\boldsymbol{x})\mathcal{G}(\boldsymbol{x}-\boldsymbol{x}')\mathrm{j}^0(t,\boldsymbol{x}') \\
&= \frac{i\hbar}{2c}\int d\boldsymbol{x}d\boldsymbol{x}'\mathbf{j}(t,\boldsymbol{x})\cdot\frac{\partial((z-z')\mathcal{G}(\boldsymbol{x}-\boldsymbol{x}'))}{\partial\boldsymbol{x}}\mathrm{j}^0(t,\boldsymbol{x}')
\end{aligned}
$$

$$+\frac{i\hbar}{2c^2}\frac{\partial}{\partial t}\int d\boldsymbol{x}d\boldsymbol{x}'\mathfrak{j}^0(t,\boldsymbol{x})z\mathcal{G}(\boldsymbol{x}-\boldsymbol{x}')\mathfrak{j}^0(t,\boldsymbol{x}')$$
$$-\frac{i\hbar}{c}\int d\boldsymbol{x}d\boldsymbol{x}'\mathfrak{j}^z(t,\boldsymbol{x})\mathcal{G}(\boldsymbol{x}-\boldsymbol{x}')\mathfrak{j}^0(t,\boldsymbol{x}'). \qquad \text{(E.16)}$$

With the help of (E.6) we get the third term on the left-hand side of (E.8), i. e.,

$$i\hbar\frac{\partial}{\partial t}Z_z^n(t) = -\frac{i\hbar}{c^2}\frac{\partial}{\partial t}\int d\boldsymbol{x}z\boldsymbol{\mathfrak{j}}(t,\boldsymbol{x})\cdot\boldsymbol{\mathcal{A}}(t,\boldsymbol{x}) - \frac{i\hbar}{c^2}\frac{\partial}{\partial t}\int d\boldsymbol{x}\mathfrak{j}^0(t,\boldsymbol{x})C_z(t,\boldsymbol{x})$$
$$-\frac{i\hbar}{2c^2}\frac{\partial}{\partial t}\int d\boldsymbol{x}d\boldsymbol{y}\mathfrak{j}^0(t,\boldsymbol{x})z\mathcal{G}(\boldsymbol{x}-\boldsymbol{y})\mathfrak{j}^0(t,\boldsymbol{y}). \qquad \text{(E.17)}$$

To calculate the last term in (E.8), we note that the only part in $\boldsymbol{Z}^n(t)$ not commuting with $V^n(t)$ is the one containing $\boldsymbol{C}$. Therefore

$$-[V^n(t),Z_z^n(t)] = -\frac{1}{c^2}\int d\boldsymbol{x}d\boldsymbol{x}'\boldsymbol{\mathfrak{j}}(t,\boldsymbol{x})\cdot[\boldsymbol{\mathcal{A}}(t,\boldsymbol{x}),C_z(t,\boldsymbol{x}')]\mathfrak{j}^0(t,\boldsymbol{x}').$$

Let us calculate this commutator at $t = 0$, using equation (C.20) as well as the integrals (A.3) and

$$\int\frac{d\boldsymbol{k}}{k^4}e^{\frac{i}{\hbar}\boldsymbol{k}\cdot\boldsymbol{r}} = \mathcal{E} - \frac{(2\pi)^3 r}{8\pi\hbar},$$

where $\mathcal{E}$ is an infinite constant.[4] We have

$$[\mathcal{A}_k(0,\boldsymbol{x}),C_z(0,\boldsymbol{x}')] = \frac{i\hbar^3 c}{(2\pi\hbar)^3}\int\frac{d\boldsymbol{p}d\boldsymbol{q}}{2\sqrt{q^3p}}\sum_{\sigma\tau}[(e_k(\boldsymbol{p},\sigma)c_{\boldsymbol{p}\sigma}e^{\frac{i}{\hbar}\boldsymbol{p}\cdot\boldsymbol{x}} + e_k^*(\boldsymbol{p},\sigma)c_{\boldsymbol{p}\sigma}^\dagger e^{-\frac{i}{\hbar}\boldsymbol{p}\cdot\boldsymbol{x}}),$$
$$(e_z(\boldsymbol{q},\tau)c_{\boldsymbol{q}\tau}e^{\frac{i}{\hbar}\boldsymbol{q}\cdot\boldsymbol{x}'} - e_z^*(\boldsymbol{q},\tau)c_{\boldsymbol{q}\tau}^\dagger e^{-\frac{i}{\hbar}\boldsymbol{q}\cdot\boldsymbol{x}'})]$$
$$= \frac{ic}{(2\pi)^3}\int\frac{d\boldsymbol{p}d\boldsymbol{q}}{2\sqrt{q^3p}}\sum_{\sigma\tau}(-e_k(\boldsymbol{p},\sigma)e_z^*(\boldsymbol{q},\tau)\delta(\boldsymbol{p}-\boldsymbol{q})\delta_{\sigma\tau}e^{\frac{i}{\hbar}\boldsymbol{p}\cdot\boldsymbol{x}-\frac{i}{\hbar}\boldsymbol{q}\cdot\boldsymbol{x}'}$$
$$- e_k^*(\boldsymbol{p},\sigma)e_z(\boldsymbol{q},\tau)\delta(\boldsymbol{p}-\boldsymbol{q})\delta_{\sigma\tau}e^{-\frac{i}{\hbar}\boldsymbol{p}\cdot\boldsymbol{x}+\frac{i}{\hbar}\boldsymbol{q}\cdot\boldsymbol{x}'})$$
$$= -\frac{ic}{(2\pi)^3}\int\frac{d\boldsymbol{p}}{2p^2}\sum_{\sigma}e_k(\boldsymbol{p},\sigma)e_z^*(\boldsymbol{p},\tau)(e^{\frac{i}{\hbar}\boldsymbol{p}\cdot(\boldsymbol{x}-\boldsymbol{x}')} + e^{-\frac{i}{\hbar}\boldsymbol{p}\cdot(\boldsymbol{x}-\boldsymbol{x}')})$$
$$= -\frac{ic}{(2\pi)^3}\int\frac{d\boldsymbol{p}}{2p^2}\left(\delta_{kz} - \frac{p_kp_z}{p^2}\right)(e^{\frac{i}{\hbar}\boldsymbol{p}\cdot(\boldsymbol{x}-\boldsymbol{x}')} + e^{-\frac{i}{\hbar}\boldsymbol{p}\cdot(\boldsymbol{x}-\boldsymbol{x}')})$$
$$= -\frac{ic}{(2\pi)^3}\int\frac{d\boldsymbol{p}}{p^2}\left(\delta_{kz} - \frac{p_kp_z}{p^2}\right)e^{\frac{i}{\hbar}\boldsymbol{p}\cdot(\boldsymbol{x}-\boldsymbol{x}')}$$
$$= -i\hbar c\delta_{kz}\mathcal{G}(\boldsymbol{x}-\boldsymbol{x}') - \frac{ic\hbar^2}{(2\pi)^3}\partial_k\partial_z\int\frac{d\boldsymbol{p}}{p^4}e^{\frac{i}{\hbar}\boldsymbol{p}\cdot(\boldsymbol{x}-\boldsymbol{x}')}$$

4 See formula (5.20) in [20].

$$= -i\hbar c\delta_{kz}\mathcal{G}(\mathbf{x}-\mathbf{x}') + i\hbar c\partial_k\partial_z\frac{|\mathbf{x}-\mathbf{x}'|}{8\pi}$$
$$= -i\hbar c\left(\delta_{kz}\mathcal{G}(\mathbf{x}-\mathbf{x}') - \frac{1}{2}\partial_k((z-z')\mathcal{G}(\mathbf{x}-\mathbf{x}'))\right).$$

Restoring the t-dependence, we have

$$\begin{aligned}
&-[V^n(t), Z_z^n(t)]\\
&= \frac{i\hbar}{c}\sum_{k=1}^{3}\int d\mathbf{x}d\mathbf{x}'\mathrm{j}^k(\tilde{x})\left(\delta_{kz}\mathcal{G}(\mathbf{x}-\mathbf{x}') - \frac{1}{2}\partial_k[(z-z')\mathcal{G}(\mathbf{x}-\mathbf{x}')]\right)\mathrm{j}^0(\tilde{x}')\\
&= \frac{i\hbar}{c}\int d\mathbf{x}d\mathbf{x}'\mathrm{j}^z(\tilde{x})\mathcal{G}(\mathbf{x}-\mathbf{x}')\mathrm{j}^0(\tilde{x}') - \frac{i\hbar}{2c}\int d\mathbf{x}d\mathbf{x}'\mathrm{j}(\tilde{x})\cdot\frac{\partial}{\partial\mathbf{x}}[(z-z')\mathcal{G}(\mathbf{x}-\mathbf{x}')]\mathrm{j}^0(\tilde{x}').
\end{aligned} \tag{E.18}$$

Adding together the four terms (E.15), (E.16), (E.17) and (E.18), we see that at $t = 0$ the first two terms in (E.17) cancel out with the two terms on the right-hand side of (E.15), the third term in (E.17) cancels with the second term on the right-hand side of (E.16) and (E.18) yields zero when added to the first and third terms on the right-hand side of (E.16). This proves equation (E.8).

Checking the last remaining nontrivial commutator,

$$[K_{0i}, Z_j^n] + [Z_i^n, K_{0j}] + [Z_i^n, Z_j^n] = 0,$$

is left as an exercise for the reader.

F Loop integrals in QED

F.1 Schwinger–Feynman integration trick

In QED calculations, we often meet integrals of expressions like $1/(abc\cdots)$, where $a, b, c, \ldots$ are certain functions of the loop momentum. Calculations of such integrals are greatly simplified if it is possible to replace the denominator by a function linear in $a, b, c, \ldots$. This can be achieved with the help of a witty trick [4].

The simplest example of such a trick is given by the integral representation of the expression $1/(ab)$, i. e.,

$$\int_0^1 \frac{dx}{(ax + b(1-x))^2} = \frac{1}{(b-a)(ax + b(1-x))}\bigg|_0^1 = \frac{1}{(b-a)a} - \frac{1}{(b-a)b} = \frac{1}{ab}. \tag{F.1}$$

The denominator of the left-hand side is a square of a function linear in a and b. Despite the introduction of additional integration (with respect to x), the problem of calculating the loop integral of the expression $(ab)^{-1}$ is substantially simplified, as we shall see below.

Using equation (F.1), we can linearize more complex denominators. For example,

$$\frac{1}{a^2 b} = -\frac{d}{da}\left(\frac{1}{ab}\right) = -\frac{d}{da}\int_0^1 \frac{dx}{(ax + b(1-x))^2} = \int_0^1 \frac{2x dx}{(ax + b(1-x))^3}. \tag{F.2}$$

From these two results we obtain an integral representation of $1/(abc)$, i. e.,

$$\begin{aligned}
\frac{1}{abc} &= \left(\frac{1}{bc}\right)\frac{1}{a} = \left(\int_0^1 \frac{dy}{(by + c(1-y))^2}\right)\frac{1}{a} \\
&= \int_0^1 dy \int_0^1 2x dx \frac{1}{[(by + c(1-y))x + a(1-x)]^3} \\
&= 2\int_0^1 x dx \int_0^1 \frac{dy}{[byx + cx(1-y) + a(1-x)]^3}.
\end{aligned} \tag{F.3}$$

One more useful formula is equation (131.2) in [1], i. e.,

$$\begin{aligned}
\frac{1}{abc} &= 2\int_0^1 dx \int_0^1 dy \int_0^1 dz \frac{\delta(x + y + z - 1)}{[ax + by + cz]^3} \\
&= 2\int_0^1 dx \int_0^{1-x} dy \frac{1}{[ax + by + c(1 - x - y)]^3}.
\end{aligned} \tag{F.4}$$

https://doi.org/10.1515/9783110493207-010

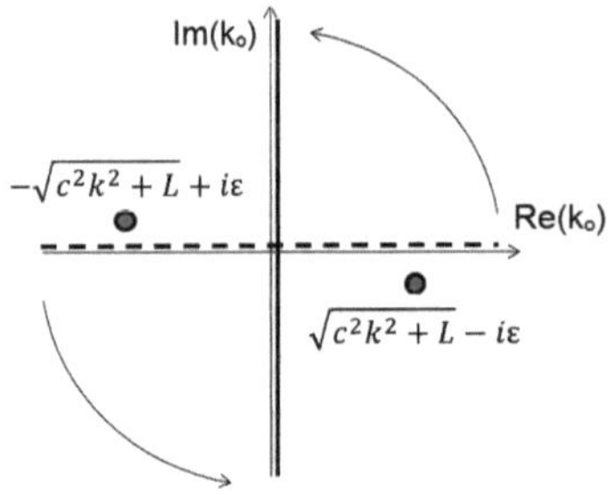

Figure F.1: Wick rotation in the integral (F.6) with respect to k_0. Dashed line, the original integration path $k_0 \in (-\infty, +\infty)$. Full line, the Wick-rotated path $k_0 \in (-i\infty, +i\infty)$.

Next we differentiate equation (F.2) with respect to a:

$$\frac{1}{a^3 d} = -\frac{1}{2} \cdot \frac{d}{da}\left(\frac{1}{a^2 d}\right) = -\frac{d}{da}\int_0^1 \frac{zdz}{[az + d(1-z)]^3} = 3\int_0^1 \frac{z^2 dz}{[az + d(1-z)]^4}.$$

This implies

$$\begin{aligned}\frac{1}{abcd} &= \left(2\int_0^1 xdx \int_0^1 dy \frac{1}{[a(1-x) + bxy + cx(1-y)]^3}\right)\frac{1}{d} \\ &= 6\int_0^1 xdx \int_0^1 dy \int_0^1 \frac{z^2 dz}{[az(1-x) + bxyz + cxz(1-y) + d(1-z)]^4}.\end{aligned} \qquad \text{(F.5)}$$

Obviously, such derivations can be continued for expressions with more factors in denominators; see, for example, the last formula on page 520 in [14] and equation (11.A.1) in [21].

F.2 Some basic four-dimensional integrals

In our study of loop integrals we will stick to the original Feynman approach [4]. Let us start from the following simple 4D integral:

$$K = \int \frac{d^4 k}{(\tilde{k}^2 - L)^3} \equiv \int \frac{dk_0 d\mathbf{k}}{(k_0^2 - c^2 k^2 - L + i\epsilon)^3}, \qquad \text{(F.6)}$$

where $L > 0$ and ϵ is a small positive number. The original path of integration with respect to k_0 goes from $-\infty$ to $+\infty$, as shown by the dashed line in Figure F.1. The integrand has two third-order poles at $k_0 = \pm\sqrt{c^2k^2 + L} \mp i\epsilon$. Without crossing these poles, we can turn the integration contour, so that it goes along the imaginary axis (from $-i\infty$ to $+i\infty$, shown by the solid line[1]), and then change the integration variables

1 This step is known as the *Wick rotation* [10].

$ik_0 = z_4$ and $\boldsymbol{ck} = \boldsymbol{z}$. This leads to the integral

$$K = \frac{1}{c^3}\int_{-i\infty}^{i\infty} dk_0 \int \frac{d\boldsymbol{z}}{(k_0^2 - z^2 - L)^3} = \frac{i}{c^3}\int_{-\infty}^{\infty} dz_4 \int \frac{d\boldsymbol{z}}{(-z_4^2 - z^2 - L)^3}.$$

Next we introduce four-dimensional spherical coordinates,[2] where $r^2 = z_4^2 + z^2$ and the "area" of the unit "sphere" is equal to $\int d\Omega = 2\pi^2$, so that

$$\begin{aligned} K &= -\frac{2\pi^2 i}{c^3}\int_0^{\infty} \frac{r^3 dr}{(r^2 + L)^3} = -\frac{\pi^2 i}{c^3}\int_L^{\infty} \frac{(t - L)dt}{t^3} = -\frac{\pi^2 i}{c^3}\left(-\frac{1}{t} + \frac{L}{2t^2}\right)\Bigg|_{t=L}^{t=\infty} \\ &= \frac{\pi^2}{2ic^3 L}. \end{aligned} \tag{F.7}$$

From symmetry properties we also obtain

$$\int d^4k \frac{k_\sigma}{(\tilde{k}^2 - L)^3} = 0. \tag{F.8}$$

Replacing $\tilde{k} \to \tilde{k} - \tilde{p}$ in (F.6) and denoting $L - \tilde{p}^2 = \Delta$, we have

$$\begin{aligned} \frac{\pi^2}{2i(\tilde{p}^2 + \Delta)c^3} &= \int \frac{d^4k}{((\tilde{k} - \tilde{p})^2 - L)^3} = \int \frac{d^4k}{(\tilde{k}^2 - 2\tilde{p}\cdot\tilde{k} + \tilde{p}^2 - L)^3} \\ &= \int \frac{d^4k}{(\tilde{k}^2 - 2\tilde{p}\cdot\tilde{k} - \Delta)^3}. \end{aligned} \tag{F.9}$$

After making the same replacements in (F.8) we get

$$0 = \int \frac{d^4k(k_\sigma - p_\sigma)}{((\tilde{k} - \tilde{p})^2 - L)^3} = \int \frac{d^4k(k_\sigma - p_\sigma)}{(\tilde{k}^2 - 2\tilde{p}\cdot\tilde{k} - \Delta)^3}.$$

Then

$$\int \frac{d^4k k_\sigma}{(\tilde{k}^2 - 2\tilde{p}\cdot\tilde{k} - \Delta)^3} = \int \frac{d^4k p_\sigma}{(\tilde{k}^2 - 2\tilde{p}\cdot\tilde{k} - \Delta)^3} = \frac{\pi^2 p_\sigma}{2i(\tilde{p}^2 + \Delta)c^3}. \tag{F.10}$$

Differentiating both sides of (F.9) first with respect to Δ and then with respect to p_σ, we obtain

$$\begin{aligned} \int \frac{d^4k}{(\tilde{k}^2 - 2\tilde{p}\cdot\tilde{k} - \Delta)^4} &= -\frac{\pi^2}{6i(\tilde{p}^2 + \Delta)^2 c^3}, \\ \int \frac{d^4k k_\sigma}{(\tilde{k}^2 - 2\tilde{p}\cdot\tilde{k} - \Delta)^4} &= -\frac{\pi^2 p_\sigma}{6i(\tilde{p}^2 + \Delta)^2 c^3}. \end{aligned} \tag{F.11}$$

2 See [3] as well as equations (6.49) and (7.81) in [10].

Next, differentiate both sides of (F.10) with respect to p_τ. If $\tau \neq \sigma$, then

$$\int \frac{d^4 k k_\sigma k_\tau}{(\tilde{k}^2 - 2\tilde{p}\cdot\tilde{k} - \Delta)^4} = -\frac{\pi^2 p_\sigma p_\tau}{6i(\tilde{p}^2 + \Delta)^2 c^3}. \tag{F.12}$$

Otherwise ($\tau = \sigma$),

$$\int \frac{d^4 k k_\sigma k_\sigma}{(\tilde{k}^2 - 2\tilde{p}\cdot\tilde{k} - \Delta)^4} = -\frac{\pi^2 p_\sigma p_\sigma}{6i(\tilde{p}^2 + \Delta)^2 c^3} + \frac{\pi^2}{12i(\tilde{p}^2 + \Delta)c^3}. \tag{F.13}$$

Joining (F.12) and (F.13) yields

$$\int \frac{d^4 k k_\sigma k_\tau}{(\tilde{k}^2 - 2\tilde{p}\cdot\tilde{k} - \Delta)^4} = -\frac{\pi^2 (p_\sigma p_\tau - \frac{1}{2}\delta_{\sigma\tau}(\tilde{p}^2 + \Delta))}{6i(\tilde{p}^2 + \Delta)^2 c^3}.$$

Then we use (F.2) and (F.9) to calculate[3]

$$\begin{aligned}
&\int \frac{d^4 k}{(\tilde{k}^2 - 2\tilde{p}_1\cdot\tilde{k} - \Delta_1)^2(\tilde{k}^2 - 2\tilde{p}_2\cdot\tilde{k} - \Delta_2)} \\
&= \int_0^1 2x dx \int \frac{d^4 k}{[(\tilde{k}^2 - 2\tilde{p}_1\cdot\tilde{k} - \Delta_1)x + (\tilde{k}^2 - 2\tilde{p}_2\cdot\tilde{k} - \Delta_2)(1-x)]^3} \\
&= \int_0^1 2x dx \int d^4 k \\
&\quad \times \frac{1}{[\tilde{k}^2 x - 2\tilde{p}_1\cdot\tilde{k}x - \Delta_1 x + \tilde{k}^2 - 2\tilde{p}_2\cdot\tilde{k} - \Delta_2 - \tilde{k}^2 x + 2\tilde{p}_2\cdot\tilde{k}x + \Delta_2 x]^3} \\
&= \int_0^1 2x dx \int \frac{d^4 k}{[\tilde{k}^2 - 2\tilde{p}_x\cdot\tilde{k} - \Delta_x]^3} = \frac{\pi^2}{ic^3}\int_0^1 \frac{x dx}{\tilde{p}_x^2 + \Delta_x},
\end{aligned} \tag{F.14}$$

where

$$\begin{aligned}
\tilde{p}_x &= x\tilde{p}_1 + (1-x)\tilde{p}_2, \\
\Delta_x &= x\Delta_1 + (1-x)\Delta_2.
\end{aligned}$$

Similarly, we use (F.2) and (F.10) to obtain

$$\begin{aligned}
&\int \frac{d^4 k k_\sigma}{(\tilde{k}^2 - 2\tilde{p}_1\cdot\tilde{k} - \Delta_1)^2(\tilde{k}^2 - 2\tilde{p}_2\cdot\tilde{k} - \Delta_2)} \\
&= \int_0^1 2x dx \int \frac{d^4 k k_\sigma}{[\tilde{k}^2 - 2\tilde{p}_x\cdot\tilde{k} - \Delta_x]^3} = \frac{\pi^2}{ic^3}\int_0^1 \frac{p_{x\sigma} x dx}{\tilde{p}_x^2 + \Delta_x}.
\end{aligned} \tag{F.15}$$

3 Here $\tilde{p}_1, \tilde{p}_2$ are two arbitrary 4-vectors and Δ_1, Δ_2 are numerical constants.

Three more integrals can be obtained: the derivatives of (F.14) with respect to Δ_2 and $p_{2\tau}$ and the derivative of (F.15) with respect to $p_{2\tau}$, i. e.,

$$\int \frac{d^4k}{(\tilde{k}^2 - 2\tilde{p}_1 \cdot \tilde{k} - \Delta_1)^2(\tilde{k}^2 - 2\tilde{p}_2 \cdot \tilde{k} - \Delta_2)^2} = -\frac{\pi^2}{ic^3}\int_0^1 \frac{x(1-x)dx}{(\tilde{p}_x^2 + \Delta_x)^2}, \tag{F.16}$$

$$\int \frac{d^4k k_\tau}{(\tilde{k}^2 - 2\tilde{p}_1 \cdot \tilde{k} - \Delta_1)^2(\tilde{k}^2 - 2\tilde{p}_2 \cdot \tilde{k} - \Delta_2)^2} = -\frac{\pi^2}{ic^3}\int_0^1 \frac{p_{x\tau}x(1-x)dx}{(\tilde{p}_x^2 + \Delta_x)^2}, \tag{F.17}$$

$$\int \frac{d^4k k_\sigma k_\tau}{(\tilde{k}^2 - 2\tilde{p}_1 \cdot \tilde{k} - \Delta_1)^2(\tilde{k}^2 - 2\tilde{p}_2 \cdot \tilde{k} - \Delta_2)^2}$$
$$= -\frac{\pi^2}{ic^3}\int_0^1 \frac{(p_{x\sigma}p_{x\tau} - (1/2)\delta_{\sigma\tau}(\tilde{p}_x^2 + \Delta_x))x(1-x)dx}{(\tilde{p}_x^2 + \Delta_x)^2}. \tag{F.18}$$

F.3 Electron self-energy integral

Using equations (B.13)–(B.15), the loop integral in square brackets in (4.15) can be rewritten as

$$J_{ad}(\not{p}) = \gamma_\mu(\not{p} + m_e c^2)I\gamma^\mu - \gamma_\mu\gamma^\nu I_\nu\gamma^\mu = (-2\not{p} + 4m_e c^2)I + 2\gamma^\nu I_\nu, \tag{F.19}$$

where

$$I \equiv \int \frac{d^4k}{[(\tilde{p} - \tilde{k})^2 - m_e^2c^4]\tilde{k}^2},$$
$$I_\nu \equiv \int \frac{d^4k k_\nu}{[(\tilde{p} - \tilde{k})^2 - m_e^2c^4]\tilde{k}^2}.$$

The factors $1/\tilde{k}^2$ in the integrands are sources of both ultraviolet and infrared divergences. Therefore, these integrals should be regularized, as explained in Subsection 4.1.5. To do this, we introduce two cutoff parameters: the *ultraviolet* cutoff Λ and the *infrared* cutoff λ.[4] Then we can replace the problematic factor $1/\tilde{k}^2$ by the following integral:

$$1/\tilde{k}^2 \to -\int_{\lambda^2c^4}^{\Lambda^2c^4} \frac{dL}{(\tilde{k}^2 - L)^2}. \tag{F.20}$$

4 The parameters $\Lambda > 0$ and $\lambda > 0$ have a physical dimension of mass. The nonzero value of λ is equivalent to introducing a fictitious photon mass $\lambda \ll m_e \ll \Lambda$.

At the end of the calculations, we must go to the limits $\Lambda \to \infty$ and $\lambda \to 0$.[5] Then, as expected, the integral reduces to $1/\tilde{k}^2$, as follows:

$$-\int_0^\infty \frac{dL}{(\tilde{k}^2 - L)^2} = -\int_{-\tilde{k}^2}^\infty \frac{dx}{x^2} = \frac{1}{\tilde{k}^2}.$$

We use equations (F.14) and (F.15) with parameters

$$\Delta_1 = L, \quad \tilde{p}_1 = 0, \quad \Delta_2 = m_e^2c^4 - \tilde{p}^2, \quad \tilde{p}_2 = \tilde{p}, \tag{F.21}$$

$$\tilde{p}_x = (1-x)\tilde{p}, \quad \Delta_x = xL + (1-x)(m_e^2c^4 - \tilde{p}^2) \tag{F.22}$$

to rewrite our integrals as follows:

$$\begin{aligned}
I &= -\int_{\lambda^2c^4}^{\Lambda^2c^4} dL \int \frac{d^4k}{(\tilde{k}^2 - 2\tilde{p}\cdot\tilde{k} + \tilde{p}^2 - m_e^2c^4)(\tilde{k}^2 - L)^2} = -\frac{\pi^2}{ic^3}\int_{\lambda^2c^4}^{\Lambda^2c^4} dL \int_0^1 \frac{xdx}{\tilde{p}_x^2 + \Delta_x} \\
&= -\frac{\pi^2}{ic^3}\int_{\lambda^2c^4}^{\Lambda^2c^4} dL \int_0^1 \frac{xdx}{(1-x)^2\tilde{p}^2 + xL + (1-x)(m_e^2c^4 - \tilde{p}^2)} \\
&= -\frac{\pi^2}{ic^3}\int_0^1 dx \ln[(1-x)^2\tilde{p}^2 + xL + (1-x)(m_e^2c^4 - \tilde{p}^2)]\Big|_{L=\lambda^2c^4}^{L=\Lambda^2c^4} \\
&= -\frac{\pi^2}{ic^3}\int_0^1 dx \ln\frac{(1-x)^2\tilde{p}^2 + x\Lambda^2c^4 + (1-x)(m_e^2c^4 - \tilde{p}^2)}{(1-x)^2\tilde{p}^2 + x\lambda^2c^4 + (1-x)(m_e^2c^4 - \tilde{p}^2)} \\
&\approx -\frac{\pi^2}{ic^3}\int_0^1 dx \ln\frac{x\Lambda^2c^4}{(1-x)^2\tilde{p}^2 + x\lambda^2c^4 + (1-x)(m_e^2c^4 - \tilde{p}^2)},
\end{aligned} \tag{F.23}$$

$$\begin{aligned}
I_\nu &= -\int_{\lambda^2c^4}^{\Lambda^2c^4} dL \int d^4k \frac{k_\nu}{(\tilde{k}^2 - 2\tilde{p}\cdot\tilde{k} + \tilde{p}^2 - m_e^2c^4)(\tilde{k}^2 - L)^2} \\
&= -\frac{\pi^2}{ic^3}\int_{\lambda^2c^4}^{\Lambda^2c^4} dL \int_0^1 \frac{p_\nu x(1-x)dx}{(1-x)^2\tilde{p}^2 + xL + (1-x)(m_e^2c^4 - \tilde{p}^2)} \\
&= -\frac{\pi^2}{ic^3}\int_0^1 dx(1-x)p_\nu \ln\frac{(1-x)^2\tilde{p}^2 + x\Lambda^2c^4 + (1-x)(m_e^2c^4 - \tilde{p}^2)}{(1-x)^2\tilde{p}^2 + x\lambda^2c^4 + (1-x)(m_e^2c^4 - \tilde{p}^2)}
\end{aligned}$$

5 In a physically acceptable theory, in these limits, all measurable quantities should tend to well-defined finite constants.

$$\approx -\frac{\pi^2}{ic^3}\int_0^1 dx(1-x)p_\nu \ln \frac{x\Lambda^2 c^4}{(1-x)^2\tilde{p}^2 + x\lambda^2 c^4 + (1-x)(m_e^2 c^4 - \tilde{p}^2)}. \tag{F.24}$$

Substituting (F.23) and (F.24) in (F.19), we obtain

$$\begin{aligned} J(\not{p}) &\approx -\frac{\pi^2}{ic^3}(-2\not{p} + 4m_e c^2)\int_0^1 dx \ln \frac{x\Lambda^2 c^4}{(1-x)^2\tilde{p}^2 + x\lambda^2 c^4 + (1-x)(m_e^2 c^4 - \tilde{p}^2)} \\ &\quad - \frac{2\pi^2 \not{p}}{ic^3}\int_0^1 dx(1-x)\ln \frac{x\Lambda^2 c^4}{(1-x)^2\tilde{p}^2 + x\lambda^2 c^4 + (1-x)(m_e^2 c^4 - \tilde{p}^2)} \\ &= -\frac{\pi^2}{ic^3}\int_0^1 dx(4m_e c^2 - 2\not{p}x)\ln \frac{x\Lambda^2 c^4}{(1-x)^2\tilde{p}^2 + x\lambda^2 c^4 + (1-x)(m_e^2 c^4 - \tilde{p}^2)}. \end{aligned} \tag{F.25}$$

For our discussions in Subsections 4.2.1 and 4.2.2 it will be convenient to represent $J(\not{p})$ as a Taylor expansion near the mass shell, where $\not{p} = m_e c^2$, i. e.,

$$J_{ad}(\not{p}) = C^{(0)}\delta_{ad} + C^{(1)}(\not{p} - m_e c^2)_{ad} + R_{ad}(\not{p}).$$

Here $C^{(0)}$ is a constant (independent on $\tilde{p}$), the term $C^{(1)}(\not{p} - mc^2)$ is proportional to $\not{p} - m_e c^2$ and $R(\not{p})$ unites all other terms (quadratic in $(\not{p} - m_e c^2)$, cubic, etc). To calculate $C^{(0)}$ we simply set $\not{p} = m_e c^2$ and $\tilde{p}^2 = m_e^2 c^4$ in (F.25), then

$$\begin{aligned} C^{(0)} &\approx -\frac{2\pi^2 m_e c^2}{ic^3}\int_0^1 dx(2-x)\ln \frac{x\Lambda^2 c^4}{(1-x)^2 m_e^2 c^4} \\ &= -\frac{2\pi^2 m_e c^2}{4ic^3}\left(-2(x^2 - 4x + 3)\ln \frac{x\Lambda^2 c^4}{(1-x)^2 m_e^2 c^4} - x^2 + 4x + 5 + 6\ln x\right)\Bigg|_{x=0}^{x=1} \\ &= -\frac{\pi^2 m_e c^2}{2ic^3}\left(8 - \left(-12\ln \frac{\Lambda}{m_e} + 5\right)\right) \\ &= -\frac{3\pi^2 m_e c^2}{2ic^3}\left(1 + 4\ln \frac{\Lambda}{m_e}\right). \end{aligned} \tag{F.26}$$

To calculate the coefficient $C^{(1)}$ we use the integral $\int_0^1 dx x \ln(1-x)^{-2} = 5/4$, so

$$\begin{aligned} C^{(1)} &= \frac{dJ}{d\not{p}}\Bigg|_{\not{p}=m_e c^2} \\ &= \frac{2\pi^2}{ic^3}\int_0^1 x dx \ln \frac{x\Lambda^2 c^4}{(1-x)^2\tilde{p}^2 + x\lambda^2 c^4 + (1-x)(m_e^2 c^4 - \tilde{p}^2)}\Bigg|_{\not{p}=m_e c^2} \\ &\quad + \frac{2\pi^2}{ic^3} m_e c^2 \int_0^1 (2-x)dx \frac{(1-x)^2\tilde{p}^2 + x\lambda^2 c^4 + (1-x)(m_e^2 c^4 - \tilde{p}^2)}{x\Lambda^2 c^4} \end{aligned}$$

$$
\begin{aligned}
&\times \frac{x\Lambda^2c^4(2(1-x)^2\not p - 2(1-x)\not p)}{[(1-x)^2\tilde{p}^2 + x\lambda^2c^4 + (1-x)(m_e^2c^4 - \tilde{p}^2)]^2}\Bigg|_{\not p = m_ec^2} \\
&= \frac{2\pi^2}{ic^3}\int_0^1 xdx \ln\frac{x\Lambda^2}{(1-x)^2m_e^2} \\
&+ \frac{2\pi^2 m_ec^2}{ic^3}\int_0^1 dx(2-x)\frac{2(1-x)^2m_ec^2 - 2(1-x)m_ec^2}{(1-x)^2m_e^2c^4 + x\lambda^2c^4} \\
&= \frac{2\pi^2}{ic^3}\int_0^1 xdx \ln\frac{x\Lambda^2}{(1-x)^2m_e^2} + \frac{4\pi^2}{ic^3}\int_0^1 dx\frac{(2-x)(x^2-x)}{(1-x)^2 + x\lambda^2/m_e^2} \\
&\approx \frac{2\pi^2}{4ic^3}\left((x-1)\left[2(x+1)\ln\frac{x\Lambda^2}{(1-x)^2m_e^2} + x + 5\right] + 2\ln x\right)\Bigg|_{x=0}^{x=1} + \frac{2\pi^2}{ic^3}\left(\ln\frac{\lambda^2}{m_e^2} + 1\right) \\
&= \frac{2\pi^2}{ic^3}\left(\ln\frac{\Lambda}{m_e} + 2\ln\frac{\lambda}{m_e} + \frac{9}{4}\right).
\end{aligned} \tag{F.27}
$$

Then the residual term

$$R_{ad}(\not p) = J_{ad}(\not p) - C^{(0)}\delta_{ad} - C^{(1)}(\not p - m_ec^2)_{ad}$$

is finite in the ultraviolet limit, because all Λ-dependent contributions cancel out there. Indeed,[6]

$$
\begin{aligned}
&-\frac{\pi^2}{ic^3}\ln\Lambda^2\int_0^1 dx(4m_ec^2 - 2\not p\, x) + \frac{2\pi^2 m_ec^2}{ic^3}\ln\Lambda^2\int_0^1 dx(2-x) \\
&-(\not p - m_ec^2)\frac{2\pi^2}{ic^3}\ln\Lambda^2\int_0^1 xdx = 0.
\end{aligned}
$$

We see that $C^{(0)}$ has an ultraviolet divergence, while $C^{(1)}$ diverges in both ultraviolet and infrared limits. In other words, the integral $J_{bc}(\not p)$, as a function of $\not p$, is infinite on the mass shell $\not p = m_ec^2$ and has an infinite derivative there. However, the second and higher derivatives are all finite.

F.4 Vertex integral

Let us calculate the integral in square brackets in equation (4.39).[7] We have

$$N_\kappa(\tilde{q},\tilde{q}') = \int d^4h\gamma_\mu\frac{-\not h + \not q' + m_ec^2}{(\tilde{h}-\tilde{q}')^2 - m_e^2c^4}\gamma_\kappa\frac{-\not h + \not q + m_ec^2}{(\tilde{h}-\tilde{q})^2 - m_e^2c^4}\gamma^\mu\frac{1}{\tilde{h}^2}$$

6 On the left-hand side we collected all terms proportional to $\ln\Lambda^2$ from (F.25), (F.26) and (F.27).

7 Here we used equation (F.20) and took into account that $\tilde{q}$ and $\tilde{q}'$ are on the mass shell, so that $\tilde{q}^2 = (\tilde{q}')^2 = m_e^2c^4$.

$$\approx -\int_{\lambda^2c^4}^{\Lambda^2c^4} dL \int d^4h \frac{\gamma_\mu(-\not{h}+\not{q}'+m_ec^2)\gamma_\kappa(-\not{h}+\not{q}+m_ec^2)\gamma^\mu}{(\tilde{h}^2-2\tilde{q}'\cdot\tilde{h})(\tilde{h}^2-2\tilde{q}\cdot\tilde{h})(\tilde{h}^2-L)^2}.$$

We can rewrite the numerator

$$\begin{aligned}
&\gamma_\mu(-\not{h}+\not{q}'+m_ec^2)\gamma_\kappa(-\not{h}+\not{q}+m_ec^2)\gamma^\mu \\
&\quad=\gamma_\mu(\not{q}'+m_ec^2)\gamma_\kappa(\not{q}+m_ec^2)\gamma^\mu-\gamma_\mu\not{h}\gamma_\kappa(\not{q}+m_ec^2)\gamma^\mu \\
&\quad-\gamma_\mu(\not{q}'+m_ec^2)\gamma_\kappa\not{h}\gamma^\mu+\gamma_\mu\not{h}\gamma_\kappa\not{h}\gamma^\mu.
\end{aligned}$$

Then the desired integral is

$$\begin{aligned}
N_\kappa(\tilde{q},\tilde{q}') &= \gamma_\mu(\not{q}'+m_ec^2)\gamma_\kappa(\not{q}+m_ec^2)\gamma^\mu\mathfrak{M}-\gamma_\mu\gamma^\sigma\gamma_\kappa(\not{q}+m_ec^2)\gamma^\mu\mathfrak{M}_\sigma \\
&-\gamma_\mu(\not{q}'+m_ec^2)\gamma_\kappa\gamma^\sigma\gamma^\mu\mathfrak{M}_\sigma+\gamma_\mu\gamma^\sigma\gamma_\kappa\gamma^\tau\gamma^\mu\mathfrak{M}_{\sigma\tau},
\end{aligned} \tag{F.28}$$

where[8]

$$\mathfrak{M} \equiv -\int_0^1 dy \int_{\lambda^2c^4}^{\Lambda^2c^4} dL \int \frac{d^4h}{[\tilde{h}^2-2\tilde{h}\cdot\tilde{q}_y]^2[\tilde{h}^2-L]^2}, \tag{F.29}$$

$$\mathfrak{M}_\sigma \equiv -\int_0^1 dy \int_{\lambda^2c^4}^{\Lambda^2c^4} dL \int \frac{d^4h h_\sigma}{[\tilde{h}^2-2\tilde{h}\cdot\tilde{q}_y]^2[\tilde{h}^2-L]^2}, \tag{F.30}$$

$$\mathfrak{M}_{\sigma\tau} \equiv -\int_0^1 dy \int_{\lambda^2c^4}^{\Lambda^2c^4} dL \int \frac{d^4h h_\sigma h_\tau}{[\tilde{h}^2-2\tilde{h}\cdot\tilde{q}_y]^2[\tilde{h}^2-L]^2}. \tag{F.31}$$

These are particular cases of the integrals (F.16)–(F.18) with parameters $\tilde{p}_1=\tilde{q}_y, \Delta_1=0$, $\tilde{p}_2=0, \Delta_2=L, \tilde{p}_x=x\tilde{q}_y$ and $\Delta_x=(1-x)L$. We have

$$\mathfrak{M} = \frac{\pi^2}{ic^3}\int_0^1 dy \int_{\lambda^2c^4}^{\Lambda^2c^4} dL \int_0^1 \frac{x(1-x)dx}{(x^2\tilde{q}_y^2+(1-x)L)^2}, \tag{F.32}$$

$$\mathfrak{M}_\sigma = \frac{\pi^2}{ic^3}\int_0^1 dy \int_{\lambda^2c^4}^{\Lambda^2c^4} dL \int_0^1 \frac{q_{y\sigma}x^2(1-x)dx}{(x^2\tilde{q}_y^2+(1-x)L)^2}, \tag{F.33}$$

$$\mathfrak{M}_{\sigma\tau} = \frac{\pi^2}{ic^3}\int_0^1 dy \int_{\lambda^2c^4}^{\Lambda^2c^4} dL \int_0^1 \frac{[x^2q_{y\sigma}q_{y\tau}-(1/2)\delta_{\sigma\tau}(x^2\tilde{q}_y^2+(1-x)L)]x(1-x)dx}{(x^2\tilde{q}_y^2+(1-x)L)^2}. \tag{F.34}$$

8 We combined the denominators by using (F.1) and $\tilde{q}_y \equiv y\tilde{q}+(1-y)\tilde{q}'$.

F.4.1 Calculation of $\mathfrak{M}$

In the limit $\Lambda \to \infty$ we obtain for (F.32)

$$\mathfrak{M} = -\frac{\pi^2}{ic^3}\int_0^1 dy \int_0^1 dx \frac{x}{x^2\tilde{q}_y^2 + (1-x)L}\Bigg|_{L=\lambda^2c^4}^{L=\infty} = \frac{\pi^2}{ic^3}\int_0^1 dy \int_0^1 dx \frac{x}{x^2\tilde{q}_y^2 + (1-x)\lambda^2c^4}$$
$$\approx \frac{\pi^2}{2ic^3}\int_0^1 \frac{dy}{\tilde{q}_y^2} \ln(-x^2\tilde{q}_y^2 - (1-x)\lambda^2c^4)\Big|_{x=0}^{x=1} = \frac{\pi^2}{2ic^3}\int_0^1 \frac{dy}{\tilde{q}_y^2} \ln\frac{\tilde{q}_y^2}{\lambda^2c^4}. \tag{F.35}$$

Next we introduce the 4-vector of the transferred energy–momentum

$$\tilde{k} \equiv \tilde{q}' - \tilde{q}. \tag{F.36}$$

Then from $(\tilde{q}')^2 = (\tilde{q} + \tilde{k})^2$ and $(\tilde{q}')^2 = \tilde{q}^2 = m_e^2c^4$ it follows that

$$(\tilde{q}\cdot\tilde{k}) = -\tilde{k}^2/2, \tag{F.37}$$
$$(\tilde{p}\cdot\tilde{k}) = \tilde{k}^2/2, \tag{F.38}$$
$$\tilde{q}_y = \tilde{q} + (1-y)\tilde{k},$$
$$\tilde{q}_y^2 = \tilde{q}^2 + 2\tilde{q}\cdot\tilde{k}(1-y) + \tilde{k}^2(1-y)^2 = m_e^2c^4 - (1-y)y\tilde{k}^2.$$

Instead of $\tilde{k}^2$ and the integration variable y it will be convenient to introduce two new variables θ and α, such that[9]

$$\tilde{k}^2 \equiv 4m_e^2c^4\sin^2\theta, \tag{F.39}$$
$$y \equiv \frac{1}{2}\left(1 + \frac{\tan\alpha}{\tan\theta}\right),$$
$$1-y = \frac{1}{2}\left(1 - \frac{\tan\alpha}{\tan\theta}\right),$$
$$\begin{aligned}\tilde{q}_y^2 &= m_e^2c^4 - 4m_e^2c^4\sin^2\theta\cdot\frac{1}{2}\left(1+\frac{\tan\alpha}{\tan\theta}\right)\cdot\frac{1}{2}\left(1-\frac{\tan\alpha}{\tan\theta}\right)\\ &= m_e^2c^4 - m_e^2c^4\cos^2\theta(\tan^2\theta - \tan^2\alpha)\\ &= m_e^2c^4(1-\cos^2\theta\tan^2\theta + \cos^2\theta\tan^2\alpha) = m_e^2c^4\cos^2\theta(1+\tan^2\alpha)\\ &= m_e^2c^4\frac{\cos^2\theta}{\cos^2\alpha},\end{aligned}$$
$$dy = \frac{d\alpha}{2\tan\theta}\frac{d}{d\alpha}\left(\frac{\sin\alpha}{\cos\alpha}\right) = \frac{d\alpha}{2\tan\theta\cos^2\alpha},$$
$$\frac{dy}{\tilde{q}_y^2} = \frac{d\alpha}{2m_e^2c^4\cos^2\theta\tan\theta} = \frac{d\alpha}{m_e^2c^4\sin(2\theta)}.$$

9 Note that, by definition, $0 \le \tilde{k}^2 \le (2\tilde{q})^2 = 4m_e^2c^4$.

The integral $\mathfrak{M}$ is infrared divergent, i. e.,[10]

$$
\begin{aligned}
\mathfrak{M} &= \frac{\pi^2}{2ic^3} \int\limits_{-\theta}^{\theta} \frac{d\alpha}{m_e^2 c^4 \sin(2\theta)} \ln\left(\frac{m_e^2 \cos^2\theta}{\lambda^2 \cos^2\alpha}\right) \\
&= \frac{2\pi^2\theta}{ic^3 m_e^2 c^4 \sin(2\theta)} \ln \frac{m_e \cos\theta}{\lambda} - \frac{2\pi^2}{ic^3 m_e^2 c^4 \sin(2\theta)} \int\limits_0^{\theta} d\alpha \ln(\cos\alpha) \\
&= \frac{2\pi^2}{ic^3 m_e^2 c^4 \sin(2\theta)} \left(\theta \ln \frac{m_e \cos\theta}{\lambda} - \alpha \ln(\cos\alpha)\Big|_{\alpha=0}^{\alpha=\theta} - \int\limits_0^{\theta} \alpha \tan\alpha d\alpha\right) \\
&= 2\mathfrak{A}\left(\theta \ln \frac{m_e}{\lambda} - \int\limits_0^{\theta} \alpha \tan\alpha d\alpha\right),
\end{aligned}
\tag{F.40}
$$

where we denoted

$$
\mathfrak{A} \equiv \frac{\pi^2}{i m_e^2 c^7 \sin(2\theta)}.
$$

F.4.2 Calculation of $\mathfrak{M}_\sigma$

Next we calculate (F.33), using the variables θ and α introduced above. Taking the limits $\lambda \to 0$ and $\Lambda \to \infty$, we obtain a result where both infrared and ultraviolet divergences are absent. We have

$$
\begin{aligned}
\mathfrak{M}_\sigma &= -\frac{\pi^2}{ic^3} \int\limits_0^1 dx \int\limits_0^1 dy \frac{x^2 q_{y\sigma}}{x^2 \tilde{q}_y^2 + (1-x)L}\Bigg|_{L=0}^{L=\infty} \approx \frac{\pi^2}{ic^3} \int\limits_0^1 dy \frac{q_{y\sigma}}{\tilde{q}_y^2} \\
&= \frac{\pi^2}{ic^3} \int\limits_{-\theta}^{\theta} \frac{d\alpha}{m_e^2 c^4 \sin(2\theta)} \left(q_\sigma + \frac{k_\sigma}{2}\left(1 - \frac{\tan\alpha}{\tan\theta}\right)\right) \\
&= \frac{\pi^2}{ic^3} \frac{2\theta}{m_e^2 c^4 \sin(2\theta)} \left(q_\sigma + \frac{k_\sigma}{2}\right) - \frac{\pi^2 k_\sigma}{2ic^3 m_e^2 c^4 \sin(2\theta) \tan\theta} \int\limits_{-\theta}^{\theta} d\alpha \tan\alpha \\
&= \mathfrak{A}\theta(q_\sigma + q_\sigma').
\end{aligned}
\tag{F.41}
$$

10 We took by parts the following integral: $\int_0^\theta d\alpha \ln(\cos\alpha) = \alpha \ln(\cos\alpha)|_{\alpha=0}^{\alpha=\theta} + \int_0^\theta \alpha \tan\alpha d\alpha$.

F.4.3 Calculation of $\mathfrak{M}_{\sigma\tau}$

To calculate (F.34), we assume that $\Lambda^2c^4 \gg \tilde{q}_y^2 \gg \lambda^2c^4$ and use integral $\int_0^1 dxx\ln[(1-x)/x^2] = -1/4$. Then

$$\begin{aligned}
\mathfrak{M}_{\sigma\tau} &= \frac{\pi^2}{ic^3}\int\limits_{\lambda^2c^4}^{\Lambda^2c^4} dL \int\limits_0^1 dx \int\limits_0^1 dy \frac{x^3(1-x)q_{y\sigma}q_{y\tau}}{(x^2\tilde{q}_y^2+(1-x)L)^2} \\
&\quad - \frac{\pi^2}{2ic^3}\int\limits_{\lambda^2c^4}^{\Lambda^2c^4} dL \int\limits_0^1 dx \int\limits_0^1 dy \frac{\delta_{\sigma\tau}x(1-x)}{x^2\tilde{q}_y^2+(1-x)L} \\
&\approx -\frac{\pi^2}{ic^3}\int\limits_0^1 dx \int\limits_0^1 dy \left(\frac{x^3 q_{y\sigma}q_{y\tau}}{x^2\tilde{q}_y^2+(1-x)\Lambda^2c^4} - \frac{xq_{y\sigma}q_{y\tau}}{\tilde{q}_y^2}\right) \\
&\quad - \frac{\pi^2}{2ic^3}\int\limits_0^1 dx \int\limits_0^1 dy \delta_{\sigma\tau}x[\ln((1-x)\Lambda^2c^4) - \ln(x^2\tilde{q}_y^2)] \\
&\approx \frac{\pi^2}{2ic^3}\int\limits_0^1 dy \frac{q_{y\sigma}q_{y\tau}}{\tilde{q}_y^2} - \frac{\pi^2\delta_{\sigma\tau}}{2ic^3}\int\limits_0^1 xdx \int\limits_0^1 dy \ln\frac{(1-x)\Lambda^2c^4}{x^2\tilde{q}_y^2} \\
&= \frac{\pi^2}{2ic^3}\int\limits_0^1 dy \frac{q_{y\sigma}q_{y\tau}}{\tilde{q}_y^2} - \frac{\pi^2\delta_{\sigma\tau}}{2ic^3}\int\limits_0^1 xdx \ln\frac{(1-x)}{x^2} - \frac{\pi^2\delta_{\sigma\tau}}{4ic^3}\int\limits_0^1 dy \ln\frac{\Lambda^2c^4}{\tilde{q}_y^2} \\
&= \frac{\pi^2}{2ic^3}\int\limits_0^1 dy \frac{q_{y\sigma}q_{y\tau}}{q_y^2} - \frac{\pi^2\delta_{\sigma\tau}}{4ic^3}\int\limits_0^1 dy \ln\frac{\Lambda^2c^4}{\tilde{q}_y^2} + \frac{\pi^2\delta_{\sigma\tau}}{8ic^3}.
\end{aligned}$$

The integrations with respect to y are carried out with the help of the variables θ, α and the table integrals

$$\begin{aligned}
\int \tan^2 x dx &= \tan x - x + C, \\
\int \frac{dx}{\cos^2 x} &= \tan x + C, \\
\int \frac{\ln\cos^2 x}{\cos^2 x}dx &= -2x + 2\tan x + \tan x \ln(\cos^2 x) + C,
\end{aligned}$$

$$\begin{aligned}
\int\limits_0^1 dy \frac{q_{y\sigma}q_{y\tau}}{\tilde{q}_y^2} &= \int\limits_{-\theta}^{\theta}\left(q_\sigma + \frac{1}{2}k_\sigma - \frac{k_\sigma \tan\alpha}{2\tan\theta}\right)\left(q_\tau + \frac{1}{2}k_\tau - \frac{k_\tau \tan\alpha}{2\tan\theta}\right)\frac{d\alpha}{m_e^2c^4\sin(2\theta)} \\
&= \int\limits_{-\theta}^{\theta}\left(q_\sigma + \frac{1}{2}k_\sigma\right)\left(q_\tau + \frac{1}{2}k_\tau\right)\frac{d\alpha}{m_e^2c^4\sin(2\theta)} + \int\limits_{-\theta}^{\theta}\frac{k_\sigma k_\tau \tan^2\alpha}{4\tan^2\theta}\frac{d\alpha}{m_e^2c^4\sin(2\theta)}
\end{aligned}$$

$$= \frac{\theta}{2m_e^2c^4\sin(2\theta)}(q_\sigma + q'_\sigma)(q_\tau + q'_\tau) + \frac{k_\sigma k_\tau \cos\theta}{4m_e^2c^4\sin^3\theta}\int_0^\theta \tan^2\alpha d\alpha$$

$$= \frac{\theta}{2m_e^2c^4\sin(2\theta)}(q_\sigma + q'_\sigma)(q_\tau + q'_\tau) + \frac{k_\sigma k_\tau}{\tilde{k}^2}(1 - \theta\cot\theta),$$

$$\int_0^1 dy \ln\frac{\Lambda^2c^4}{\tilde{q}_y^2} = \int_0^\theta \frac{d\alpha}{\tan\theta\cos^2\alpha}\ln\frac{\Lambda^2\cos^2\alpha}{m_e^2\cos^2\theta}$$

$$= \ln\frac{\Lambda^2}{m_e^2\cos^2\theta} + \frac{1}{\tan\theta}(-2\theta + \ln(\cos^2\theta)\,\tan\theta + 2\,\tan\theta)$$

$$= 2\ln\frac{\Lambda}{m_e} + 2(1 - \theta\cot\theta).$$

Now we see that $\mathfrak{M}_{\sigma\tau}$ diverges in the ultraviolet, i. e.,

$$\mathfrak{M}_{\sigma\tau} = \frac{\mathfrak{A}\theta}{4}(q_\sigma + q'_\sigma)(q_\tau + q'_\tau) + \mathfrak{D}k_\sigma k_\tau + \mathfrak{E}\delta_{\sigma\tau}, \tag{F.42}$$

where we introduced the notation

$$\mathfrak{D} \equiv \frac{\pi^2(1 - \theta\cot\theta)}{2ic^3\tilde{k}^2},$$

$$\mathfrak{E} \equiv -\frac{\pi^2}{2ic^3}\left(\ln\frac{\Lambda}{m_e} + \frac{3}{4} - \theta\cot\theta\right).$$

F.4.4 Complete integral

Using results (F.40), (F.41) and (F.42), we obtain the full integral (F.28). We have

$$\begin{aligned} N_\kappa(\tilde{q}, \tilde{q}') &= \mathfrak{M}\gamma_\mu(\not{q}' + m_ec^2)\gamma_\kappa(\not{q} + m_ec^2)\gamma^\mu \\ &\quad - \mathfrak{A}\theta\gamma_\mu(\not{q} + \not{q}')\gamma_\kappa(\not{q} + m_ec^2)\gamma^\mu - \mathfrak{A}\theta\gamma_\mu(\not{q}' + m_ec^2)\gamma_\kappa(\not{q} + \not{q}')\gamma^\mu \\ &\quad + \frac{\mathfrak{A}\theta}{4}\gamma_\mu(\not{q} + \not{q}')\gamma_\kappa(\not{q} + \not{q}')\gamma^\mu + \mathfrak{D}\gamma_\mu\not{k}\gamma_\kappa\not{k}\gamma^\mu + 4\mathfrak{E}\gamma_\kappa \\ &\equiv \mathfrak{M}T_\kappa^{(1)} - \mathfrak{A}\theta T_\kappa^{(2)} - \mathfrak{A}\theta T_\kappa^{(3)} + \frac{\mathfrak{A}\theta}{4}T_\kappa^{(4)} + \mathfrak{D}\gamma_\mu\not{k}\gamma_\kappa\not{k}\gamma^\mu + 4\mathfrak{E}\gamma_\kappa. \end{aligned} \tag{F.43}$$

First, we notice that according to (4.39), the integral $N_\kappa(\tilde{q}, \tilde{q}')$ is multiplied by $\overline{u}(\boldsymbol{q}', \sigma')$ from the left and by $u(\boldsymbol{q}, \sigma)$ from the right. Then from (B.93)–(B.94) it follows that if in some term the factor $\not{q}'$ is in the leftmost position, then it can be replaced with m_ec^2. Similarly, the factor $\not{q}$ in the rightmost position is replaceable by m_ec^2. So, our plan is to employ formulas from Appendix B.2 to move factors $\not{q}'$ to the left and factors $\not{q}$ to the right. For example, in the environment $\overline{u}(\boldsymbol{q}', \sigma')\cdots u(\boldsymbol{q}, \sigma)$ it is true that

$$\gamma_\mu\not{q}' = -\not{q}'\gamma_\mu + 2q'_\mu = -m_ec^2\gamma_\mu + 2q'_\mu,$$

$$\not{q}\gamma^\mu = -\gamma^\mu\not{q} + 2q^\mu = -m_ec^2\gamma^\mu + 2q^\mu.$$

Then, using (F.36), (F.37), we obtain

$$\begin{aligned}
T_\kappa^{(1)} &\equiv \gamma_\mu(\not{q}' + m_e c^2)\gamma_\kappa(\not{q} + m_e c^2)\gamma^\mu \\
&= \gamma_\mu \not{q}'\gamma_\kappa \not{q}\gamma^\mu + m_e c^2 \gamma_\mu\gamma_\kappa\not{q}\gamma^\mu + m_e c^2\gamma_\mu\not{q}'\gamma_\kappa\gamma^\mu + m_e^2c^4\gamma_\mu\gamma_\kappa\gamma^\mu \\
&= (-\not{q}'\gamma_\mu + 2q'_\mu)\gamma_\kappa(-\gamma^\mu\not{q} + 2q^\mu) + m_e c^2\gamma_\mu\gamma_\kappa(-\gamma^\mu\not{q} + 2q^\mu) \\
&\quad + m_e c^2(-\not{q}'\gamma_\mu + 2q'_\mu)\gamma_\kappa\gamma^\mu - 2m_e^2c^4\gamma_\kappa \\
&= m_e^2c^4\gamma_\mu\gamma_\kappa\gamma^\mu - 2m_e c^2\gamma_\mu\gamma_\kappa q^\mu - 2m_e c^2 q'_\mu\gamma_\kappa\gamma^\mu + 4q'_\mu\gamma_\kappa q^\mu \\
&\quad - m_e^2c^4\gamma_\mu\gamma_\kappa\gamma^\mu + 2m_e c^2\gamma_\mu\gamma_\kappa q^\mu - m_e^2c^4\gamma_\mu\gamma_\kappa\gamma^\mu + 2m_e c^2 q'_\mu\gamma_\kappa\gamma^\mu - 2m_e^2c^4\gamma_\kappa \\
&= 4\gamma_\kappa(\tilde{q}'\cdot\tilde{q}) = (-2\tilde{k}^2 + 4m_e^2c^4)\gamma_\kappa.
\end{aligned}$$

For further calculations we will need some simple results:

$$\not{k}\gamma_\kappa\not{k} = -\gamma_\kappa\not{k}^2 + 2\not{k}\,k_\kappa = -\gamma_\kappa\tilde{k}^2 + 2(\not{q}' - \not{q})k_\kappa = -\tilde{k}^2\gamma_\kappa,$$

$$\begin{aligned}
\gamma_\kappa\not{k} - \not{k}\gamma_\kappa &= \gamma_\kappa(\not{q}' - \not{q}) - (\not{q}' - \not{q})\gamma_\kappa = \gamma_\kappa(\not{q}' - m_e c^2) - (m_e c^2 - \not{q})\gamma_\kappa \\
&= -2\gamma_\kappa m_e c^2 + \gamma_\kappa\not{q}' + \not{q}\gamma_\kappa = -2\gamma_\kappa m_e c^2 - \not{q}'\gamma_\kappa + 2q'_\kappa - \gamma_\kappa\not{q} + 2q_\kappa \\
&= -2\gamma_\kappa m_e c^2 - m_e c^2\gamma_\kappa + 2q'_\kappa - m_e c^2\gamma_\kappa + 2q_\kappa \\
&= -4m_e c^2\gamma_\kappa + 2(q_\kappa + q'_\kappa),
\end{aligned}$$

$$\begin{aligned}
\not{q}\gamma_\kappa\not{q}' &= (\not{q}' - \not{k})\gamma_\kappa(\not{q} + \not{k}) = \not{q}'\gamma_\kappa\not{q} - \not{k}\gamma_\kappa\not{q} + \not{q}'\gamma_\kappa\not{k} - \not{k}\gamma_\kappa\not{k} \\
&= m_e^2c^4\gamma_\kappa - m_e c^2\not{k}\gamma_\kappa + m_e c^2\gamma_\kappa\not{k} + \tilde{k}^2\gamma_\kappa \\
&= -3m_e^2c^4\gamma_\kappa + 2m_e c^2(q_\kappa + q'_\kappa) + \tilde{k}^2\gamma_\kappa.
\end{aligned}$$

Then

$$\begin{aligned}
T_\kappa^{(2)} &\equiv \gamma_\mu(\not{q} + \not{q}')\gamma_\kappa(\not{q} + m_e c^2)\gamma^\mu \\
&= \gamma_\mu\not{q}\gamma_\kappa\not{q}\gamma^\mu + \gamma_\mu\not{q}'\gamma_\kappa\not{q}\gamma^\mu + m_e c^2\gamma_\mu\not{q}\gamma_\kappa\gamma^\mu + m_e c^2\gamma_\mu\not{q}'\gamma_\kappa\gamma^\mu \\
&= -2\not{q}\gamma_\kappa\not{q} - 2\not{q}\gamma_\kappa\not{q}' + 2m_e c^2\not{q}\gamma_\kappa + 2m_e c^2\gamma_\kappa\not{q} + 2m_e c^2\not{q}'\gamma_\kappa + 2m_e c^2\gamma_\kappa\not{q}' \\
&= -2\not{q}\gamma_\kappa\not{q} - 2\not{q}\gamma_\kappa\not{q}' + 4m_e c^2 q_\kappa + 4m_e c^2 q'_\kappa,
\end{aligned}$$

$$\begin{aligned}
T_\kappa^{(3)} &\equiv \gamma_\mu(\not{q}' + m_e c^2)\gamma_\kappa(\not{q} + \not{q}')\gamma^\mu \\
&= \gamma_\mu\not{q}'\gamma_\kappa\not{q}\gamma^\mu + m_e c^2\gamma_\mu\gamma_\kappa\not{q}\gamma^\mu + \gamma_\mu\not{q}'\gamma_\kappa\not{q}'\gamma^\mu + m_e c^2\gamma_\mu\gamma_\kappa\not{q}'\gamma^\mu \\
&= -2\not{q}\gamma_\kappa\not{q}' - 2\not{q}'\gamma_\kappa\not{q}' + 2m_e c^2\gamma_\kappa\not{q} + 2m_e c^2\not{q}\gamma_\kappa + 2m_e c^2\gamma_\kappa\not{q}' + 2m_e c^2\not{q}'\gamma_\kappa \\
&= -2\not{q}\gamma_\kappa\not{q}' - 2\not{q}'\gamma_\kappa\not{q}' + 4m_e c^2 q_\kappa + 4m_e c^2 q'_\kappa,
\end{aligned}$$

$$\begin{aligned}
T_\kappa^{(2)} + T_\kappa^{(3)} &= -2m_e c^2\not{q}\gamma_\kappa - 2m_e c^2\gamma_\kappa\not{q}' - 4\not{q}\gamma_\kappa\not{q}' + 8m_e c^2(q_\kappa + q'_\kappa) \\
&= 2m_e c^2\gamma_\kappa\not{q} - 2m_e c^2 q_\kappa + 2m_e c^2\not{q}'\gamma_\kappa - 2m_e c^2 q'_\kappa \\
&\quad - 4(-3m_e^2c^4\gamma_\kappa + 2m_e c^2(q_\kappa + q'_\kappa) + \tilde{k}^2\gamma_\kappa) + 8m_e c^2(q_\kappa + q'_\kappa) \\
&= 16m_e^2c^4\gamma_\kappa - 4m_e c^2(q_\kappa + q'_\kappa) - 4\tilde{k}^2\gamma_\kappa,
\end{aligned}$$

$$
\begin{aligned}
T_\kappa^{(4)} &\equiv \gamma_\mu(\not{q} + \not{q}')\gamma_\kappa(\not{q} + \not{q}')\gamma^\mu \\
&= \gamma_\mu \not{q}\gamma_\kappa \not{q}\gamma^\mu + \gamma_\mu \not{q}'\gamma_\kappa \not{q}\gamma^\mu + \gamma_\mu \not{q}\gamma_\kappa \not{q}'\gamma^\mu + \gamma_\mu \not{q}'\gamma_\kappa \not{q}'\gamma^\mu \\
&= -2\not{q}\gamma_\kappa \not{q} - 2\not{q}\gamma_\kappa \not{q}' - 2\not{q}'\gamma_\kappa \not{q} - 2\not{q}'\gamma_\kappa \not{q}' \\
&= -2m_e c^2 \not{q}\gamma_\kappa - 2(-3m_e^2 c^4 \gamma_\kappa + 2m_e c^2(q_\kappa + q'_\kappa) + \tilde{k}^2\gamma_\kappa) \\
&\quad - 2m_e^2 c^4 \gamma_\kappa - 2m_e c^2 \gamma_\kappa \not{q}' \\
&= 8m_e^2 c^4 \gamma_\kappa - 8m_e c^2(q_\kappa + q'_\kappa) - 2\tilde{k}^2\gamma_\kappa.
\end{aligned}
$$

Adding all terms together, we get the full integral:

$$
\begin{aligned}
N_\kappa(q, q') &= \mathfrak{M}(-2\tilde{k}^2 + 4m_e^2 c^4)\gamma_\kappa - \mathfrak{A}\theta[16m_e^2 c^4 \gamma_\kappa - 4m_e c^2(q_\kappa + q'_\kappa) - 4\tilde{k}^2\gamma_\kappa] \\
&\quad + \mathfrak{A}\theta\left(2m_e^2 c^4 \gamma_\kappa - 2m_e c^2(q_\kappa + q'_\kappa) - \frac{1}{2}\tilde{k}^2\gamma_\kappa\right) + 2\mathfrak{D}\tilde{k}^2\gamma_\kappa + 4\mathfrak{E}\gamma_\kappa \\
&= \left(-2\mathfrak{M}\tilde{k}^2 + 2\mathfrak{D}\tilde{k}^2 + 4\mathfrak{M}m_e^2 c^4 + 4\mathfrak{E} - 14\mathfrak{A}\theta m_e^2 c^4 + \frac{7}{2}\mathfrak{A}\theta\tilde{k}^2\right)\gamma_\kappa \\
&\quad + 2\mathfrak{A}\theta m_e c^2(q_k + q'_k).
\end{aligned}
$$

The coefficient in front of γ_κ is

$$
\begin{aligned}
&\frac{2\pi^2(-2\tilde{k}^2 + 4m_e^2 c^4)}{ic^3 m_e^2 c^4 \sin(2\theta)}\left[\theta \ln\frac{m_e}{\lambda} - \int_0^\theta x \tan x dx\right] + \frac{\pi^2(1 - \theta\cot\theta)}{ic^3} \\
&\quad - \frac{2\pi^2}{ic^3}\left(\ln\frac{\Lambda}{m_e} + (1 - \theta\cot\theta) - \frac{1}{4}\right) - \frac{14\pi^2\theta}{ic^3\sin(2\theta)} + \frac{7\pi^2\tilde{k}^2\theta}{2ic^3 m_e^2 c^4 \sin(2\theta)} \\
&= \frac{2\pi^2(-8m_e^2 c^4\sin^2\theta + 4m_e^2 c^4)}{ic^3 m_e^2 c^4\sin(2\theta)}\left[\theta\ln\frac{m_e}{\lambda} - \int_0^\theta x\tan x dx\right] \\
&\quad - \frac{\pi^2(1 - \theta\cot\theta)}{ic^3} - \frac{2\pi^2}{ic^3}\ln\frac{\Lambda}{m_e} + \frac{\pi^2}{2ic^3} - \frac{14\pi^2\theta}{ic^3\sin(2\theta)} + \frac{14\pi^2\theta\sin^2\theta}{ic^3\sin(2\theta)} \\
&= \frac{8\pi^2}{ic^3\tan(2\theta)}\left[-\theta\ln\frac{\lambda}{m_e} - \int_0^\theta x\tan x dx\right] \\
&\quad - \frac{\pi^2(1 - \theta\cot\theta)}{ic^3} - \frac{2\pi^2}{ic^3}\ln\frac{\Lambda}{m_e} + \frac{\pi^2}{2ic^3} - \frac{7\pi^2\theta\cot\theta}{ic^3}.
\end{aligned}
$$

Therefore, finally we have

$$
\begin{aligned}
N_\kappa(\tilde{q}, \tilde{q}') &= -\frac{\pi^2\gamma_\kappa}{ic^3}\left(\frac{8\theta}{\tan(2\theta)}\ln\frac{\lambda}{m_e} + \frac{8}{\tan(2\theta)}\int_0^\theta x\tan x dx + \frac{1}{2} + 6\theta\cot\theta + 2\ln\frac{\Lambda}{m_e}\right) \\
&\quad + \frac{2\pi^2\theta(q + q')_\kappa}{im_e c^5\sin(2\theta)}.
\end{aligned}
\tag{F.44}
$$

F.5 Integral for ladder diagram

For the integral (4.50)

$$b(\boldsymbol{p},\boldsymbol{q},\boldsymbol{k}) = \int d^4h[\tilde{h}^2 - 2(\tilde{q}\cdot\tilde{h})]^{-1}[\tilde{h}^2 + 2(\tilde{p}\cdot\tilde{h})]^{-1}[\tilde{h}^2 - \lambda^2c^4]^{-1} \\ \times [\tilde{h}^2 + 2(\tilde{h}\cdot\tilde{k}) + \tilde{k}^2 - \lambda^2c^4]^{-1},$$

we follow the calculation method from [13] (see also [11]). First we use equation (F.5) with parameters

$$\begin{aligned} a &= \tilde{h}^2 + 2(\tilde{k}\cdot\tilde{h}) + \tilde{k}^2 - \lambda^2c^4, \\ b &= \tilde{h}^2 - 2(\tilde{q}\cdot\tilde{h}), \\ c &= \tilde{h}^2 + 2(\tilde{p}\cdot\tilde{h}), \\ d &= \tilde{h}^2 - \lambda^2c^4, \end{aligned}$$

to obtain

$$\begin{aligned} b(\boldsymbol{p},\boldsymbol{q},\boldsymbol{k}) &= 6\int d^4h\int_0^1 dx\int_0^1 dy\int_0^1 xz^2dz \\ &\quad \times [(\tilde{h}^2 + 2(\tilde{k}\cdot\tilde{h}) + k^2 - \lambda^2c^4)z(1-x) \\ &\qquad + (\tilde{h}^2 - 2(\tilde{q}\cdot\tilde{h}))xyz + (\tilde{h}^2 + 2(\tilde{p}\cdot\tilde{h}))xz(1-y) + (\tilde{h}^2 - \lambda^2c^4)(1-z)]^{-4} \\ &= 6\int d^4h\int_0^1 dx\int_0^1 dy\int_0^1 xz^2dz \\ &\quad \times [\tilde{h}^2 - 2\tilde{h}\cdot(-\tilde{k}z(1-x) + \tilde{q}xyz - \tilde{p}xz(1-y)) + \tilde{k}^2z(1-x) + \lambda^2c^4(zx-1)]^{-4} \\ &= 6\int d^4h\int_0^1 dx\int_0^1 dy\int_0^1 \frac{xz^2dz}{[\tilde{h}^2 - 2(\tilde{h}\cdot\tilde{p}_x)z - \Delta]^4}, \end{aligned}$$

where

$$\begin{aligned} \Delta &\equiv \lambda^2c^4(1-zx) - \tilde{k}^2z(1-x), \\ \tilde{p}_x &= -\tilde{k}(1-x) - \tilde{p}x(1-y) + \tilde{q}xy = -\tilde{k}(1-x) - x\tilde{p}_y, \\ \tilde{p}_y &= \tilde{p}(1-y) - \tilde{q}y. \end{aligned}$$

Equation (F.11) permits us to take the integral with respect to $\tilde{h}$, i. e.,

$$b(\boldsymbol{p},\boldsymbol{q},\boldsymbol{k}) = -\frac{\pi^2}{ic^3}\int_0^1 dx\int_0^1 dy\int_0^1 \frac{xz^2dz}{(z^2\tilde{p}_x^2 + \Delta)^2}.$$

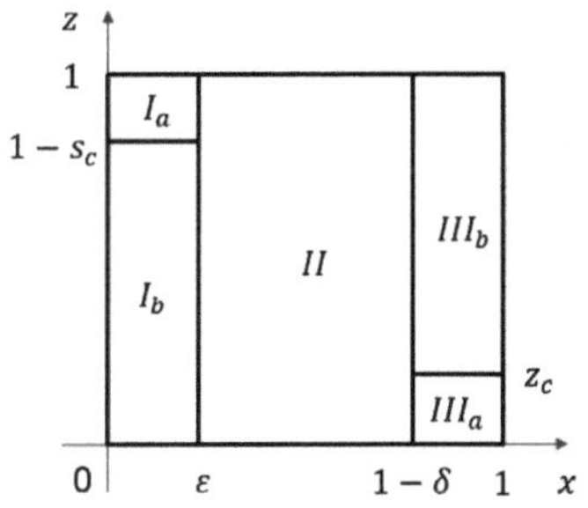

Figure F.2: Splitting of the integration area in the x-z plane.

Using (F.37) and (F.38), we obtain

$$(\tilde{k}\cdot\tilde{p}_y) = (\tilde{k}\cdot\tilde{p})(1-y) - (\tilde{k}\cdot\tilde{q})y = \frac{\tilde{k}^2}{2}(1-y) + \frac{\tilde{k}^2}{2}y = \frac{\tilde{k}^2}{2}.$$

$$\begin{aligned}\tilde{p}_x^2 &= (x\tilde{p}_y + \tilde{k}(1-x))^2 = x^2\tilde{p}_y^2 + \tilde{k}^2(1-x)^2 + 2x(1-x)(\tilde{p}_y\cdot\tilde{k})\\ &= x^2\tilde{p}_y^2 + \tilde{k}^2 - 2\tilde{k}^2x + \tilde{k}^2x^2 + \tilde{k}^2x - \tilde{k}^2x^2 = x^2\tilde{p}_y^2 + \tilde{k}^2(1-x),\end{aligned}$$

$$b(\boldsymbol{p},\boldsymbol{q},\boldsymbol{k}) = -\frac{\pi^2}{ic^3}\int_0^1 dx\int_0^1 dy\int_0^1 \frac{xz^2dz}{[z^2(x^2\tilde{p}_y^2 + \tilde{k}^2(1-x)) + \lambda^2c^4(1-zx) - \tilde{k}^2z(1-x)]^2}. \tag{F.45}$$

Even if the parameter λ is small, the term $\lambda^2c^4(1-zx)$ cannot be neglected[11] when $x\to 0$, $z\to 0$, when $x\to 1$, $z\to 0$ and when $x\to 0$, $z\to 1$. Therefore, we are going to break the segment of integration with respect to x into three regions: $0<x<\epsilon$, $\epsilon<x<1-\delta$ and $1-\delta<x<1$, where ϵ and δ are small, but large enough so that in the interval $\epsilon<x<1-\delta$ one can ignore the term $\lambda^2c^4(1-zx)$ (see Figure F.2). So, our integral consists of three parts, i. e.,

$$b(\boldsymbol{p},\boldsymbol{q},\boldsymbol{k}) = L_{\mathrm{I}} + L_{\mathrm{II}} + L_{\mathrm{III}} = \int_0^{\epsilon}\cdots dx + \int_{\epsilon}^{1-\delta}\cdots dx + \int_{1-\delta}^{1}\cdots dx.$$

F.5.1 Calculation of L_{I}

In the integral L_{I} we make the substitution $s = 1-z$ to obtain

$$\begin{aligned}L_{\mathrm{I}} &= \frac{\pi^2}{ic^3}\int_0^{\epsilon} dx\int_0^1 dy\int_0^1 dsx(1-s)^2\\ &\quad\times\left[(1-s)^2(x^2\tilde{p}_y^2 + \tilde{k}^2(1-x)) + \lambda^2c^4(1-(1-s)x) - \tilde{k}^2(1-s)(1-x)\right]^{-2}\end{aligned}$$

11 Because other terms in the denominator can be even smaller.

and we break the s-integration segment into two regions $0 \le s < s_c$ and $s_c \le s \le 1$, where the parameter s_c is small, but sufficiently large for us to neglect the λ-term in the latter region. Then

$$L_{\mathrm{I}} = L_{\mathrm{Ia}} + L_{\mathrm{Ib}} = \int_0^{s_c} \cdots ds + \int_{s_c}^{1} \cdots ds.$$

In the first integral we take into account that $s^2 \ll s \ll 1$ and $x^2 \ll x \ll 1$. Then we have

$$\begin{aligned}
L_{\mathrm{Ia}} &\approx -\frac{\pi^2}{ic^3}\int_0^{\epsilon} dx \int_0^1 dy \int_0^{s_c} ds \\
&\times \frac{x}{[(1-2s)(x^2\tilde{p}_y^2 + \tilde{k}^2(1-x)) + \lambda^2c^4 - \tilde{k}^2(1-s)(1-x)]^2} \\
&\approx -\frac{\pi^2}{ic^3}\int_0^{\epsilon} dx \int_0^1 dy \int_0^{s_c} \frac{xds}{[(x^2\tilde{p}_y^2 + \lambda^2c^4) - (2x^2\tilde{p}_y^2 + \tilde{k}^2(1-x))s]^2} \\
&= -\frac{\pi^2}{ic^3}\int_0^{\epsilon} xdx \int_0^1 dy \\
&\times \left.\frac{1}{[2x^2\tilde{p}_y^2 + \tilde{k}^2(1-x)][(x^2\tilde{p}_y^2 + \lambda^2c^4) - (2x^2\tilde{p}_y^2 + \tilde{k}^2(1-x))s]}\right|_{s=0}^{s=s_c} \\
&= -\frac{\pi^2}{ic^3}\int_0^{\epsilon} xdx \int_0^1 dy \frac{1}{2x^2\tilde{p}_y^2 + \tilde{k}^2(1-x)} \\
&\times \left(\frac{1}{(x^2\tilde{p}_y^2 + \lambda^2c^4) - (2x^2\tilde{p}_y^2 + \tilde{k}^2(1-x))s_c} - \frac{1}{x^2\tilde{p}_y^2 + \lambda^2c^4}\right) \\
&= -\frac{\pi^2}{ic^3}\int_0^{\epsilon} xdx \int_0^1 dy \frac{1}{2x^2\tilde{p}_y^2 + \tilde{k}^2(1-x)} \\
&\times \frac{(2x^2\tilde{p}_y^2 + \tilde{k}^2(1-x))s_c}{[(x^2\tilde{p}_y^2 + \lambda^2c^4) - (2x^2\tilde{p}_y^2 + \tilde{k}^2(1-x))s_c][x^2\tilde{p}_y^2 + \lambda^2c^4]} \\
&= -\frac{\pi^2}{ic^3}\int_0^{\epsilon} xdx \int_0^1 dy \frac{s_c}{[(x^2\tilde{p}_y^2 + \lambda^2c^4) - (2x^2\tilde{p}_y^2 + \tilde{k}^2(1-x))s_c][x^2\tilde{p}_y^2 + \lambda^2c^4]} \\
&\approx -\frac{\pi^2}{ic^3}\int_0^{\epsilon} xdx \int_0^1 dy \frac{s_c}{[(x^2\tilde{p}_y^2 + \lambda^2c^4) - \tilde{k}^2s_c][x^2\tilde{p}_y^2 + \lambda^2c^4]} \\
&\approx \frac{\pi^2}{ic^3\tilde{k}^2}\int_0^1 dy \int_0^{\epsilon} \frac{xdx}{x^2\tilde{p}_y^2 + \lambda^2c^4} - \frac{\pi^2}{ic^3}\int_0^1 dy \int_0^{\epsilon} \frac{xdx}{\tilde{k}^4 s_c}
\end{aligned}$$

$$
\begin{aligned}
&= \frac{\pi^2}{ic^3\tilde{k}^2}\int_0^1 \frac{dy}{2\tilde{p}_y^2}\ln(x^2\tilde{p}_y^2+\lambda^2c^4)\Big|_{x=0}^{x=\epsilon} - \frac{\pi^2}{ic^3}\int_0^1 dy\int_0^\epsilon \frac{xdx}{\tilde{k}^4 s_c}\\
&= \frac{\pi^2}{ic^3\tilde{k}^2}\int_0^1 \frac{dy}{2\tilde{p}_y^2}\ln\frac{\epsilon^2\tilde{p}_y^2}{\lambda^2c^4} - \frac{\pi^2}{ic^3}\int_0^1 dy\int_0^\epsilon \frac{xdx}{\tilde{k}^4 s_c}.
\end{aligned}
$$

In the second part, L_{Ib}, we drop the λ-term and obtain

$$
\begin{aligned}
L_{\mathrm{Ib}} &\approx -\frac{\pi^2}{ic^3}\int_0^\epsilon dx\int_0^1 dy\int_{s_c}^1 ds\frac{x(1-s)^2}{[(1-s)^2(x^2\tilde{p}_y^2+\tilde{k}^2(1-x))-\tilde{k}^2(1-s)(1-x)]^2}\\
&= -\frac{\pi^2}{ic^3}\int_0^\epsilon dx\int_0^1 dy\int_{s_c}^1 \frac{xds}{[x^2\tilde{p}_y^2-(x^2\tilde{p}_y^2+\tilde{k}^2(1-x))s]^2}\\
&= -\frac{\pi^2}{ic^3}\int_0^\epsilon xdx\int_0^1 dy\frac{1}{[x^2\tilde{p}_y^2+\tilde{k}^2(1-x)][x^2\tilde{p}_y^2-(x^2p_y^2+\tilde{k}^2(1-x))s]}\Bigg|_{s=s_c}^{s=1}\\
&= \frac{\pi^2}{ic^3}\int_0^\epsilon dx\int_0^1 dy\frac{x}{x^2\tilde{p}_y^2+\tilde{k}^2(1-x)}\left(\frac{1}{\tilde{k}^2(1-x)}+\frac{1}{x^2\tilde{p}_y^2-(x^2\tilde{p}_y^2+\tilde{k}^2(1-x))s_c}\right)\\
&= \frac{\pi^2}{ic^3}\int_0^\epsilon dx\int_0^1 dy\frac{x}{x^2\tilde{p}_y^2+\tilde{k}^2(1-x)}\cdot\frac{x^2\tilde{p}_y^2-(x^2\tilde{p}_y^2+\tilde{k}^2(1-x))s_c+\tilde{k}^2(1-x)}{\tilde{k}^2(1-x)[x^2\tilde{p}_y^2-(x^2\tilde{p}_y^2+\tilde{k}^2(1-x))s_c]}\\
&= \frac{\pi^2}{ic^3}\int_0^1 dy\int_0^\epsilon \frac{xdx}{\tilde{k}^2(1-x)[x^2\tilde{p}_y^2-(x^2\tilde{p}_y^2+\tilde{k}^2(1-x))s_c]}\\
&\approx \frac{\pi^2}{ic^3}\int_0^1 dy\int_0^\epsilon \frac{xdx}{\tilde{k}^4 s_c}.
\end{aligned}
$$

The sum of both parts

$$
L_{\mathrm{I}} = L_{\mathrm{Ia}} + L_{\mathrm{Ib}} \approx \frac{\pi^2}{ic^3\tilde{k}^2}\int_0^1 \frac{dy}{2\tilde{p}_y^2}\ln\left(\frac{\epsilon^2\tilde{p}_y^2}{\lambda^2c^4}\right)
$$

does not depend on s_c, as expected.

F.5.2 Calculation of L_{II}

In the second integration region we neglect the λ-term, so

$$
L_{\mathrm{II}} \approx -\frac{\pi^2}{ic^3}\int_\epsilon^{1-\delta} dx\int_0^1 dy\int_0^1 \frac{xdz}{[z(x^2\tilde{p}_y^2+\tilde{k}^2(1-x))-\tilde{k}^2(1-x)]^2}.
$$

We use the table integrals

$$\int \frac{dz}{(az+b)^2} = -\frac{1}{a(az+b)} + \text{const},$$
$$\int \frac{dx}{x(1-x)} = \ln(x) - \ln(x-1) + \text{const}$$

and obtain

$$\begin{aligned}
L_{\text{II}} &= \frac{\pi^2}{ic^3}\int_{\epsilon}^{1-\delta} dx \int_0^1 dy \frac{x}{[x^2\tilde{p}_y^2 + \tilde{k}^2(1-x)][z(x^2\tilde{p}_y^2 + \tilde{k}^2(1-x)) - \tilde{k}^2(1-x)]}\Bigg|_{z=0}^{z=1} \\
&= \frac{\pi^2}{ic^3}\int_{\epsilon}^{1-\delta} dx \int_0^1 dy \frac{x}{x^2\tilde{p}_y^2 + \tilde{k}^2(1-x)}\left(\frac{1}{x^2\tilde{p}_y^2} + \frac{1}{\tilde{k}^2(1-x)}\right) \\
&= \frac{\pi^2}{ic^3\tilde{k}^2}\int_{\epsilon}^{1-\delta} dx \int_0^1 dy \frac{1}{x\tilde{p}_y^2(1-x)} = \frac{\pi^2}{ic^3\tilde{k}^2}\int_0^1 \frac{dy}{\tilde{p}_y^2}[\ln(x) - \ln(x-1)]\Big|_{x=\epsilon}^{x=1-\delta} \\
&\approx \frac{\pi^2}{ic^3\tilde{k}^2}\int_0^1 \frac{dy}{\tilde{p}_y^2}[-\ln(\delta) - \ln(-1) - \ln(\epsilon) + \ln(-1)] \approx -\frac{\pi^2\ln(\delta\epsilon)}{ic^3\tilde{k}^2}\int_0^1 \frac{dy}{\tilde{p}_y^2}.
\end{aligned}$$

F.5.3 Calculation of L_{III}

In the third integral we substitute $x \to 1-x$ and obtain

$$\begin{aligned}
L_{\text{III}} &= \frac{\pi^2}{ic^3}\int_{\delta}^{0} dx \int_0^1 dy \int_0^1 dz \frac{(1-x)z^2}{[z^2((1-x)^2\tilde{p}_y^2 + \tilde{k}^2x) + \lambda^2c^4(1-z(1-x)) - \tilde{k}^2zx]^2} \\
&\approx \frac{\pi^2}{ic^3}\int_{\delta}^{0} dx \int_0^1 dy \int_0^1 \frac{z^2dz}{[z^2\tilde{p}_y^2 + z^2\tilde{k}^2x + \lambda^2c^4(1-z) - \tilde{k}^2zx]^2} \\
&= -\frac{\pi^2}{ic^3\tilde{k}^2}\int_0^1 dy \int_0^1 \frac{zdz}{(z-1)}\left(\frac{1}{z^2\tilde{p}_y^2 + z^2\tilde{k}^2x + \lambda^2c^4(1-z) - \tilde{k}^2zx}\right)\Bigg|_{x=\delta}^{x=0} \\
&= -\frac{\pi^2}{ic^3\tilde{k}^2}\int_0^1 dy \int_0^1 \frac{zdz}{(z-1)}\left(\frac{1}{z^2\tilde{p}_y^2 + \lambda^2c^4(1-z)} - \frac{1}{z^2\tilde{p}_y^2 + z^2\tilde{k}^2\delta + \lambda^2c^4(1-z) - \tilde{k}^2z\delta}\right) \\
&= -\frac{\pi^2}{ic^3\tilde{k}^2}\int_0^1 dy \int_0^1 \frac{zdz}{(z-1)}\frac{z^2\tilde{p}_y^2 + z^2\tilde{k}^2\delta + \lambda^2c^4(1-z) - \tilde{k}^2z\delta - z^2\tilde{p}_y^2 - \lambda^2c^4(1-z)}{[z^2\tilde{p}_y^2 + \lambda^2c^4(1-z)][z^2\tilde{p}_y^2 + z^2\tilde{k}^2\delta + \lambda^2c^4(1-z) - \tilde{k}^2z\delta]} \\
&= -\frac{\pi^2\delta}{ic^3}\int_0^1 dy \int_0^1 dz \frac{z^2}{[z^2\tilde{p}_y^2 + \lambda^2c^4(1-z)][z^2\tilde{p}_y^2 + z^2\tilde{k}^2\delta + \lambda^2c^4(1-z) - \tilde{k}^2z\delta]}.
\end{aligned}$$

Then we break the region of z-integration into two subregions $0 \le z < z_c$ and $z_c \le z < 1$, where the boundary z_c is selected so that $\lambda^2 c^4 \ll z_c^2 \tilde{p}_y^2 \ll \tilde{k}^2 z_c \delta$. Then

$$L_{\text{III}} = L_{\text{IIIa}} + L_{\text{IIIb}},$$

where

$$\begin{aligned}
L_{\text{IIIa}} &= -\frac{\pi^2\delta}{ic^3}\int_0^1 dy \int_0^{z_c} dz \frac{z^2}{[z^2\tilde{p}_y^2 + \lambda^2 c^4(1-z)][z^2\tilde{p}_y^2 + z^2\tilde{k}^2\delta + \lambda^2 c^4(1-z) - \tilde{k}^2 z\delta]} \\
&\approx -\frac{\pi^2\delta}{ic^3}\int_0^1 dy \int_0^{z_c} \frac{z^2 dz}{(z^2\tilde{p}_y^2 + \lambda^2 c^4)(\lambda^2 c^4 - \tilde{k}^2 z\delta)} \\
&= \frac{\pi^2\delta}{ic^3}\int_0^1 dy \int_0^{z_c} \frac{dz}{\tilde{p}_y^2\lambda^2 c^4 + \tilde{k}^4\delta^2}\left(\frac{\lambda^2 c^4 + \tilde{k}^2 z\delta}{z^2\tilde{p}_y^2 + \lambda^2 c^4} - \frac{\lambda^2 c^4}{\lambda^2 c^4 - \tilde{k}^2 z\delta}\right) \\
&= \frac{\pi^2\delta}{ic^3}\int_0^1 \frac{dy}{\tilde{p}_y^2\lambda^2 c^4 + \tilde{k}^4\delta^2}\cdot\frac{\tilde{k}^2\delta}{2\tilde{p}_y^2}\ln(\lambda^2 c^4 + z^2\tilde{p}_y^2)\Big|_{z=0}^{z=z_c} \\
&= \frac{\pi^2\delta}{ic^3}\int_0^1 \frac{dy}{\tilde{p}_y^2\lambda^2 c^4 + \tilde{k}^4\delta^2}\left(\frac{\tilde{k}^2\delta}{2\tilde{p}_y^2}\ln(\lambda^2 c^4 + z_c^2\tilde{p}_y^2) - \frac{k^2\delta}{2\tilde{p}_y^2}\ln(\lambda^2 c^4)\right) \\
&\approx \frac{\pi^2}{2ic^3\tilde{k}^2}\int_0^1 \frac{dy}{\tilde{p}_y^2}\ln\left(\frac{z_c^2\tilde{p}_y^2}{\lambda^2 c^4}\right),
\end{aligned}$$

$$\begin{aligned}
L_{\text{IIIb}} &= -\frac{\pi^2\delta}{ic^3}\int_0^1 dy \int_{z_c}^1 dz \frac{z^2}{[z^2\tilde{p}_y^2 + \lambda^2 c^4(1-z)][z^2\tilde{p}_y^2 + z^2\tilde{k}^2\delta + \lambda^2 c^4(1-z) - \tilde{k}^2 z\delta]} \\
&\approx -\frac{\pi^2\delta}{ic^3}\int_0^1 dy \int_{z_c}^1 \frac{dz}{\tilde{p}_y^2 z[z(\tilde{p}_y^2 + \tilde{k}^2\delta) - \tilde{k}^2\delta]} \\
&= \frac{\pi^2\delta}{ic^3}\int_0^1 \frac{dy}{\tilde{p}_y^2}\frac{1}{\tilde{k}^2\delta}[\ln(z) - \ln(z(\tilde{p}_y^2 + \tilde{k}^2\delta) - \tilde{k}^2\delta)]\Big|_{z=z_c}^{z=1} \\
&= \frac{\pi^2}{ic^3}\int_0^1 \frac{dy}{\tilde{p}_y^2}\frac{1}{\tilde{k}^2}[-\ln(\tilde{p}_y^2) - \ln(z_c) + \ln(z_c(\tilde{p}_y^2 + \tilde{k}^2\delta) - \tilde{k}^2\delta)] \\
&\approx \frac{\pi^2}{ic^3\tilde{k}^2}\int_0^1 \frac{dy}{\tilde{p}_y^2}\ln\left(\frac{-\tilde{k}^2\delta}{\tilde{p}_y^2 z_c}\right).
\end{aligned}$$

Adding together these two contributions, we obtain

$$L_{\text{III}} = L_{\text{IIIa}} + L_{\text{IIIb}}$$

$$= \frac{\pi^2}{ic^3\tilde{k}^2}\int_0^1 \frac{dy}{\tilde{p}_y^2}\ln\left(\frac{-\tilde{k}^2\delta}{\tilde{p}_y^2 z_c}\right) + \frac{\pi^2}{ic^3\tilde{k}^2}\int_0^1 \frac{dy}{\tilde{p}_y^2}\ln\left(\frac{z_c\tilde{p}_y}{\lambda c^2}\right)$$

$$= \frac{\pi^2}{ic^3\tilde{k}^2}\int_0^1 \frac{dy}{\tilde{p}_y^2}\ln\left(-\frac{\tilde{k}^2\delta}{\tilde{p}_y\lambda c^2}\right).$$

F.5.4 Complete integral

Collecting all the above terms, we obtain equation (A20) from [13], i. e.,

$$\begin{aligned} b(\boldsymbol{p},\boldsymbol{q},\boldsymbol{k}) &= L_{\mathrm{I}} + L_{\mathrm{II}} + L_{\mathrm{III}} \\ &\approx \frac{\pi^2}{2ic^3\tilde{k}^2}\int_0^1 \frac{dy}{\tilde{p}_y^2}\ln\left(\frac{\epsilon^2\tilde{p}_y^2}{\lambda^2c^4}\right) - \frac{\pi^2}{2ic^3\tilde{k}^2}\int_0^1 \frac{dy}{\tilde{p}_y^2}\ln(\delta^2\epsilon^2) + \frac{\pi^2}{2ic^3\tilde{k}^2}\int_0^1 \frac{dy}{\tilde{p}_y^2}\ln\left(\frac{\tilde{k}^4\delta^2}{\tilde{p}_y^2\lambda^2c^4}\right) \\ &= \frac{\pi^2}{2ic^3\tilde{k}^2}\int_0^1 \frac{dy}{\tilde{p}_y^2}\ln\left(\frac{\epsilon^2\tilde{p}_y^2}{\lambda^2c^4}\cdot\frac{1}{\delta^2\epsilon^2}\cdot\frac{\tilde{k}^4\delta^2}{\tilde{p}_y^2\lambda^2c^4}\right) = \frac{\pi^2}{ic^3\tilde{k}^2}\ln\left(\frac{\tilde{k}^2}{\lambda^2c^4}\right)\int_0^1 \frac{dy}{\tilde{p}_y^2}. \end{aligned} \tag{F.46}$$

G Scattering matrix in $(v/c)^2$ approximation

In formulas (3.25) and (4.47) we expressed contributions to the S-operator through functions $\mathcal{U}^\mu$ and $\mathcal{W}^\mu$. For further analysis, it would be convenient to rewrite these formulas using the $(v/c)^2$ approximation from Appendix B.9. In Appendix G.1 we will do this in the second perturbation order (3.25), and in Appendix G.2 we will consider the fourth-order formula (4.47).

G.1 Second perturbation order

The coefficient function (3.25) of the scattering phase operator is composed of three components, i. e.,

$$\phi_2 = \phi_2^A + \phi_2^B + \phi_2^C. \tag{G.1}$$

Using properties of the Pauli matrices from Appendix B.1, the formula

$$\begin{aligned}
[\boldsymbol{\sigma}_{\mathrm{pr}} \times \boldsymbol{k}] \cdot [\boldsymbol{\sigma}_{\mathrm{el}} \times \boldsymbol{k}] &= [[\boldsymbol{\sigma}_{\mathrm{el}} \times \boldsymbol{k}] \times \boldsymbol{\sigma}_{\mathrm{pr}}] \cdot \boldsymbol{k} \\
&= (\boldsymbol{k}(\boldsymbol{\sigma}_{\mathrm{el}} \cdot \boldsymbol{\sigma}_{\mathrm{pr}}) - \boldsymbol{\sigma}_{\mathrm{pr}}(\boldsymbol{\sigma}_{\mathrm{el}} \cdot \boldsymbol{k})) \cdot \boldsymbol{k} \\
&= k^2(\boldsymbol{\sigma}_{\mathrm{el}} \cdot \boldsymbol{\sigma}_{\mathrm{pr}}) - (\boldsymbol{\sigma}_{\mathrm{pr}} \cdot \boldsymbol{k})(\boldsymbol{\sigma}_{\mathrm{el}} \cdot \boldsymbol{k})
\end{aligned}$$

and approximations (B.74)–(B.78), we obtain

$$\begin{aligned}
\phi_2^A &\equiv -\frac{e^2\hbar^2}{(2\pi\hbar)^3} \frac{m_p m_e c^4}{\sqrt{\omega_{\boldsymbol{q}}\omega_{\boldsymbol{q}+\boldsymbol{k}}\Omega_{\boldsymbol{p}}\Omega_{\boldsymbol{p}-\boldsymbol{k}}}} \frac{1}{k^2} \mathcal{U}^0((\boldsymbol{q}+\boldsymbol{k})\sigma', \boldsymbol{q}\sigma) \mathcal{W}^0((\boldsymbol{p}-\boldsymbol{k})\tau', \boldsymbol{p}\tau) \\
&\approx -\frac{e^2\hbar^2}{(2\pi\hbar)^3 k^2}\left(1 - \frac{p^2}{2m_p^2c^2} + \frac{\boldsymbol{p}\cdot\boldsymbol{k}}{2m_p^2c^2} - \frac{k^2}{4m_p^2c^2} - \frac{q^2}{2m_e^2c^2} - \frac{\boldsymbol{q}\cdot\boldsymbol{k}}{2m_e^2c^2} - \frac{k^2}{4m_e^2c^2}\right) \\
&\quad \times \chi_{\sigma'}^{(\mathrm{el})\dagger}\left(1 + \frac{(2\boldsymbol{q}+\boldsymbol{k})^2 + 2i\boldsymbol{\sigma}_{\mathrm{el}} \cdot [\boldsymbol{k} \times \boldsymbol{q}]}{8m_e^2c^2}\right)\chi_\sigma^{(\mathrm{el})} \\
&\quad \times \chi_{\tau'}^{(\mathrm{pr})\dagger}\left(1 + \frac{(2\boldsymbol{p}-\boldsymbol{k})^2 - 2i\boldsymbol{\sigma}_{\mathrm{pr}} \cdot [\boldsymbol{k} \times \boldsymbol{p}]}{8m_p^2c^2}\right)\chi_\tau^{(\mathrm{pr})} \\
&\approx -\frac{e^2\hbar^2}{(2\pi\hbar)^3}\chi_{\sigma'}^{(\mathrm{el})\dagger}\chi_{\tau'}^{(\mathrm{pr})\dagger} \\
&\quad \times \left(\frac{1}{k^2} - \frac{1}{8m_p^2c^2} - \frac{1}{8m_e^2c^2} - \frac{i\boldsymbol{\sigma}_{\mathrm{pr}} \cdot [\boldsymbol{k} \times \boldsymbol{p}]}{4m_p^2c^2k^2} + \frac{i\boldsymbol{\sigma}_{\mathrm{el}} \cdot [\boldsymbol{k} \times \boldsymbol{q}]}{4m_e^2c^2k^2}\right)\chi_\sigma^{(\mathrm{el})}\chi_\tau^{(\mathrm{pr})},
\end{aligned}$$

$$\begin{aligned}
\phi_2^B &\equiv -\frac{e^2\hbar^2c^2}{(2\pi\hbar)^3} \frac{m_p m_e c^4}{\sqrt{\omega_{\boldsymbol{q}}\omega_{\boldsymbol{q}+\boldsymbol{k}}\Omega_{\boldsymbol{p}}\Omega_{\boldsymbol{p}-\boldsymbol{k}}}} \frac{\boldsymbol{\mathcal{U}}((\boldsymbol{q}+\boldsymbol{k})\sigma', \boldsymbol{q}\sigma) \cdot \boldsymbol{\mathcal{W}}((\boldsymbol{p}-\boldsymbol{k})\tau', \boldsymbol{p}\tau)}{((\omega_{\boldsymbol{q}+\boldsymbol{k}} - \omega_{\boldsymbol{q}})^2 - c^2k^2)} \\
&\approx -\frac{e^2\hbar^2}{(2\pi\hbar)^3(-k^2)}\chi_{\sigma'}^{(\mathrm{el})\dagger}\chi_{\tau'}^{(\mathrm{pr})\dagger}
\end{aligned}$$

https://doi.org/10.1515/9783110493207-011

$$\times (2\boldsymbol{q} + \boldsymbol{k} + i[\boldsymbol{\sigma}_{\text{el}} \times \boldsymbol{k}])(2\boldsymbol{p} - \boldsymbol{k} - i[\boldsymbol{\sigma}_{\text{pr}} \times \boldsymbol{k}])\frac{\chi_{\sigma}^{(\text{el})}\chi_{\tau}^{(\text{pr})}}{4m_e m_p c^2}$$

$$\approx \frac{e^2\hbar^2}{(2\pi\hbar)^3}\chi_{\sigma'}^{(\text{el})\dagger}\chi_{\tau'}^{(\text{pr})\dagger}$$

$$\times\left(\frac{\boldsymbol{p}\cdot\boldsymbol{q}}{m_p m_e c^2 k^2} - \frac{\boldsymbol{k}\cdot\boldsymbol{q}}{2m_p m_e c^2 k^2} + \frac{\boldsymbol{p}\cdot\boldsymbol{k}}{2m_p m_e c^2 k^2} - \frac{1}{4m_p m_e c^2} - \frac{i[\boldsymbol{\sigma}_{\text{pr}} \times \boldsymbol{k}]\cdot\boldsymbol{q}}{2m_p m_e c^2 k^2}\right.$$

$$\left. + \frac{i\boldsymbol{p}\cdot[\boldsymbol{\sigma}_{\text{el}} \times \boldsymbol{k}]}{2m_p m_e c^2 k^2} + \frac{(\boldsymbol{\sigma}_{\text{pr}}\cdot\boldsymbol{\sigma}_{\text{el}})}{4m_p m_e c^2} - \frac{(\boldsymbol{\sigma}_{\text{pr}}\cdot\boldsymbol{k})(\boldsymbol{\sigma}_{\text{el}}\cdot\boldsymbol{k})}{4m_p m_e c^2 k^2}\right)\chi_{\sigma}^{(\text{el})}\chi_{\tau}^{(\text{pr})},$$

$$\phi_2^C \equiv \frac{e^2\hbar^2 c^2}{(2\pi\hbar)^3}\frac{m_p m_e c^4}{\sqrt{\omega_{\boldsymbol{q}}\omega_{\boldsymbol{q}+\boldsymbol{k}}\Omega_{\boldsymbol{p}}\Omega_{\boldsymbol{p}-\boldsymbol{k}}}}\frac{1}{((\omega_{\boldsymbol{q}+\boldsymbol{k}} - \omega_{\boldsymbol{q}})^2 - c^2k^2)k^2}$$

$$\times (\boldsymbol{k}\cdot\mathcal{U}((\boldsymbol{q}+\boldsymbol{k})\sigma', \boldsymbol{q}\sigma))(\boldsymbol{k}\cdot\mathcal{W}((\boldsymbol{p}-\boldsymbol{k})\tau', \boldsymbol{p}\tau))$$

$$\approx -\frac{e^2\hbar^2}{(2\pi\hbar)^3 k^4}\frac{1}{4m_p m_e c^2}\chi_{\sigma'}^{(\text{el})\dagger}\chi_{\tau'}^{(\text{pr})\dagger}(2\boldsymbol{p}\cdot\boldsymbol{k} - k^2)(2\boldsymbol{q}\cdot\boldsymbol{k} + k^2)\chi_{\sigma}^{(\text{el})}\chi_{\tau}^{(\text{pr})}$$

$$= -\frac{e^2\hbar^2}{(2\pi\hbar)^3}\chi_{\sigma'}^{(\text{el})\dagger}\chi_{\tau'}^{(\text{pr})\dagger}$$

$$\times\left(\frac{(\boldsymbol{p}\cdot\boldsymbol{k})(\boldsymbol{q}\cdot\boldsymbol{k})}{m_p m_e c^2 k^4} - \frac{\boldsymbol{q}\cdot\boldsymbol{k}}{2m_p m_e c^2 k^2} + \frac{\boldsymbol{p}\cdot\boldsymbol{k}}{2m_p m_e c^2 k^2} - \frac{1}{4m_p m_e c^2}\right)\chi_{\sigma}^{(\text{el})}\chi_{\tau}^{(\text{pr})}.$$

Adding together these three terms, we obtain the coefficient function (G.1)

$$\phi_2(\boldsymbol{p}'\tau', \boldsymbol{q}'\sigma'; \boldsymbol{p}\tau, \boldsymbol{q}\sigma)$$

$$\approx \frac{e^2\hbar^2}{(2\pi\hbar)^3}\chi_{\sigma'}^{(\text{el})\dagger}\chi_{\tau'}^{(\text{pr})\dagger}\left(-\frac{1}{k^2} + \frac{1}{8m_p^2c^2} + \frac{1}{8m_e^2c^2} + \frac{\boldsymbol{p}\cdot\boldsymbol{q}}{m_p m_e c^2 k^2}\right.$$

$$- \frac{(\boldsymbol{p}\cdot\boldsymbol{k})(\boldsymbol{q}\cdot\boldsymbol{k})}{m_p m_e c^2 k^4} + \frac{i\boldsymbol{\sigma}_{\text{pr}}\cdot[\boldsymbol{k}\times\boldsymbol{p}]}{4m_p^2c^2k^2} - \frac{i\boldsymbol{\sigma}_{\text{el}}\cdot[\boldsymbol{k}\times\boldsymbol{q}]}{4m_e^2c^2k^2} - \frac{i\boldsymbol{\sigma}_{\text{pr}}\cdot[\boldsymbol{k}\times\boldsymbol{q}]}{2m_p m_e c^2 k^2}$$

$$\left. + \frac{i\boldsymbol{\sigma}_{\text{el}}\cdot[\boldsymbol{k}\times\boldsymbol{p}]}{2m_p m_e c^2 k^2} + \frac{(\boldsymbol{\sigma}_{\text{pr}}\cdot\boldsymbol{\sigma}_{\text{el}})}{4m_p m_e c^2} - \frac{(\boldsymbol{\sigma}_{\text{pr}}\cdot\boldsymbol{k})(\boldsymbol{\sigma}_{\text{el}}\cdot\boldsymbol{k})}{4m_p m_e c^2 k^2}\right)\chi_{\sigma}^{(\text{el})}\chi_{\tau}^{(\text{pr})}, \qquad \text{(G.2)}$$

where $\boldsymbol{k} \equiv \boldsymbol{q}' - \boldsymbol{q} = \boldsymbol{p} - \boldsymbol{p}'$ is the *transferred momentum*. The proton mass is much larger than the electron mass ($m_p \gg m_e$), so we are interested primarily in the terms that do not have m_p in the denominator:

$$\phi_2(\boldsymbol{p}'\tau', \boldsymbol{q}'\sigma'; \boldsymbol{p}\tau, \boldsymbol{q}\sigma) \approx \frac{e^2\hbar^2}{(2\pi\hbar)^3}\delta_{\tau\tau'}\chi_{\sigma'}^{(\text{el})\dagger}\left(-\frac{1}{k^2} + \frac{1}{8m_e^2c^2} - i\frac{\boldsymbol{\sigma}_{\text{el}}\cdot[\boldsymbol{k}\times\boldsymbol{q}]}{4m_e^2c^2k^2}\right)\chi_{\sigma}^{(\text{el})}. \qquad \text{(G.3)}$$

These terms survive in the limit $m_p \to \infty$ and give the dominant contribution to the electron–proton scattering amplitude.

G.2 Vertex contribution in fourth order

It will be convenient to break the S-matrix element (4.47) into two parts,

$$s_4^{(e)+(h)} = s_4^{\text{fin}} + s_4^{\text{div}},$$

where s_4^{fin} remains finite in the infrared limit $\lambda \to 0$, while s_4^{div} contains the infrared-divergent logarithm $\ln(\lambda/m_e)$. As usual, we introduce the 3-vector of the transferred momentum $\boldsymbol{k} \equiv \boldsymbol{q}' - \boldsymbol{q} = \boldsymbol{p} - \boldsymbol{p}'$. Then, for the infrared-finite part we obtain

$$\begin{aligned}
s_4^{\text{fin}} &\approx -\frac{ic^3\alpha^2}{4\pi^2\tilde{k}^2}\frac{m_p m_e c^4}{\sqrt{\Omega_{\boldsymbol{p}-\boldsymbol{k}}\Omega_{\boldsymbol{p}}\omega_{\boldsymbol{q}+\boldsymbol{k}}\omega_{\boldsymbol{q}}}} \\
&\quad \times \overline{u}(\boldsymbol{q}+\boldsymbol{k},\sigma')\left[2\gamma_\kappa\left(1-\frac{\tilde{k}^2}{12m_e^2c^4}\right) - \frac{(\tilde{q}+\tilde{k})_\kappa + \tilde{q}_\kappa}{m_e c^2}\left(1+\frac{\tilde{k}^2}{6m_e^2c^4}\right)\right]u(\boldsymbol{q},\sigma) \\
&\quad \times \mathcal{W}^\kappa((\boldsymbol{p}-\boldsymbol{k})\tau',\boldsymbol{p}\tau) \\
&= -\frac{ic^3\alpha^2}{4\pi^2\tilde{k}^2}\frac{m_p m_e c^4}{\sqrt{\Omega_{\boldsymbol{p}-\boldsymbol{k}}\Omega_{\boldsymbol{p}}\omega_{\boldsymbol{q}+\boldsymbol{k}}\omega_{\boldsymbol{q}}}} \\
&\quad \times \left[2\left(1-\frac{\tilde{k}^2}{12m_e^2c^4}\right)\mathcal{U}^0((\boldsymbol{q}+\boldsymbol{k})\sigma',\boldsymbol{q}\sigma)\mathcal{W}^0((\boldsymbol{p}-\boldsymbol{k})\tau',\boldsymbol{p}\tau)\right. \\
&\quad - 2\left(1-\frac{\tilde{k}^2}{12m_e^2c^4}\right)\boldsymbol{\mathcal{U}}((\boldsymbol{q}+\boldsymbol{k})\sigma',\boldsymbol{q}\sigma)\cdot\boldsymbol{\mathcal{W}}((\boldsymbol{p}-\boldsymbol{k})\tau',\boldsymbol{p}\tau) \\
&\quad - \left(1+\frac{\tilde{k}^2}{6m_e^2c^4}\right)\frac{\omega_{\boldsymbol{q}+\boldsymbol{k}}+\omega_{\boldsymbol{q}}}{m_e c^2}\overline{u}(\boldsymbol{q}+\boldsymbol{k},\sigma')u(\boldsymbol{q},\sigma)\mathcal{W}^0((\boldsymbol{p}-\boldsymbol{k})\tau',\boldsymbol{p}\tau) \\
&\quad \left. + \left(1+\frac{\tilde{k}^2}{6m_e^2c^4}\right)\frac{(2\boldsymbol{q}+\boldsymbol{k})\cdot\boldsymbol{\mathcal{W}}((\boldsymbol{p}-\boldsymbol{k})\tau',\boldsymbol{p}\tau)}{m_e c}\overline{u}(\boldsymbol{q}+\boldsymbol{k},\sigma')u(\boldsymbol{q},\sigma)\right] \\
&\equiv -\frac{ic^3\alpha^2}{4\pi^2\tilde{k}^2}\frac{m_p m_e c^4}{\sqrt{\Omega_{\boldsymbol{p}-\boldsymbol{k}}\Omega_{\boldsymbol{p}}\omega_{\boldsymbol{q}+\boldsymbol{k}}\omega_{\boldsymbol{q}}}}(\mathfrak{s}_4^{\text{fin}(1)} + \mathfrak{s}_4^{\text{fin}(2)} + \mathfrak{s}_4^{\text{fin}(3)} + \mathfrak{s}_4^{\text{fin}(4)}).
\end{aligned}$$

Next we use formulas from Appendices B.6 and B.9[1] and calculate in turn all four contributions. As in the preceding section, we will take the limit $m_p \to \infty$, but in the part $\mathfrak{s}_4^{\text{fin}(2)}$ we also leave a term which contains both factors $\boldsymbol{\sigma}_{\text{el}}$ (= electron spin) and $\boldsymbol{p}$ (= proton momentum). Although this term vanishes in the limit $m_p \to \infty$, we will need it in the third volume to derive the so-called anomalous magnetic moment of the electron. We have

$$\mathfrak{s}_4^{\text{fin}(1)} \approx 2\left(1+\frac{k^2}{12m_e^2c^2}\right)\mathcal{U}^0((\boldsymbol{q}+\boldsymbol{k})\sigma',\boldsymbol{q}\sigma)\mathcal{W}^0((\boldsymbol{p}-\boldsymbol{k})\tau',\boldsymbol{p}\tau)$$

1 For example, in these approximations we can set $\mathcal{W}^0 \approx \delta_{\tau\tau'}$, $\boldsymbol{\mathcal{W}} \approx 0$ and $\tilde{k}^2 \approx -c^2k^2$. In particular, this means that $\mathfrak{s}_4^{\text{fin}(4)} \approx 0$.

$$\approx \chi_{\sigma'}^{(\mathrm{el})\dagger}\chi_{\tau'}^{(\mathrm{pr})\dagger}2\left(1+\frac{k^2}{12m_e^2c^2}\right)\left(1+\frac{(2\boldsymbol{q}+\boldsymbol{k})^2+2i\boldsymbol{\sigma}_{\mathrm{el}}\cdot[\boldsymbol{k}\times\boldsymbol{q}]}{8m_e^2c^2}\right)$$
$$\times\left(1+\frac{(2\boldsymbol{p}-\boldsymbol{k})^2-2i\boldsymbol{\sigma}_{\mathrm{pr}}\cdot[\boldsymbol{k}\times\boldsymbol{p}]}{8m_p^2c^2}\right)\chi_{\tau}^{(\mathrm{pr})}\chi_{\sigma}^{(\mathrm{el})}$$
$$=2\chi_{\sigma'}^{(\mathrm{el})\dagger}\chi_{\tau'}^{(\mathrm{pr})\dagger}\left(1+\frac{k^2}{12m_e^2c^2}+\frac{q^2}{2m_e^2c^2}+\frac{\boldsymbol{q}\cdot\boldsymbol{k}}{2m_e^2c^2}+\frac{k^2}{8m_e^2c^2}\right.$$
$$\left.+\frac{i\boldsymbol{\sigma}_{\mathrm{el}}\cdot[\boldsymbol{k}\times\boldsymbol{q}]}{4m_e^2c^2}+\frac{p^2}{2m_p^2c^2}-\frac{\boldsymbol{p}\cdot\boldsymbol{k}}{2m_p^2c^2}+\frac{k^2}{8m_p^2c^2}-\frac{i\boldsymbol{\sigma}_{\mathrm{pr}}\cdot[\boldsymbol{k}\times\boldsymbol{p}]}{4m_p^2c^2}\right)\chi_{\tau}^{(\mathrm{pr})}\chi_{\sigma}^{(\mathrm{el})}$$
$$\approx 2\delta_{\tau\tau'}\chi_{\sigma'}^{(\mathrm{el})\dagger}\left(1+\frac{5k^2}{24m_e^2c^2}+\frac{q^2}{2m_e^2c^2}+\frac{\boldsymbol{q}\cdot\boldsymbol{k}}{2m_e^2c^2}+\frac{i\boldsymbol{\sigma}_{\mathrm{el}}\cdot[\boldsymbol{k}\times\boldsymbol{q}]}{4m_e^2c^2}\right)\chi_{\sigma}^{(\mathrm{el})},$$

$$\mathfrak{s}_4^{\mathrm{fin}(2)}\approx-2\left(1+\frac{k^2}{12m_e^2c^2}\right)\mathcal{U}((\boldsymbol{q}+\boldsymbol{k})\sigma',\boldsymbol{q}\sigma)\cdot\mathcal{W}((\boldsymbol{p}-\boldsymbol{k})\tau',\boldsymbol{p}\tau)$$
$$\approx-\frac{2}{4m_em_pc^2}\chi_{\sigma'}^{(\mathrm{el})\dagger}\chi_{\tau'}^{(\mathrm{pr})\dagger}\left(1+\frac{k^2}{12m_e^2c^2}\right)$$
$$\times(2\boldsymbol{q}+\boldsymbol{k}+i[\boldsymbol{\sigma}_{\mathrm{el}}\times\boldsymbol{k}])\cdot(2\boldsymbol{p}-\boldsymbol{k}-i[\boldsymbol{\sigma}_{\mathrm{pr}}\times\boldsymbol{k}])\chi_{\tau}^{(\mathrm{pr})}\chi_{\sigma}^{(\mathrm{el})}$$
$$\approx-\frac{i\delta_{\tau\tau'}}{m_em_pc^2}\chi_{\sigma'}^{(\mathrm{el})\dagger}([\boldsymbol{\sigma}_{\mathrm{el}}\times\boldsymbol{k}]\cdot\boldsymbol{p})\chi_{\sigma}^{(\mathrm{el})}.$$

To calculate the third term, we also use (B.4) and formulas

$$\overline{u}(\boldsymbol{q}+\boldsymbol{k},\sigma')u(\boldsymbol{q},\sigma)$$
$$=\chi_{\sigma'}^{(\mathrm{el})\dagger}\left[\sqrt{\omega_{\boldsymbol{q}+\boldsymbol{k}}+m_ec^2},-\sqrt{\omega_{\boldsymbol{q}+\boldsymbol{k}}-m_ec^2}\left(\frac{\boldsymbol{q}+\boldsymbol{k}}{|\boldsymbol{q}+\boldsymbol{k}|}\cdot\boldsymbol{\sigma}_{\mathrm{el}}\right)\right]$$
$$\times\begin{bmatrix}\sqrt{\omega_{\boldsymbol{q}}+m_ec^2}\\ \sqrt{\omega_{\boldsymbol{q}}-m_ec^2}(\frac{\boldsymbol{q}}{q}\cdot\boldsymbol{\sigma}_{\mathrm{el}})\end{bmatrix}\frac{\chi_{\sigma}^{(\mathrm{el})}}{2m_ec^2}$$
$$=\chi_{\sigma'}^{(\mathrm{el})\dagger}\left[\frac{\sqrt{\omega_{\boldsymbol{q}+\boldsymbol{k}}+m_ec^2}\sqrt{\omega_{\boldsymbol{q}}+m_ec^2}}{2m_ec^2}\right.$$
$$\left.-\frac{\sqrt{\omega_{\boldsymbol{q}+\boldsymbol{k}}-m_ec^2}\sqrt{\omega_{\boldsymbol{q}}-m_ec^2}}{2m_ec^2}\left(\frac{\boldsymbol{q}+\boldsymbol{k}}{|\boldsymbol{q}+\boldsymbol{k}|}\cdot\boldsymbol{\sigma}_{\mathrm{el}}\right)\left(\frac{\boldsymbol{q}}{q}\cdot\boldsymbol{\sigma}_{\mathrm{el}}\right)\right]\chi_{\sigma}^{(\mathrm{el})}$$
$$\approx\chi_{\sigma'}^{(\mathrm{el})\dagger}\left[\left(1+\frac{(\boldsymbol{q}+\boldsymbol{k})^2}{8m_e^2c^2}\right)\left(1+\frac{q^2}{8m_e^2c^2}\right)-\frac{|\boldsymbol{q}+\boldsymbol{k}|q}{4m_e^2c^2}\left(\frac{\boldsymbol{q}+\boldsymbol{k}}{|\boldsymbol{q}+\boldsymbol{k}|}\cdot\boldsymbol{\sigma}_{\mathrm{el}}\right)\left(\frac{\boldsymbol{q}}{q}\cdot\boldsymbol{\sigma}_{\mathrm{el}}\right)\right]\chi_{\sigma}^{(\mathrm{el})}$$
$$\approx\chi_{\sigma'}^{(\mathrm{el})\dagger}\left(1+\frac{k^2}{8m_e^2c^2}-\frac{i\boldsymbol{\sigma}_{\mathrm{el}}\cdot[\boldsymbol{k}\times\boldsymbol{q}]}{4m_e^2c^2}\right)\chi_{\sigma}^{(\mathrm{el})},$$

$$\frac{\omega_{\boldsymbol{q}+\boldsymbol{k}}+\omega_{\boldsymbol{q}}}{m_ec^2}\approx\frac{1}{m_ec^2}\left(m_ec^2+\frac{(\boldsymbol{q}+\boldsymbol{k})^2}{2m_e}+m_ec^2+\frac{q^2}{2m_e}\right)$$
$$=2\left(1+\frac{q^2}{2m_e^2c^2}+\frac{\boldsymbol{q}\cdot\boldsymbol{k}}{2m_e^2c^2}+\frac{k^2}{4m_e^2c^2}\right)$$

to obtain

$$
\begin{aligned}
\mathfrak{s}_4^{\mathrm{fin}(3)} &\approx -\left(1-\frac{k^2}{6m_e^2c^2}\right)\frac{\omega_{\boldsymbol{q}+\boldsymbol{k}}+\omega_{\boldsymbol{q}}}{m_ec^2}\overline{u}(\boldsymbol{q}+\boldsymbol{k},\sigma')u(\boldsymbol{q},\sigma)\mathcal{W}^0((\boldsymbol{p}-\boldsymbol{k})\tau',\boldsymbol{p}\tau)\\
&= -2\chi_{\sigma'}^{(\mathrm{el})\dagger}\chi_{\tau'}^{(\mathrm{pr})\dagger}\left(1-\frac{k^2}{6m_e^2c^2}\right)\left(1+\frac{q^2}{2m_ec^2}+\frac{\boldsymbol{q}\cdot\boldsymbol{k}}{2m_ec^2}+\frac{k^2}{4m_ec^2}\right)\\
&\quad\times\left(1+\frac{k^2}{8m_e^2c^2}-\frac{i\boldsymbol{\sigma}_{\mathrm{el}}\cdot[\boldsymbol{k}\times\boldsymbol{q}]}{4m_e^2c^2}\right)\left(1+\frac{(2\boldsymbol{p}-\boldsymbol{k})^2-2i\boldsymbol{\sigma}_{\mathrm{pr}}\cdot[\boldsymbol{k}\times\boldsymbol{p}]}{8m_p^2c^2}\right)\chi_\tau^{(\mathrm{pr})}\chi_\sigma^{(\mathrm{el})}\\
&\approx -2\delta_{\tau\tau'}\chi_{\sigma'}^{(\mathrm{el})\dagger}\left(1-\frac{k^2}{6m_e^2c^2}+\frac{q^2}{2m_e^2c^2}+\frac{\boldsymbol{q}\cdot\boldsymbol{k}}{2m_e^2c^2}+\frac{k^2}{4m_e^2c^2}+\frac{k^2}{8m_e^2c^2}\right.\\
&\quad\left.-\frac{i\boldsymbol{\sigma}_{\mathrm{el}}\cdot[\boldsymbol{k}\times\boldsymbol{q}]}{4m_e^2c^2}\right)\chi_\sigma^{(\mathrm{el})}\\
&= -2\delta_{\tau\tau'}\chi_{\sigma'}^{(\mathrm{el})\dagger}\left(1+\frac{q^2}{2m_e^2c^2}+\frac{\boldsymbol{q}\cdot\boldsymbol{k}}{2m_e^2c^2}+\frac{5k^2}{24m_e^2c^2}-\frac{i\boldsymbol{\sigma}_{\mathrm{el}}\cdot[\boldsymbol{k}\times\boldsymbol{q}]}{4m_e^2c^2}\right)\chi_\sigma^{(\mathrm{el})}.
\end{aligned}
$$

So, the sum $\mathfrak{s}_4^{\mathrm{fin}(1)}+\mathfrak{s}_4^{\mathrm{fin}(3)}$ determines the part independent of m_p, i. e.,

$$
\begin{aligned}
s_4^{\mathrm{fin}(1+3)} &= \frac{ic\alpha^2}{4\pi^2k^2}\frac{m_pm_ec^4}{\sqrt{\Omega_{\boldsymbol{p}-\boldsymbol{k}}\Omega_{\boldsymbol{p}}\omega_{\boldsymbol{q}+\boldsymbol{k}}\omega_{\boldsymbol{q}}}}\left(\mathfrak{s}_4^{\mathrm{fin}(1)}+\mathfrak{s}_4^{\mathrm{fin}(3)}\right)\\
&\approx \frac{2ic\alpha^2\delta_{\tau\tau'}}{4\pi^2k^2}\frac{m_pm_ec^4}{\sqrt{\Omega_{\boldsymbol{p}-\boldsymbol{k}}\Omega_{\boldsymbol{p}}\omega_{\boldsymbol{q}+\boldsymbol{k}}\omega_{\boldsymbol{q}}}}\chi_{\sigma'}^{(\mathrm{el})\dagger}\frac{2i\boldsymbol{\sigma}_{\mathrm{el}}\cdot[\boldsymbol{k}\times\boldsymbol{q}]}{4m_e^2c^2}\chi_\sigma^{(\mathrm{el})}\\
&\approx -\frac{\alpha^2\delta_{\tau\tau'}}{4\pi^2m_e^2c}\chi_{\sigma'}^{(\mathrm{el})\dagger}\frac{(\boldsymbol{\sigma}_{\mathrm{el}}\cdot[\boldsymbol{k}\times\boldsymbol{q}])}{k^2}\chi_\sigma^{(\mathrm{el})}
\end{aligned}
\tag{G.4}
$$

and the contribution due to the electron's anomalous magnetic moment is

$$
\begin{aligned}
s_4^{\mathrm{fin}(2)} &= \frac{ic\alpha^2\delta_{\tau\tau'}}{4\pi^2k^2}\frac{m_pm_ec^4}{\sqrt{\Omega_{\boldsymbol{p}-\boldsymbol{k}}\Omega_{\boldsymbol{p}}\omega_{\boldsymbol{q}+\boldsymbol{k}}\omega_{\boldsymbol{q}}}}\mathfrak{s}_4^{\mathrm{fin}(2)}\\
&\approx \frac{ic\alpha^2\delta_{\tau\tau'}}{4\pi^2k^2}\cdot\frac{-i\chi_{\sigma'}^{(\mathrm{el})\dagger}}{m_em_pc^2}([\boldsymbol{\sigma}_{\mathrm{el}}\times\boldsymbol{k}]\cdot\boldsymbol{p})\chi_\sigma^{(\mathrm{el})}\\
&= \frac{\alpha^2\delta_{\tau\tau'}}{4\pi^2m_em_pc}\chi_{\sigma'}^{(\mathrm{el})\dagger}\frac{\boldsymbol{\sigma}_{\mathrm{el}}\cdot[\boldsymbol{k}\times\boldsymbol{p}]}{k^2}\chi_\sigma^{(\mathrm{el})}.
\end{aligned}
\tag{G.5}
$$

For the λ-dependent part in (4.47) we use the nonrelativistic approximations (B.79)–(B.83) and obtain

$$
\begin{aligned}
s_4^{\mathrm{div}} &\approx \frac{i\alpha^2m_pm_ec^4}{3\pi^2m_e^2c\sqrt{\Omega_{\boldsymbol{p}-\boldsymbol{k}}\omega_{\boldsymbol{q}+\boldsymbol{k}}\Omega_{\boldsymbol{p}}\omega_{\boldsymbol{q}}}}\ln\left(\frac{\lambda}{m_e}\right)(\tilde{\mathcal{U}}\cdot\tilde{\mathcal{W}})\\
&\approx \frac{i\alpha^2}{3\pi^2m_e^2c}\ln\left(\frac{\lambda}{m_e}\right)\delta_{\sigma\sigma'}\delta_{\tau\tau'}.
\end{aligned}
\tag{G.6}
$$

H Checks of physical dimensions

In our formulas, we explicitly indicate all fundamental constants, such as c and $\hbar$, instead of the usual practice of choosing a system of units in which $\hbar = c = 1$. This makes our formulas somewhat more cumbersome than usual, but this is offset by two important advantages. First, in our notation it is easy to sort the terms in the order of their smallness in the important classical ($\hbar \to 0$) and nonrelativistic ($c \to \infty$) limits. Second, it is much easier to monitor the physical dimensions of terms at all stages of computations. In this appendix, we will offer several rules for evaluating physical dimensions of expressions containing quantum fields.

From the familiar formula

$$\int d\boldsymbol{p}\delta(\boldsymbol{p}) = 1,$$

it follows that the dimension of the delta-function is the inverse cube of momentum

$$<\delta(\boldsymbol{p})> = \frac{1}{<p^3>}.$$

(Anti)commutators of creation and annihilation operators

$$\{a^\dagger_{\boldsymbol{p}\sigma}, a_{\boldsymbol{p}'\sigma'}\} = \delta(\boldsymbol{p} - \boldsymbol{p}')\delta_{\sigma\sigma'},$$
$$[c_{\boldsymbol{p}\tau}, c^\dagger_{\boldsymbol{p}'\tau'}] = \delta(\boldsymbol{p} - \boldsymbol{p}')\delta_{\tau\tau'}$$

suggest their dimensions

$$<a^\dagger_{\boldsymbol{p}\sigma}> = <a_{\boldsymbol{p}'\sigma'}> = <c_{\boldsymbol{p}\tau}> = <c^\dagger_{\boldsymbol{p}'\tau'}> = \frac{1}{<p^{3/2}>}. \tag{H.1}$$

In the definition of the Dirac quantum field (B.34),

$$\psi(t, \boldsymbol{x}) = \int \frac{d\boldsymbol{p}}{(2\pi\hbar)^{3/2}} \sqrt{\frac{m_e c^2}{\omega_{\boldsymbol{p}}}} \sum_\sigma (e^{-\frac{i}{\hbar}\tilde{p}\cdot\tilde{x}} u(\boldsymbol{p}, \sigma) a_{\boldsymbol{p}\sigma} + e^{\frac{i}{\hbar}\tilde{p}\cdot\tilde{x}} v(\boldsymbol{p}, \sigma) b^\dagger_{\boldsymbol{p}\sigma}),$$

the 4-vectors $\tilde{p}$ and $\tilde{x}$ have the dimensions of energy and time, respectively:

$$<\tilde{p}> = <E>,$$
$$<\tilde{x}> = <t>.$$

The Planck constant has the dimension of *action*

$$<\hbar> = <p><r> = <E><t>,$$

https://doi.org/10.1515/9783110493207-012

which implies that the arguments $(\frac{i}{\hbar}\tilde{p} \cdot \tilde{x})$ of the exponents are dimensionless, as expected. The functions u and v are dimensionless as well (see equations (B.45)–(B.48)). Then the dimension of the field operator is

$$<\psi> = \frac{<p^3>}{<\hbar^{3/2}><p^{3/2}>} = \frac{<p^{3/2}>}{<\hbar^{3/2}>} = \frac{1}{<r^{3/2}>}.$$

Similarly, we get the dimension of the photon quantum field (C.2)

$$<\mathcal{A}> = \frac{<\hbar><c^{1/2}><p^3>}{<\hbar^{3/2}><p^{1/2}><p^{3/2}>} = \frac{<p><c^{1/2}>}{<\hbar^{1/2}>} = \frac{<p^{1/2}><c^{1/2}>}{<r^{1/2}>}, \tag{H.2}$$

the current density operator (D.1),

$$<\mathrm{j}> = <e\overline{\psi}\psi> = \frac{<e>}{<r^3>},$$

and the potential energy (3.14),[1]

$$<V_1> = <r^3>\frac{<e>}{<r^3>}\frac{<p^{1/2}><c^{1/2}>}{<r^{1/2}>} = \frac{<e><\hbar^{1/2}><c^{1/2}>}{<r>} = \frac{<e^2>}{<r>}.$$

The latter result is expected from comparison with the Coulomb interaction energy $E = e^2/(4\pi r)$. In QED, the second-order potential (3.15) also has the dimension of energy,

$$<V_2> = \frac{<r^3><r^3>}{<r>}\frac{<e>}{<r^3>}\frac{<e>}{<r^3>} = \frac{<e^2>}{<r>}.$$

Following these rules, it is not difficult to confirm that all three terms in the potential boost (3.17) have the expected dimension of $<m><r>$.

Let us now illustrate the dimension check on the example of the scattering amplitude (3.31). The S-operator is a dimensionless quantity, while the dimension of creation/annihilation operators is $<p^{-3/2}>$. So, for the dimension of the matrix element $s_2 = \langle \mathrm{vac}|a_{q\tau}d_{p\sigma}S_2 d^\dagger_{\mathbf{p}'\sigma'}a^\dagger_{\mathbf{q}'\tau'}|\mathrm{vac}\rangle$ we expect to get $<p^{-6}>$. Since

$$<\delta^4(p)> = <\delta(E)\delta(\mathbf{p})> = \frac{1}{<E><p^3>},$$

the dimension of (3.33) is

$$\frac{<e^2><c^2>}{<\hbar><E><p^3><E^2>} = \frac{<c^3>}{<E^3><p^3>} = \frac{1}{<p^6>},$$

in agreement with our expectation.

1 We simplified this expression by using the formula $<e^2> = <\hbar><c>$, which follows from the fact that $\alpha \equiv e^2/(4\pi\hbar c) \approx 1/137$ is the dimensionless fine structure constant.

Bibliography

[1] Berestetskii VB, Livshitz EM, Pitaevskii LP. Quantum electrodynamics. Oxford: Elsevier; 1982.
[2] Bjorken JD, Drell SD. Relativistic quantum mechanics. New York: McGraw–Hill; 1964.
[3] Blumenson LE. A derivation of n-dimensional spherical coordinates. Am Math Mon. 1960;67:63.
[4] Feynman RP. Space–time approach to quantum electrodynamics. Phys Rev. 1949;76:769.
[5] Fiore G, Modanese G. General properties of the decay amplitudes for massless particles. Nucl Phys B. 1996;477:623.
[6] Giunti C, Laveder M. Neutrino mixing. In: Columbus FH, Krasnoholovets V, editors. Developments in quantum physics. New York: Nova Science; 2004. p. 197–254. arXiv:hep-ph/0310238v2.
[7] Goldberger ML, Watson KM. Collision theory. New York: J. Wiley & Sons; 1964.
[8] Kraus P, Griffits DJ. Renormalization of a model quantum field theory. Am J Phys. 1992;60:1013.
[9] Mutze U. Relativistic quantum mechanics of n-particle systems with cluster-separable interactions. Phys Rev D. 1984;29:2255.
[10] Peskin ME, Schroeder DV. An introduction to quantum field theory. Boulder: Westview Press; 1995.
[11] Polovin RV. Radiative corrections to the scattering of electrons by electrons and positrons. Sov Phys JETP. 1957;4:385.
[12] Polyzou WN. Relativistic quantum mechanics – particle production and cluster properties. Phys Rev C. 2003;68:015202. arXiv:nucl-th/0302023v1.
[13] Redhead MLG. Radiative corrections to the scattering of electrons and positrons by electrons. Proc R Soc A. 1953;220:219.
[14] Schweber SS. An introduction to relativistic quantum field theory. Evanston, Il: Row, Peterson & Co.; 1961.
[15] Shvarts AS. Mathematical foundations of quantum field theory. Moscow: Atomizdat; 1975 (in Russian).
[16] Stefanovich E. Elementary Particle Theory. Vol. 1: Quantum Mechanics. De Gruyter Stud Math Phys. Vol. 45. Berlin: De Gruyter; 2018.
[17] Stefanovich E. Elementary particle theory. Vol. 3: Relativistic quantum dynamics. De Gruyter Stud Math Phys. Vol. 47. Berlin: De Gruyter; 2018.
[18] Weinberg S. Photons and gravitons in S-matrix theory: Derivation of charge conservation and equality of gravitational and inertial mass. Phys Rev. 1964;135:B1049.
[19] Weinberg S. The quantum theory of massless particles. In: Deser S, Ford KW, editors. Lectures on particles and field theory, vol. 2. Englewood Cliffs: Prentice–Hall; 1964.
[20] Weinberg S. Photons and gravitons in perturbation theory Derivation of Maxwell's and Einstein's equations. Phys Rev B. 1965;138:988.
[21] Weinberg S. The quantum theory of fields, vol. 1. Cambridge: University Press; 1995.

https://doi.org/10.1515/9783110493207-013

Index

De Gruyter Studies in Mathematical Physics

Volume 47
Eugene Stefanovich
Elementary Particle Theory: Volume 3: Relativistic Quantum Dynamics, 2018
ISBN 978-3-11-049090-9, e-ISBN (PDF) 978-3-11-049322-1,
e-ISBN (EPUB) 978-3-11-049139-5

Volume 45
Eugene Stefanovich
Elementary Particle Theory: Volume 1: Quantum Mechanics, 2018
ISBN 978-3-11-049088-6, e-ISBN (PDF) 978-3-11-049213-2,
e-ISBN (EPUB) 978-3-11-049103-6

Volume 44
Vladimir V. Kiselev
Collective Effects in Condensed Matter Physics, 2018
ISBN 978-3-11-058509-4, e-ISBN (PDF) 978-3-11-058618-3,
e-ISBN (EPUB) 978-3-11-058513-1

Volume 43
Robert F. Snider
Irreducible Cartesian Tensors, 2017
ISBN 978-3-11-056363-4, e-ISBN (PDF) 978-3-11-056486-0,
e-ISBN (EPUB) 978-3-11-056373-3

Volume 42
Javier Roa
Regularization in Orbital Mechanics: Theory and Practice, 2017
ISBN 978-3-11-055855-5, e-ISBN (PDF) 978-3-11-055912-5,
e-ISBN (EPUB) 978-3-11-055862-3

Volume 41
Esra Russell, Oktay K. Pashaev
Oscillatory Models in General Relativity, 2017
ISBN 978-3-11-051495-7, e-ISBN (PDF) 978-3-11-051536-7,
e-ISBN (EPUB) 978-3-11-051522-0

Volume 40
Joachim Schröter
Minkowski Space: The Spacetime of the Special Relativity Theory Groups, 2017
ISBN 978-3-11-048457-1, e-ISBN (PDF) 978-3-11-048573-8,
e-ISBN (EPUB) 978-3-11-048461-8

Bei Fragen zur Produktsicherheit wenden Sie sich bitte an:
If you have any questions regarding product safety,
please contact:

Walter de Gruyter GmbH
Genthiner Straße 13
10785 Berlin
productsafety@degruyterbrill.com